D1218023

ADVANCES IN

CHROMATOGRAPHY

Volume 13

ADVANCES IN CHROMATOGRAPHY

Volume 13

Edited by

J. CALVIN GIDDINGS
EXECUTIVE EDITOR

DEPARTMENT OF CHEMISTRY
UNIVERSITY OF UTAH
SALT LAKE CITY, UTAH

ELI GRUSHKA
GAS CHROMATOGRAPHY

DEPARTMENT OF CHEMISTRY
STATE UNIVERSITY OF NEW YORK AT BUFFALO
BUFFALO, NEW YORK

ROY A. KELLER
LIQUID CHROMATOGRAPHY

DEPARTMENT OF CHEMISTRY
STATE UNIVERSITY OF NEW YORK
COLLEGE AT FREDONIA
FREDONIA, NEW YORK

JACK CAZES
MACROMOLECULAR CHROMATOGRAPHY

WATERS ASSOCIATES, INC.
MILFORD, MASSACHUSETTS

MARCEL DEKKER, Inc., New York

MARCEL DEKKER, INC.

270 Madison Avenue, New York, New York 10016

LIBRARY OF CONGRESS CATALOG CARD NUMBER 65-27435

ISBN 0-8247-6274-6

Current printing (last digit):
10 9 8 7 6 5 4 3 2 1

PRINTED IN THE UNITED STATES OF AMERICA

PREFACE

This volume of Advances in Chromatography marks a significant departure from earlier volumes in the series in editorial approach. Roy Keller, after shouldering for many years the major burden of general editorial work, has opted to restrict his efforts to a single field of chromatography: liquid chromatography. Accordingly, the editorial structure has been revised. We have created several "area editorships" to permit a greater editorial concentration in the major subfields of chromatography. We hope to generate a more meaningful coverage of each chromatographic area as the various area editors are able to focus their talents more specifically. The focus on liquid chromatography is provided by Roy Keller. Eli Grushka has taken responsibility for the broad and still active field of gas chromatography. Jack Cazes is focusing on large molecules; he is serving as area editor for macromolecular chromatography. Other area editors may be added in the future in order to keep in step with the ever-dynamic field of chromatography.

While emphases may shift, our goal here and in the future is the same goal we have pursued for twelve previous volumes of Advances beginning in 1965: to provide the community of chromatographers with stimulating, critical, readable, and relevant reviews of this broad and multifaceted subject. We welcome suggestions from readers to help us better reach these objectives.

J. Calvin Giddings
Executive Editor

CONTRIBUTORS TO VOLUME 13

KLAUS H. ALTGELT, Fundamental Research Division, Chevron Research Company, Richmond, California

NILS FRIIS, Department of Chemical Engineering, McMaster University, Hamilton, Ontario

MARIE-FRANCE GONNORD, Laboratoire de Chimie Analytique Physique, Ecole Polytechnique, Paris, France

T. H. GOUW, Analytical Research and Services Division, Chevron Research Company, Richmond, California

GEORGES GUIOCHON, Laboratoire de Chimie Analytique Physique, Ecole Polytechnique, Paris, France

ARCHIE HAMIELEC, Department of Chemical Engineering, McMaster University, Hamilton, Ontario

RALPH E. JENTOFT, Analytical Research and Services Division, Chevron Research Company, Richmond, California

DAVID A. LEATHARD, Department of Chemistry and Biology, Sheffield Polytechnic, Sheffield, England

CLAIRE VIDAL-MADJAR, Laboratoire de Chimie Analytique Physique, Ecole Polytechnique, Paris, France

J. M. SMITH, Department of Chemical Engineering, University of California at Davis, Davis, California

MOTOYUKI SUZUKI, Institute of Industrial Science, University of Tokyo, Tokyo, Japan

CONTENTS

CONTENTS OF OTHER VOLUMES

Volume 1

Volume 2

Volume 3

Volume 4

Volume 5

Volume 6

Volume 7

Volume 8

Volume 9

Volume 10

Volume 11

Volume 12

ADVANCES IN

CHROMATOGRAPHY

Volume 13

Chapter 1

PRACTICAL ASPECTS IN SUPERCRITICAL FLUID CHROMATOGRAPHY

T. H. Gouw and Ralph E. Jentoft
Analytical and Research Services Division
Chevron Research Company
Richmond, California

I. INTRODUCTION

When a compound is heated in a confined space of such a volume that the compound is always present as a liquid and as a gas in equilibrium with each other, the intensive properties of these two phases will converge until, at the critical point, these two phases become identical. Further heating will result in the formation of a supercritical phase.

The mobile phase in supercritical fluid chromatography is a compound that is maintained at a temperature somewhat above its critical point. Depending on the nature of the stationary phase, the technique is also called fluid-liquid or fluid-solid chromatography.

Classically, a compound is said to be in the gaseous state when it is heated to a temperature above its critical temperature. In this particular region of the phase diagram, however, the mobile phase has properties intermediate between the properties of a liquid and of a "regular" gas. Intermolecular energy levels are of the order of kT, where k is the Boltzmann constant and T is the absolute temperature. In a "regular" gas, even when it is highly compressed, intermolecular energy levels are much lower. In the liquid state, on the other hand, pronounced molecular interactions result in intermolecular energy levels much larger than kT.

The density of the mobile phase in supercritical fluid chromatography is about 200 to 500 times the density observed in gas chromatography. The effect of shorter intermolecular distances and the resultant increase in molecular interactions is an enhanced solubilizing capability of the solvent towards a wide variety of solutes. Compounds of much higher molecular weight than gas chromatography allows can therefore be chromatographed.

At a given mobile phase velocity, the speed of migration in gas chromatography is essentially determined by the vapor pressure of the solute alone. By its nature, a gas has very little influence on the concentration of the solute molecules in the moving phase. Vapor pressures can be increased by increasing the temperature of the column, thereby decreasing the observed retention times. However, the thermal stability of the solute and of the column packing sets an upper limit on the temperature at which the separation can be carried out. In supercritical fluid chromatography, the mobile phase competes actively for the solute molecules, and substantially lower partition ratios (defined as the concentration of solute in the stationary phase over the concentration in the mobile phase) are observed. Since retention times decrease almost proportionaly with a decrease in partition rations, the much lower retention time allows one to chromatograph much larger molecules in a reasonable length of time. The higher solubilizing capability also permits chromatography to be carried out at much lower temperatures, allowing many thermally less stable compounds, such as biologically active materials, to be processed by this technique. An additional advantage is that many substances that cannot be volatilized without decomposition, such as some ionic species, are also soluble in the supercritical mobile phase.

A specific phenomenon associated with the supercritical phase is the very large change in density as a function of the pressure. This phenomenon is used to practical advantage for processing samples with a wide molecular-weight range. Whereas temperature programming is used in gas chromatography, pressure programming is the most advisable route in supercritical fluid chromatography to obtain satisfactory chromatograms for mixtures of wide molecular-weight range.

High solution capabilities are also observed if a liquid is used as the mobile phase. The viscosity of a supercritical fluid is, however, more closely akin to the viscosity of a gas and is hence about a factor of 100 smaller than the value for the same substance in the liquid phase. For a given flow rate, this low viscosity results in much lower pressure drops over the chromatographic column that can be observed in liquid chromatography. Higher mobile phase velocities are therefore possible. At low mobile phase velocities, the main contribution to the HETP in liquid chromatography is resistance to mass transfer in the mobile phase. At higher flow rates, the determining factor becomes the resistance to mass transfer in the stationary phase. A supercritical fluid has the advantage of possessing a diffusion coefficient about 100 to 200 times larger than the coefficient observed in a liquid. Since this property is very important in mass transfer considerations, the superior transport properties of a supercritical fluid suggest that much higher separating speeds can be attained compared with those in liquid chromatography.

Many practical problems still have to be solved before this technique enjoys the general acceptability now accorded to gas and modern high resolution liquid chromatography. The theoretical advantages described above have, however, encouraged many investigators to study and develop this technique into a practical chromatographic method. It thus appears that supercritical fluid chromatography will not supplant the other two established chromatographic techniques; but in many special applications it will be an indispensable tool to solve problems that cannot be easily elucidated by other existing techniques.

II. HISTORICAL

Klesper, Corwin and Turner appear to be the first investigators to report a column chromatographic separation with a supercritical fluid as the mobile phase [26]. In their paper, they described the separation of Ni etioporphyrin II from Ni mesoporphyrin IX dimethyl ester with dichlorodifluoromethane or monochlorodifluoromethane as the eluent. In subsequent communications, Karayannis et al. described the behavior of a large number of porphyrins, etioporphyrin II metal chelates [21, 23], metal acetylacetonates [22], and other metal chelates [24] when they were chromatographed with dichlorodifluoromethane. The chromatograms shown were not very encouraging, since very large band spreads were observed with these compounds.

Soon after the first paper was published, several patent applications were filed in Austria for a separation process based on the enhanced solution of mixtures of organic compounds in a supercritical fluid [49]. By decrease in pressure or by increase in temperature of the system or both, a portion of the dissolved material is precipitated and separated out. It was not, however, until a few years later that the same applicants filed a patent application for a column chromatographic process with a supercritical fluid as the mobile phase [48].

The most comprehensive study on this subject was published by Sie and Rijnders in 1966 on investigations they had completed a few years earlier [42-46]. The papers are valuable not only because of the fundamental treatment of the subject but also because of the large number of analytical applications included in the text.

Parallel to these developments are the investigations Giddings and his school carried out on using gases at extremely high pressures as the mobile phase [10, 13, 14, 28]. This technique, called "dense gas chromatography," is carried out using pressures up to 2000 atm. Although there are many points of similarity, we prefer, for various reasons, to consider dense gas chromatography as a technique separate from supercritical fluid chromatography. Intermolecular energy levels are lower in gas chromatography than they are when a supercritical fluid is used as the mobile phase. The high pressures applied in the dense gas technique will also necessitate specialized equipment that will require more precautions and costs.

Many investigators have since reported on newer developments in this field. Supercritical fluid chromatography has been used to chromatograph paraffins [2]; halogenated paraffins [19]; mono- and polynuclear aromatic hydrocarbons [46]; their alkyl derivatives; ionic species such as the phospholipids and the metalloorganic compounds [47, 19] polymers such as polystyrenes [20] and epoxy resins [47]; pharmaceuticals [47]; natural products, such as cinchonine and brucine [47]; vitamins, steroids; and complex, undefined heavy substances such as tar [35, 45]. This chapter concerns some of the practical aspects and problems that have been reported in the last few years; it is an extension of a review published earlier [15].

III. RANGE OF OPERATIONS

Many people tend to associate critical phenomena with very high pressures and temperatures, probably because water has a critical temperature of 374.4 °C, and a critical pressure of 226.8 atm. Actually, critical pressures of most hydrocarbons are less than 40 atm; in a homologous series, critical temperatures increase with increasing molecular weight. For ethane, the critical temperature T_c is 32.4 °C; for propane, T_c is 98.6 °C; and for n-heptane, T_c is 267.0 °C. The fluids in general use are those with relatively low critical temperatures. A very popular mobile phase, carbon dioxide, shows a critical temperature of only 31.3 °C. Pressures in supercritical fluid chromatography usually range from 30 atm to 300 atm. Higher pressures are generally not necessary since the critical pressure of the mobile phase is generally only in the 40-60 atm range.

Figure 1 shows a phase diagram with the reduced pressure and the reduced density on the two axes [14]. Lines of constant reduced temperatures are shown. The region of dense gas chromatography for which McLaren, Myers, and Giddings have carried out their experiments is crosshatched in the figure; the region for which supercritical fluid chromatography is generally carried out by current investigators is dotted. Excursions outside this region are possible.

Note the slight slopes of the lines of constant T_r in the region close to and above the critical point. This indicates that in this area there is a sharp increase in density with only a slight increase in pressure. At higher T_r's, this effect is not so pronounced. Chromatographic separations are generally carried out about 3 °C or more above the critical temperature to reduce the very strong dependence of the operational parameters on the temperature.

Table 1 lists some compounds that can be used as the mobile phase. Many more compounds can obviously be included in this list. Theoretically, one can use any compound that is thermally stable at the operating conditions. In practice, in choosing the fluid, one may want to consider the solvent selectivity, product availability, cost, toxicity, and chemical reactivity.

The first reported use of supercritical fluid chromatography by Klesper, Corwin, and Turner [26] described using dichlorodifluoromethane and monochlorodifluoromethane as the eluent. Isopropanol has been used by Sie and Rijnders [44, 45] and Jentoft and Gouw [18]. Rijnders has also reported on diethyl ether as the supercritical fluid [35]. The most popular mobile phases have been CO_2 and pentane, however.

An important consideration in choosing the mobile phase is related to the type of detector used in the chromatographic system. If a flame ionization detector is used, the choice of compounds is limited to carbon

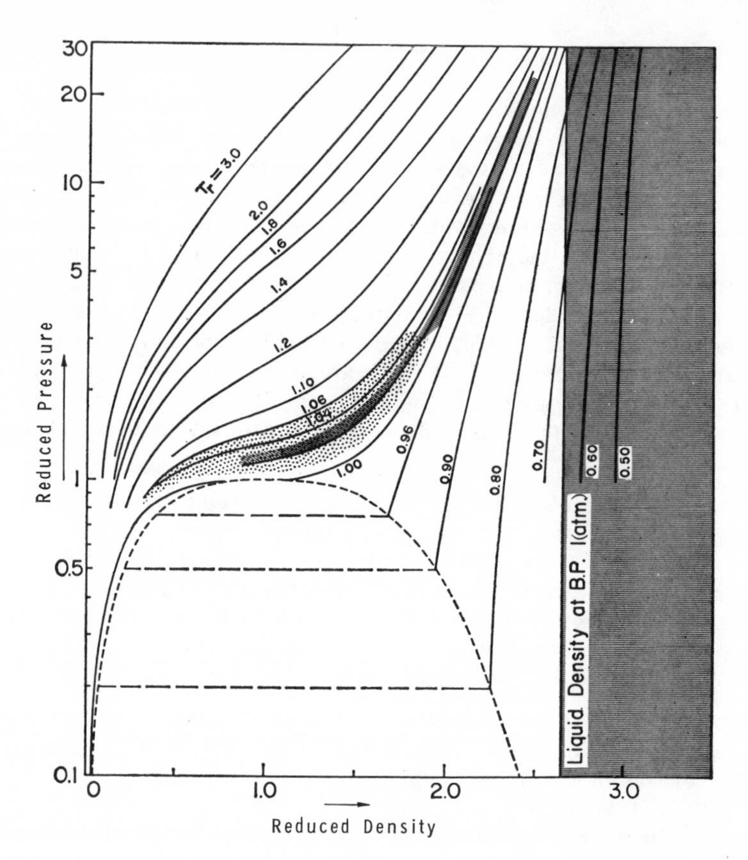

FIG. 1. Operating regions of "dense gas" (crosshatched field) and supercritical fluid chromatography (dotted field). Excursions outside the indicated regions are possible. (Adapted from Ref. 14. Courtesy Dr. J. C. Giddings. Copyright 1968 by the American Association for the Advancement of Science.)

dioxide or nitrous oxide because of the relative insensitivity of the flame to these compounds. If other eluents are used, an additional stage is necessary to separate the solutes from the mobile phase prior to detection. This is the case if a moving wire detector is used. Moving wire detectors are difficult to operate, however.

A UV absorption detector necessitates mobile phases that are transparent in the wavelength region where detection is carried out. In general, this does not create any problems. Essentially all applicable mobile phases are of sufficiently low molecular weight and do not possess UV chromophoric groups.

TABLE 1

Properties of Possible Mobile Phases for Supercritical Fluid Chromatography

Compound	Atm b.p. (°C)	Critical point data		
		T_c (°C)	P_c (atm)	d_c (g/ml)
Nitrous oxide	-89	36.5	71.4	0.457
Carbon dioxide	-78.5[a]	31.3	72.9	0.448
Sulfur dioxide	-10	157.5	77.6	0.524
Sulfur hexafluoride	-63.8[a]	45.6	37.1	0.752
Ammonia	-33.4	132.3	111.3	0.24
Water	100	374.4	226.8	0.344
Methanol	64.7	240.5	78.9	0.272
Ethanol	78.4	243.4	63.0	0.276
Isopropanol	82.5	235.3	47.0	0.273
Ethane	-88	32.4	48.3	0.203
n-Propane	-44.5	96.8	42.0	0.220
n-Butane	-0.5	152.0	37.5	0.228
n-Pentane	36.3	196.6	33.3	0.232
n-Hexane	69.0	234.2	29.6	0.234
n-Heptane	98.4	267.0	27.0	0.235
2,3-Dimethylbutane	58.0	226.8	31.0	0.241
Benzene	80.1	288.9	48.3	0.302
Diethyl ether	34.6	193.6	36.3	0.267
Methyl ethyl ether	7.6	164.7	43.4	0.272
Dichlorodifluoromethane	-29.8	111.7	39.4	0.558
Dichlorofluoromethane	8.9	178.5	51.0	0.522
Trichlorofluoromethane	23.7	196.6	41.7	0.554
Dichlorotetrafluoroethane	3.5	146.1	35.5	0.582

[a] Sublimation point

Another aspect that might be important is whether the mobile phase is liquid at ambient conditions. From the relation known as Guldberg's rule [31], the boiling point of a compound at atmospheric pressure has been observed empirically to be about two-thirds the critical temperature, both taken on an absolute temperature scale. This indicates that for a compound to be liquid at room temperature and atmospheric pressure, its critical temperature should be at least about 165 °C. The substance n-pentane has a critical temperature of 196.6 °C; n-butane has a critical temperature of 152.0 °C. Compounds with low critical temperatures, such as carbon dioxide and nitrous oxide, are obviously gaseous at ambient conditions.

A. Solubility of Solute in Mobile Phase

Increased solubility in a supercritical phase has already been reported, the first instance several decades ago. Hannay and Hogarth observed that cobalt and ferric chlorides dissolve in much higher proportions in supercritical ethanol than would be predicted from the molecular vapor pressure of the salts alone [17]. This behavior is a special case of the general observation that the partial pressure of a substance with low vapor pressure is increased when it is brought into contact with a gas at elevated pressure. This increase in vapor pressure is especially pronounced if the second phase is a substance close to its critical point. Many investigators have since confirmed this rule. Diepen and Scheffer [11] studied the solubility of naphthalene in compressed ethylene and observed that at 200 atm the solubility of the aromatic hydrocarbon was about 10^4 times greater than the solubility computed on the basis of the ideal gas laws. Van Nieuwenburg and van Zon heated water and quartz in a closed container to a temperature above the critical temperature of water [29]. When the apparatus was cooled and dismantled, crystals of quartz were observed on the walls of the container, indicating extensive solubilization of the quartz during the heating cycle. Recent investigations [36] have shown that at 600 °C and 1000 atm the supercritical water phase can contain as much as 50% NaCl. This observation parallels the well-known phenomenon of the precipitation of salts, amorphous silica, and quartz on the convex sides of steam turbine blades during decompression. This phenomenon has led to the use of deionized water to circumvent these problems.

Sie, van Beersum, and Rijnders [42] described the pressure-induced increase in solvent power of dense gases by considering the changes in the second virial coefficient to describe the initial departure from ideality. This approach was found to be valid in their systems for pressures up to 50 atm. For pressures higher than 100 atm, however, this approach becomes quite laborious because of the consideration that must be given to the higher terms of the series and to series convergence problems [39]. A more recent quantitative treatment of this phenomenon has been given by Giddings et al. [10, 13, 14], who based their approach on the Hildebrand solubility parameter. Experimental data covering very high pressures show a rough accord with the predictions based on this theory.

Bartmann and Schneider [5, 6] indicate, however, that the treatment of these solubility phenomena in the critical range by the solubility parameter alone is often not adequate except for approximative purposes. Their fundamental approach is to consider the form of the phase diagram for a multicomponent system in the critical region [2, 5]. Since this area can be extremely complex, we shall only summarize the phase behavior of some binary mixtures in this region. Rowlinson [38], Ricci [34], Zernike [50], Schneider [40], Prausnitz [32, 33], and many others have described this subject in detail.

For a single compound, we observe a single critical point in the p-(T) phase diagram. In a binary system, the critical points (= extreme values in the isothermal p-(x) or isobaric T-(x) cross sections) lie on a line, called the critical locus curve. This curve, which is anchored by the critical points of the two components, does not always have to have a continuous shape; in some cases the curve can also be interrupted [5]. Phase equilibria in binary mixtures are best discussed with the aid of the p-(T) projections of these critical curves. In this region of the phase diagram, one can observe three types of two-phase equilibria, i.e., liquid-liquid, liquid-gas, and gas-gas equilibria. The latter is also referred to as the region of "immiscibility of gases." Van der Waals had predicted the existence of this phenomenon as early as 1894, but only in the last decade has this phenomenon been extensively studied in different laboratories all over the world. Figure 2 shows the critical p-(T) curve projection for a number of n-alkane-CO_2 mixtures [41]. The critical points of CO_2 and of the n-alkanes under consideration are shown as small solid circles in this figure. The solid lines going from the critical points of the n-alkanes down almost horizontally to the left are the liquid-gas equilibrium lines, or vapor pressure curves, for the pure n-hydrocarbons. The dashed lines are the critical locus curves of the binary CO_2-n-alkane mixtures. The symbol LG stands for the liquid-gas boundary of the critical curve; LL stands for the line that separates the region of two immiscible liquids and the supercritical phase.

The critical locus curve for mixtures of CO_2 with methane, butane, and the higher-molecular-weight n-alkanes up to C_{13} shows a pressure maximum that increases with increasing molecular weight. Ethane shows a temperature minimum, and propane shows a p-(T) critical locus with a monotonic shape between the critical points of the pure components of the mixture. For propane to tridecane above the critical temperature of CO_2,

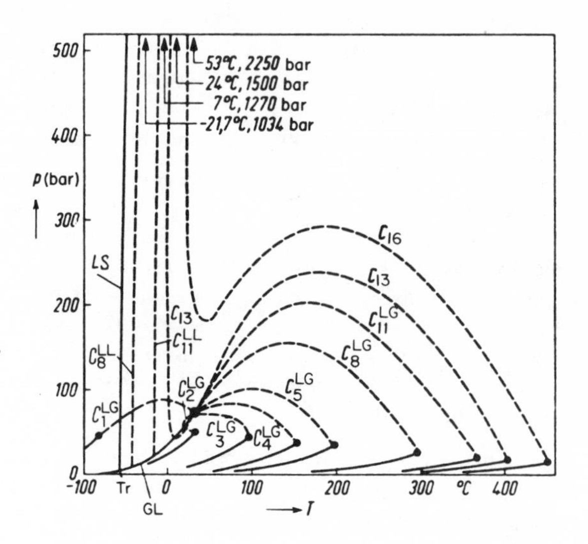

FIG. 2. p-(T) projections of binary systems of CO_2 and n-alkanes. Tr is the triple point; LS is the fusion curve; and GL is the vapor pressure curve of CO_2. The filled dots indicate the critical points of CO_2 and the various n-alkanes. Dashed lines are the critical locus curves separating the region of multiphase equilibrium from the single-phase supercritical region. (From Ref. 41. Courtesy Verlag Chemie GmbH.)

one finds a mixture of gas and liquid that will coalesce during isothermal compression into a single phase. Beyond the critical locus curve, no two-phase equilibrium can exist any longer. At lower temperatures, the lines drawn for C_8 to C_{13} indicate the condition at which the two immiscible liquids coalesce into one single phase. The symbol LG stands for the vapor pressure curve, and LS stands for the solid-liquid, or the fusion, curve for pure CO_2.

It is interesting to note the more complicated shape of the critical locus line of CO_2 and n-hexadecane. Although one observes a pressure maximum, the critical locus line does not go through the critical point of CO_2. The number of immiscible phases below this curve depends on the value of x and T and can be as many as three, i.e., two liquid phases and one gas phase.

p-(T) projections with a maximum are characteristic not only for mixtures of CO_2 and n-alkanes but also for binary mixtures of CO_2 with nonpolar compounds of low-to-medium molecular weight, such as cyclohexane, toluene, and the unsaturated linear hydrocarbons. Binary mixtures of CO_2

and several low-molecular-weight substances, such as alcohols, acetic acid, and acetaldehyde, also show this behavior. For some systems, complete mutual miscibility by isothermal compresseion can be obtained only above a certain temperature.

Phase diagrams of other mixtures can be much more complicated. We have already seen a three-phase line, which defines a region in which two liquid phases and one gas phase can be in equilibrium with each other. In many cases the critical locus line is not continuous. For mixtures of CO_2 and the aromatic amines, the phase diagram is further complicated by the appearance of solid phases and the formation of solid compounds. In many of these diagrams, complete miscibility is possible only in certain temperature regions.

These phase diagrams can be used effectively to describe some of the phenomena observed in supercritical fluid chromatography. Most of the papers on this technique have, for instance, stressed the sharp increase in the solubility of the solute in the mobile phase. Bartmann and Schneider warned, however, that even then extremely low solubilities can occur [5]. Figure 3 shows the isothermal p-x diagram for the system CO_2-n-decane at 37.8 °C (T_c for CO_2 is 31.3 °C) based on data determined by Reamer and Sage [37]. One notes that for pressures between 20 and 80 atm, the solubility of n-decane in CO_2 is only 10^{-4} to 10^{-3} g/cm^3. The line drawn at 72.9 atm denotes the critical pressure of CO_2.

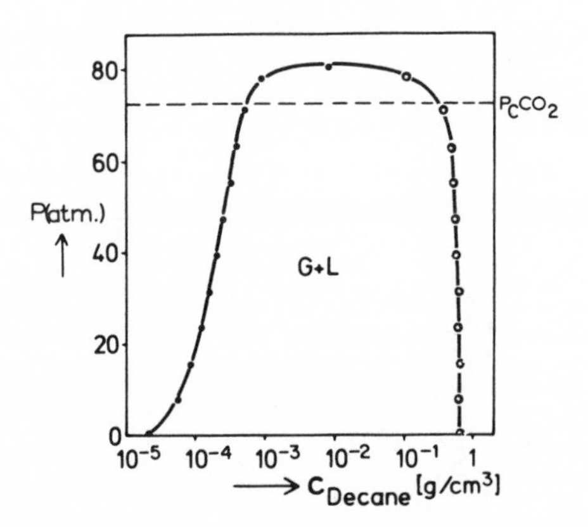

FIG. 3. Solubility of n-decane in CO_2, based on data from Reamer and Sage. (From Ref. 37. Courtesy Dr. G. M. Schneider and Elsevier Publishing Company.)

Since the molecular weight of n-decane is only 142, we can expect the solubility of many higher-molecular-weight compounds to be substantially lower; for the higher-molecular-weight n-paraffins it is possible to observe solubilities one or two magnitudes lower. It is therefore important to realize that in supercritical fluid chromatography very low solubilities can also occur and that this circumstance will limit the applicability of the technique for a particular problem.

Isothermal compression in this region generally results in a sharp "increase" in solubilities at the intersection with the solubility curve because of the transition into the region of mutual solubility. In some systems, this phenomenon occurs irrespective of the operating temperatures; in other systems, it occurs only above a certain maximum temperature.

Giddings and coworkers have reported this sudden increase in solubility for many high-molecular-weight compounds. They note that these compounds remain essentially stationary in a chromatographic column until the inlet pressure is increased to above a certain value. Migration is then observed as this pressure is exceeded. Giddings calls this the "threshold pressure" [13, 14]. Threshold values obviously depend on the composition of both the mobile phase and the solute. Some threshold values with CO_2 as the carrier fluid are shown in Table 2.

TABLE 2

Threshold Pressure with CO_2 at 40 °C

Compound	P_t (atm)
n-Octadecane	87.5
n-Eicosane	89.5
n-Hexatriacontane	104.8
1-Undecanol	61.2
1-Octadecanol	98.7
Carbowax 400	90
Carbowax 4000	115
Carbowax 20M	190
Silicone gum rubber SE-30	770
Dinonylphthalate	165

The solubility of the solute in the supercritical phase can normally also be influenced by adding modifiers to the mobile phase. The use of modifiers has been reported by Jentoft and Gouw [18] and by Novotny et al. [30]. The latter authors showed that adding 0.1% isopropanol to n-pentane as the mobile phase decreases the observed K values for many polynuclear aromatic hydrocarbons by 20-35%. This indicates that solvent programming may be useful in some applications. All reported work on the technique has been carried out under isocratic (constant solvent composition) conditions, however.

B. Selectivity

The high density and the resultant large intermolecular interactions in a supercritical fluid allow separations to be carried out on a more selective basis than gas chromatography allows. High selectivities are also observed in liquid chromatography, but this capability can be much more pronounced in supercritical fluid chromatography. In addition, selectivity can be adjusted by varying the pressure or the temperature or both of the system, allowing a greater operational flexibility than liquid chromatography allows.

In supercritical fluid chromatography, the use of a nonpolar mobile phase or a polar stationary substrate or both favors separations according to group type. Tremendously selective separations can be attained by this approach. Sie and Rijnders observed that with n-pentane as the mobile phase and alumina as the substrate, the k' value (k' = the partition ratio multiplied by the phase ratio) of 2-phenylnaphthalene can be as high as the k' value of hexacosylnaphthalene, which has 20 additional methylene units [45]. A still more extreme example was observed in the separation of di-n-triacontylphthalate from diphenylphthalate. With n-pentane as the mobile phase and 23% polyethylene glycol 6000 on 120/140 mesh Sil-O-Cel as the stationary substrate, the alkylphthalate was eluted well before the arylphthalate, notwithstanding that the former had 48 more C atoms [44].

To obtain separations according to molecular weight, the preferred approach would be to use a polar eluent such as isopropanol, or a nonpolar substrate, such as polyethylene or a lower-molecular-weight paraffin, or both. An excellent way to illustrate this point is summarized in Fig. 4, in which the capacity factor k' has been plotted against the carbon number for various members of the homologous series of the di-n-alkylphthalates [46]. If we take the middle curve as reference, we can note that by replacing the mobile phase with a more polar solvent, the difference in the elution times between the members of the homologous series is increased. Separation according to molecular weight is thereby favored by this approach. Replacing the nonpolar stationary phase with a more

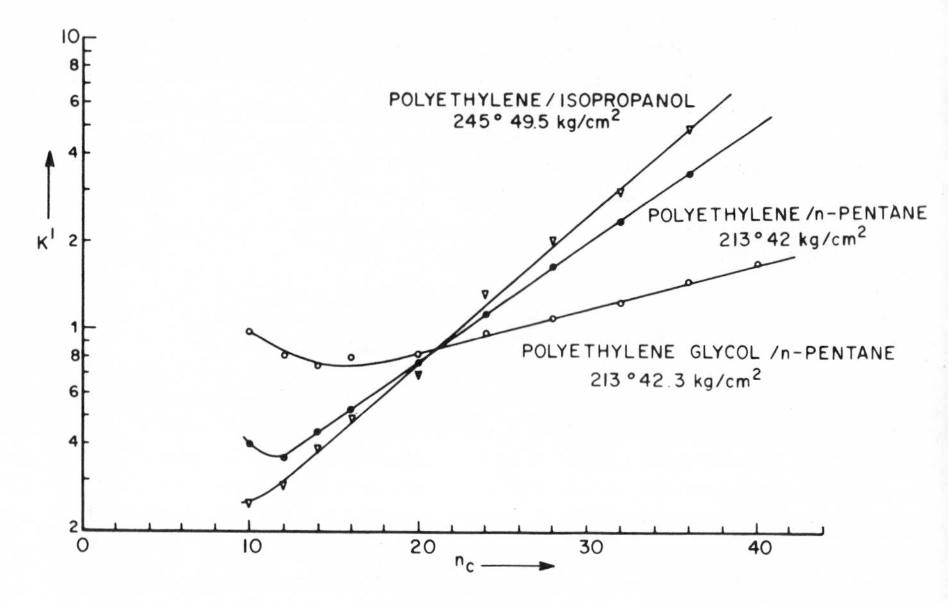

FIG. 4. Retention of di-n-alkylphthalates as function of carbon atom number under different operating conditions. Increasing the polarity of the mobile phase or decreasing the polarity of the stationary phase increases k'. For separations according to group type, one should use a polar substrate and a nonpolar mobile phase. (From Ref. 46. Courtesy of the authors and of Elsevier Publishing Company.)

polar substrate, i.e., polyethylene glycol, reduces the difference in elution time between consecutive members of the series. For the lower-molecular-weight compounds, we even note a reversal in the trend, indicating the enhanced influence of the aromatic moiety of the molecule. Under these conditions, separation according to type is therefore preferred.

Selectivity can in some cases also be influenced by the pressure of the system. Rijnders [36] finds, e.g., with n-pentane as the mobile fluid and 23% polyethylene glycol 6000 on 40-50 mesh Sil-O-Cel as the stationary substrate, below 40 atm and at 210 °C 1,3-dimethylnaphthalene (b.p. 263 °C) is eluted well before 2-n-decylnaphthalene (b.p. 387 °C). Under these conditions, volatility considerations dominate; and the compounds are eluted according to molecular weight. At higher pressures, however, the dominating factor is the interaction of the aromatic moiety with the polar substrate. Type separation takes place, and 2-n-decylnaphthalene is eluted well before dimethylnaphthalene.

Another example of the selectivity capabilities in supercritical fluid chromatography is shown in Fig. 5 and 6. Figure 5 shows a chromatogram of a synthetic mixture of polynuclear aromatic hydrocarbons on a column 3.5 m long and of 1/8-in. OD, packed with "Permaphase" ETH, an ether polymer substrate chemically bonded to superficially porous microbeads of 30 μm nominal diameter. The separation was carried out isothermally at 39 °C at a pressure of 1800 psi. Carbon dioxide was used as the mobile phase. Detection was carried out in a high pressure UV detector at 265 nm.

Note that under these conditions type selectivity takes place. Pyrene, the two methylpyrenes, and 1,3-dimethylpyrene elute together. Benzo(c)-phenathrene, a catacondensed molecule compared to the pericondensed pyrene, is eluted well after the pyrenes. The benz(a)anthracenes elute close to each other. It is especially interesting to note that 7,12-dimethylbenz(a)anthracene is eluted before the 7-methyl derivative.

Figure 6 shows essentially the same mixture chromatographed under the same operating conditions on a 3-m by 1/8-in. column packed with VYDAC reversed phase, a 30-44-μm superficially porous bead packing to which (nonpolar) octadecyl groups are bonded. Both the Permaphase and VYDAC packing have a fluid-impermeable glass core with a thin outer layer of a porous material.

Under these conditions, separations are carried out more or less according to molecular weight. Good separation is observed between each parent polynuclear aromatic hydrocarbon and its alkyl derivatives. Since recovery and the reinjection of solutes can be carried out almost quantitatively in supercritical fluid chromatography without too much effort, it is generally possible to recover a particular compound in reasonably high purity from a mixture of polynuclear aromatic hydrocarbons. The mixture is first separated under type-selective conditions, and the isolated compounds are subsequently rechromatographed under conditions in which separations are carried out according to molecular weight.

The possibility of obtaining noteworthy separations of polynuclear aromatic hydrocarbons has already been reported in one of the earliest papers on this technique. Sie and Rijnders [45] published chromatograms demonstrating complete baseline resolution of the isomeric condensed polynuclear aromatic hydrocarbons phenanthrene and anthracene, and also of 1,2- and 3,4-benzpyrene on a column showing only 200-250 theoretical plates. The separation of these isomers by gas chromatography can be carried out only in columns with a substantially larger number of theoretical plates, because on most common isotropic phases, the relative retention for each pair is close to unity.

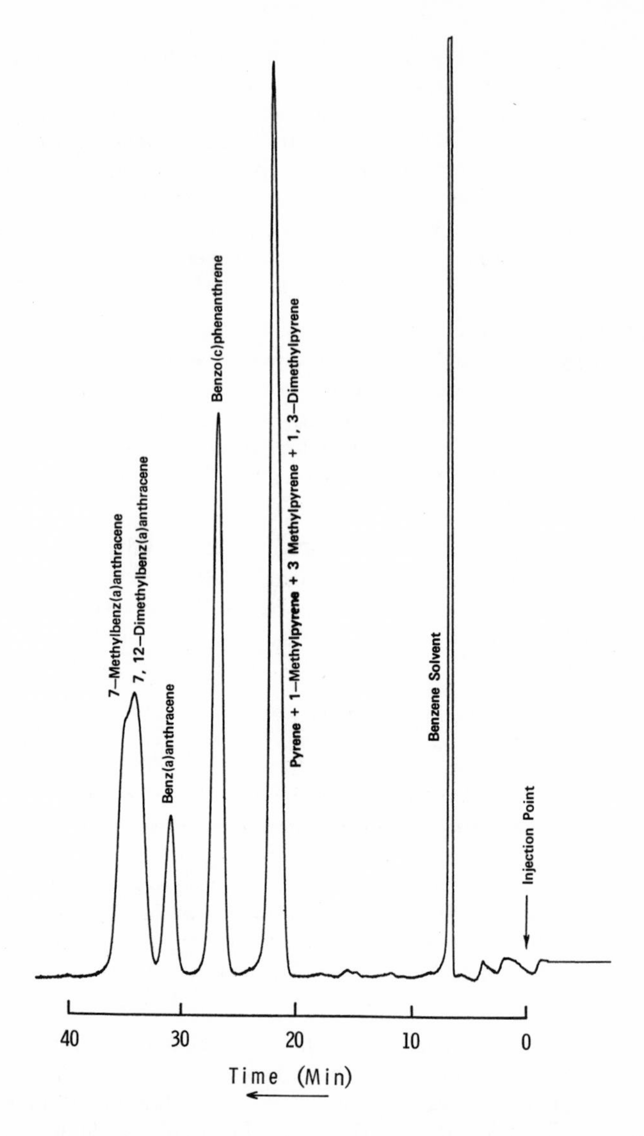

FIG. 5. Separation of polynuclear aromatic hydrocarbons on "Permaphase" ETH. Separations carried out according to group type. The parent hydrocrabon and its alkyl derivatives are eluted close together. Column size: 3.5 m by 1/8-in. OD. CO_2 was used as the mobile phase. Separation carried out at 39° C at pressure of 1800 psi.

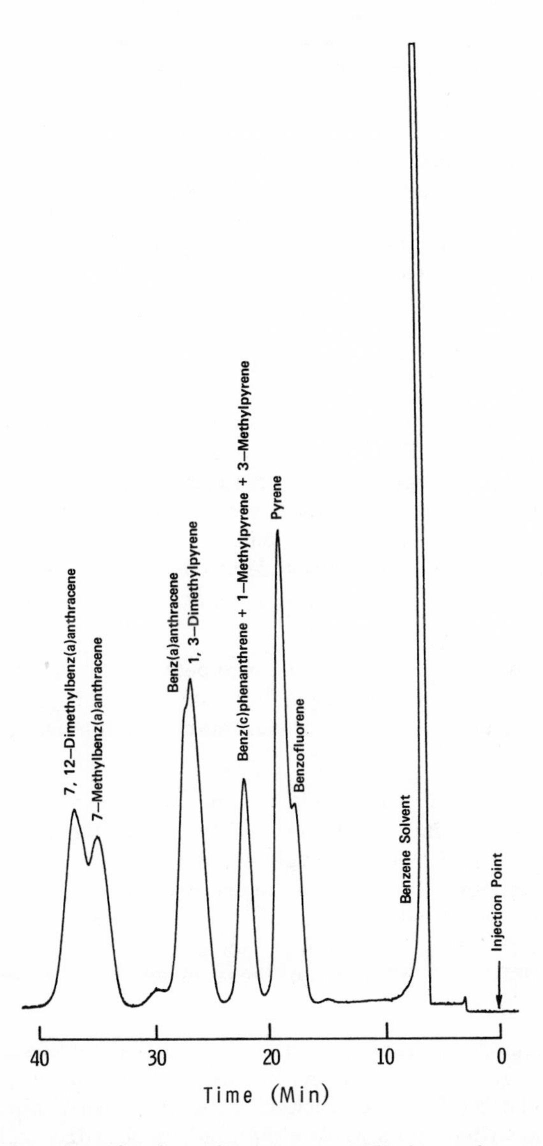

FIG. 6. Separation of polynuclear aromatic hydrocarbons on "VYDAC" reverse phase. Operating conditions are similar to condition of previous chromatogram. Separations carried out more according to molecular weight.

Selectivity can also be influenced by the temperature at which the separation is carried out. With n-pentane as the mobile phase on an alumina column at a temperature of 300° C, 1,1'-binaphthyl is eluted well after anthracene. At this high temperature, volatility considerations dominate and elution takes place according to molecular weight. At 200° C, however, the vapor pressure of these polynuclear aromatic hydrocarbons is considerably reduced; and the effect of the adsorption forces of the solid substrate results in the elution of the binaphthyl compound well before anthracene [45]. Both runs were carried out at approximately 40 atm.

C. Pressure

Pressure is the most important operating parameter in supercritical fluid chromatography. The strong dependence of the density as a function of the pressure in the region close to the critical point has already been pointed out. Because the solubility of a solute is a function of the density of the mobile state, partition coefficients, and hence retention times, can be adjusted by varying the pressure of the system. Thermodynamic descriptions of these phenomena have been presented by Giddings [12], Sie and Rijnders [42], and others [2].

Practical pressure requirements range between 30 and 300 atm, which is comparable to what is currently being used in modern high-resolution liquid chromatography. Higher pressures are possible but generally not necessary.

From the economical point of view, at higher pressures substantially higher costs would be involved because of the necessity for more safety features and the use of heavier-gauge tubing and vessels. From the physical point of view, a too-high pressure may even be deleterious to separation. Sie and Rijnders find, e.g., that with increasing pressure, the increase in HETP as a function of flow velocity becomes steeper with a concurrent shift of the minimum of these curves to lower flow velocities [43].

A more thorough discussion on the influence of the pressure on the HETP has been given by Bartmann [2]. With the special properties of the supercritical phase taken into account, the well-known Van Deemter equation was modified into a more simplified, generalized relation with a smaller number of variables:

$$\text{HETP} = 0.1\ \text{Re}\cdot d_p + \frac{1.3 d_p}{\text{Pe}} + \frac{1.5\ k/g^*\text{Pe}}{(1 + k')^2} \tag{1}$$

where

Re = Reynolds number based on average particle diameter

Pe (vd_p/D_m) = dimensionless Peclèt number for mass transfer in the mobile phase

v = mobile phase velocity

d_p = average particle diameter of column packing

k' = capacity factor

D_m = diffusivity

g* = ratio between volume of stationary phase and active surface area available for mass transfer

In the practical region of interest, the second term on the right is generally negligibly small relative to the first term; and Eq. (1) can be rewritten

$$\text{HETP} = \left[\frac{0.1\, d_p}{\text{Pr}} + \frac{1.5k'g^*}{(1 + k')^2}\right] \text{Pe} \tag{2}$$

where Pr is the dimensionless Prandtl number. The Reynolds, Prandtl, and Peclèt numbers are related to each other according to

$$\text{Pe} = \text{Re} \cdot \text{Pr} \tag{3}$$

Since k' will decrease with increasing pressure, it is necessary that Pe does not increase appreciably with increasing pressure. This is necessary to control the increase in the HETP. This is quite difficult to accomplish since the diffusion coefficient in the mobile phase is inversely proportional to the pressure of the system. For pressure-programmed operation, it is therefore imperative that the linear velocity at the outlet of the column be maintained at an approximately constant rate and not allowed to increase significantly in value.

HETPs as a function of the pressure for several capacity factors are shown in Figure 7. These graphs have been derived using Eq. (2) for a binary system of CO_2 and a compound exhibiting a phase behavior comparable to that observed for CO_2 and the lower-molecular-weight n-alkanes (Fig. 2). A value of 10^{-2} cm was chosen for g*.

The upper dotted line indicates the actual behavior in the column of a substance such as n-decane. Here the solubility of the solute in the mobile phase increases, and k' is observed to decrease when the pressure is

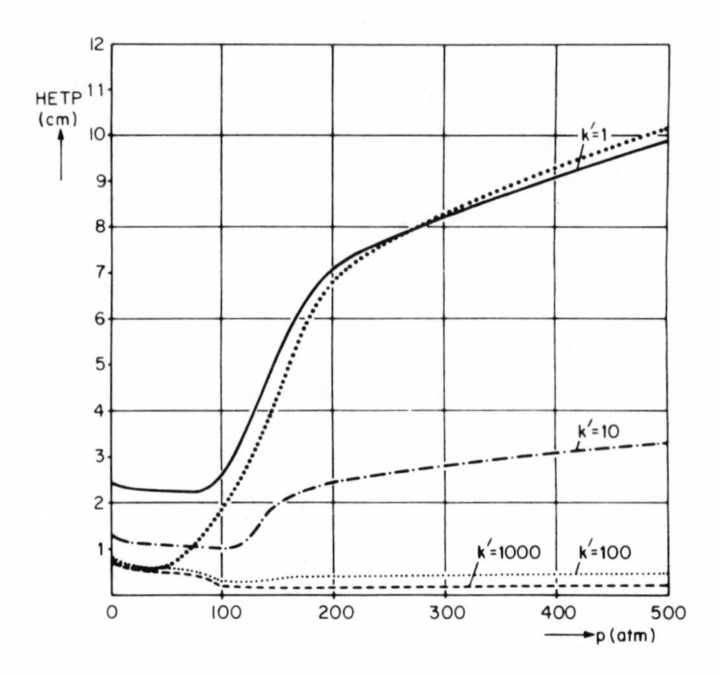

FIG. 7. HETP as a function of pressure. Upper dotted line constructed for the actual behavior of a substance such as n-decane, in which solubility of the solute in the mobile phase increases and k' is observed to decrease with increasing pressure. Effect of pressure is more extreme on compounds with low partition coefficient. (From Ref. 2.)

increased in a range encompassing several orders of magnitude. The pressure drop over the column in question is actually quite small and has been disregarded. Although this diagram has been derived for a special case, it is noteworthy for its general implication on the change in the HETP with change in the operating pressure.

1. Pressure Drop

An equally important consideration in supercritical fluid chromatography is the pressure drop over the chromatographic column. Because of the lower viscosity of the mobile phase, a lower pressure drop can be expected compared with what is observed in liquid chromatography. This is true under normal operating conditions. Fig. 8 shows the pressure drop over a 1.5-m by a 2-mm column with n-pentane as the mobile phase [30]. The three top curves have been obtained with Corasil I (37-50 μ m) as the

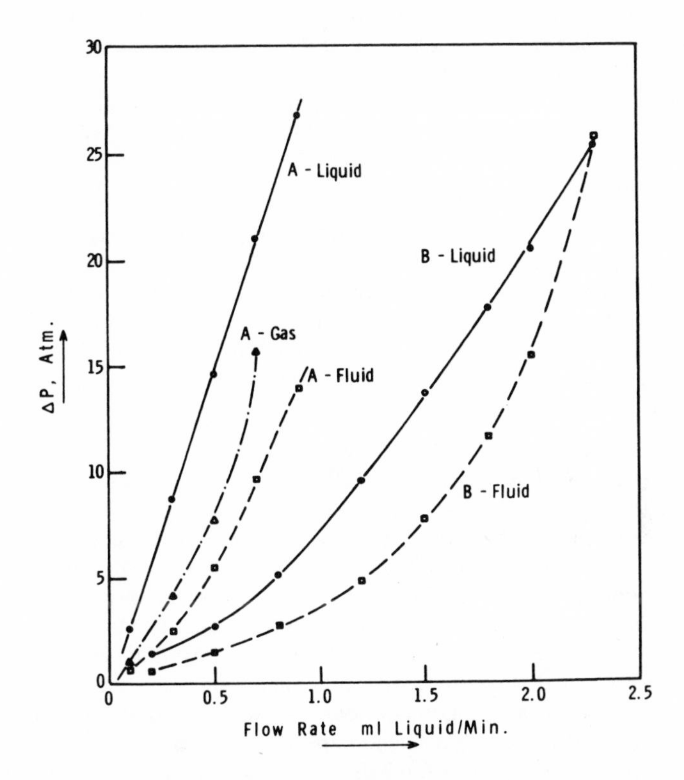

FIG. 8. Pressure drop as a function of the flow rate. Based on data from Novotny et al. Data have been obtained on a 1.5-m by 2-mm column packed with 37-50 μm Corasil I (A) and packed with 100-125 μm Porasil C (B). Mobile phase is n-pentane. (From Ref. 30.)

column packing. These are the curves that are labeled "A." The two lower curves labeled "B" were obtained on the same-sized column packed with Porasil C, which has a much larger average particle diameter (100-125 μm). Since pressure drops are roughly inversely proportional to particle size, the curves for Column B are displaced to the right relative to the comparable curves for Column A. The measurements with liquid pentane as the mobile phase have been carried out at room temperature. The data observed for the mobile phase in the gaseous state were obtained at 27.2 atm and 220° C. So that pentane could be maintained in the supercritical state, the column was held at 40.8-atm pressure and 220° C.

Except at the higher velocities, where incipient turbulence can be expected to occur, the pressure drop in supercritical fluid chromatography is generally about one-half to one-third the drop observed in liquid chromatography for the same mobile-phase velocity. At high-flow velocities, the pressure drop becomes approximately the same for liquid and for supercritical chromatography [8, 30]. The pressure drop for a gas as the mobile phase is, in general, higher than the pressure drop observed if a supercritical fluid is used. In the case of the gas, the excessive pressure drop is due to the very high gas velocities needed to obtain equivalent mass flow rates.

For the same linear velocities, the pressure drop in supercritical fluid chromatography remains essentially constant with increasing inlet pressure. The percentage pressure drop therefore decreases with increasing pressure. Comparable measurements with the mobile phase as a gas and as a liquid show an increase in pressure drop with increasing inlet pressure; one actually observes an increase in the percentage pressure drop with an increase in pressure [8, 9, 30].

Pressure drops in supercritical chromatography are important from the physical point of view. Because of the strong dependence of the solubility of the solute on the system pressure, partition coefficients can increase significantly as the solute travels through the column. In extreme cases, this phenomenon will result in an appreciable deterioration in the quality of the obtained chromatogram. The increase in partition coefficients results in a "holding back" effect for the solute, which is the opposite to what is observed in temperature programmed gas chromatography or pressure-programmed supercritical fluid chromatography. The result is an increase in band broadening and a decrease in resolution. In addition, there is also the possibility that phase separation occurs with decreasing pressure. This can be derived from Figures 2 and 3 as the case for which an isothermal decrease in pressure would lead to an intersection with the critical locus curve.

The effect of pressure is less pronounced at larger distances from the critical point; and in some cases, an increase in temperature may be advisable to counteract the effect of excessive pressure drop. It is therefore often desirable to operate at lower flow velocities to decrease the magnitude of the observed pressure drop.

2. Pressure Programming

Since partition coefficients can be readily modified by changing the pressure of the system, pressure programming is an indispensable tool for chromatographing a mixture of compounds present in a wide molecular-weight range. By increasing the pressure during the chromatographic separation process, one can increase both the flow rate and the solubility of the solute in the mobile phase so that high-molecular-weight compounds can be chromatographed in much shorter times without loss of resolution for lower-molecular-weight components in the sample.

Under isothermal conditions, the relative change in partition coefficients as a function of the pressure is about the same for compounds of a given molecular type. The effect of pressure is therefore quite predictable.

An example [18] of the power of pressure-programmed supercritical fluid chromatography is shown in Figure 9. This chromatogram was obtained on a polystyrene mixture with a nominal molecular weight of 900. This sample had been prepared by anionic polymerization and obtained commerically (Pressure Chemical Company, Pittsburgh, Pennsylvania). The measured viscosity average molecular weight was given as $1220 \pm 7\%$. The separation was carried out on a 4-m by 1/8-in. OD column packed with 120-150 mesh Porasil C, to which normal octane groups are bonded. The temperature was maintained at 215 °C; the column inlet pressure was programmed from 650 to 1000 psi. The mobile phase was 5% methanol in n-pentane. Detection was carried out at 260 nm in a UV absorption flow-through cell after the pressure in the system was decompressed to atmospheric levels. Each peak in the chromatogram corresponds to one polystyrene oligomer, whose general formula is shown in the figure and whose molecular weight is given by

$$MW = 104n + 58 \tag{4}$$

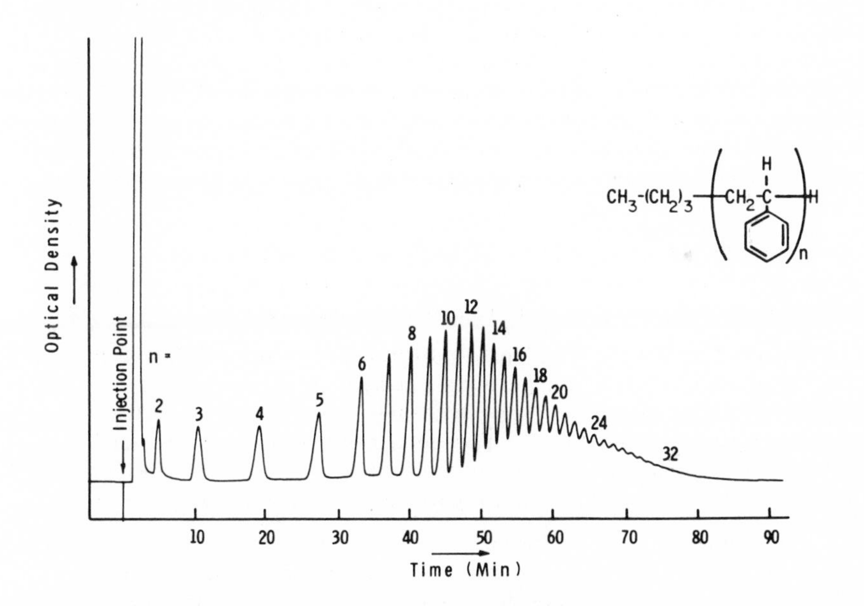

FIG. 9. Chromatogram of polystyrene oligomers obtained by pressure-programmed supercritical fluid chromatography. Molecular-weight range of this sample is from 266 to about 3300. (From Ref. 18. Courtesy of Preston Technical Abstracts Company.)

where n is an integer number denoting the number of styrene molecules in the compound. This chromatogram shows, therefore, a molecular-weight range from 266 to about 3300.

Another example [2] of a pressure-programmed operation is shown in Figure 10. This chromatogram was obtained on a synthetic mixture of oxygen-containing compounds on a 3-m by 1.7-mm ID column packed with 100-120 mesh Porasil C-Carbowax 400. Carbon dioxide was used as the mobile phase. The separation was carried out isothermally at 40 °C with the inlet pressure programmed from 55 to 117 atm. The flow rate at the outlet of the column was maintained at 1.0 liter/min (STP).

D. Temperature

Because pressure is important in supercritical fluid chromatography, there may be a tendency to deemphasize the role of temperature as an operational parameter. Temperatures can, however, have a pronounced effect on separation. In the region close to the critical point, a change of a few degrees can result in a change in the observed partition ratios by a factor of 2 or more.

In gas chromatography and in the temperature region below the critical point, increasing the temperature of the system results in a shorter elution time. In the region close to and above the critical point, partition coefficients are increased with an increase in temperature. This effect is thus the opposite to what is observed in gas chromatography. Negative temperature programming can theoretically be used to achieve the same effect as positive temperature programming in gas chromatography.

Sie and Rijnders have observed, however, that the lines denoting partition coefficients as a function of the temperature often intersect each other, even for compounds from a single molecular type [44]. This makes the results of negative temperature programming somewhat unpredictable.

Close to the critical point, partition ratios increase with increasing temperature. At higher temperature, partition ratios can remain constant, increase slightly, or even decrease because of increased volatility considerations. The maxima, if present, occur at different temperatures for different compounds. The position of these maxima also depends on the eluent and to a minor extent on the properties of the stationary phase.

Temperatures can therefore be used judiciously to increase the resolution between two compounds of interest. Selectivity can be influenced by modifying the temperature of the column as mentioned earlier in the separation of 1,1'-binaphthyl and anthracene on an alumina column. For the analysis of mixtures with a wide boiling range, pressure programming is therefore preferred because it has a more predictable effect on the separation than changing the temperature during the run.

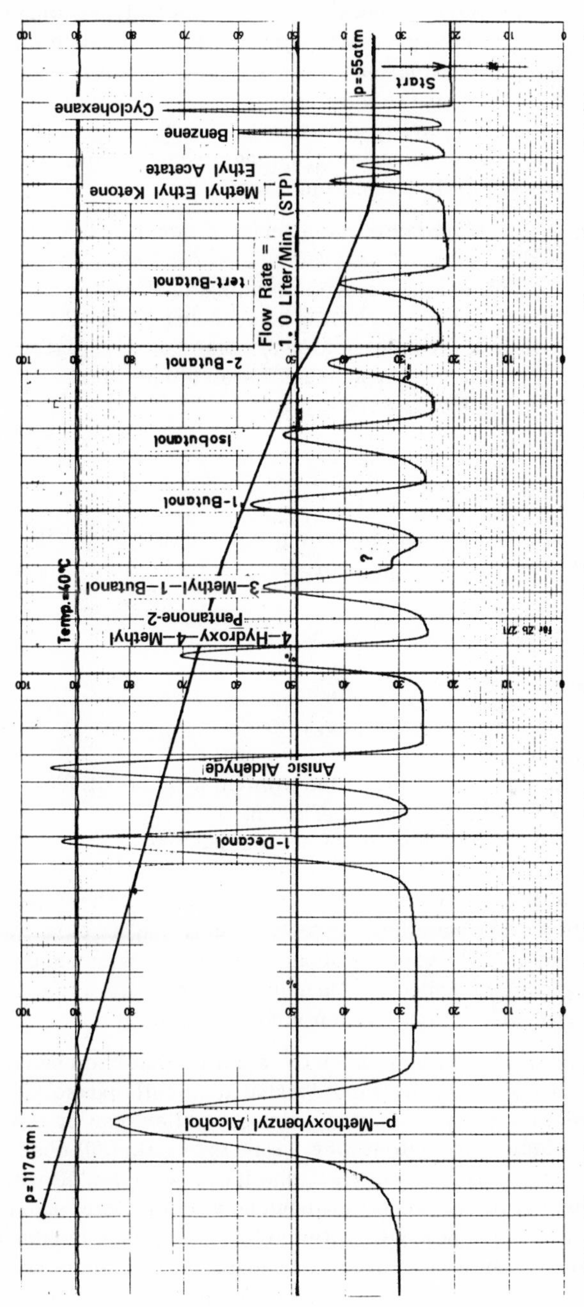

FIG. 10. Chromatogram of some oxygenated compounds. Column: 3 m by 1.7-mm ID Carbowax 400/Porasil C, 100-120 mesh. Mobile phase is CO_2 at 40° C. Sample size is 1.5 μl. Pressure programmed from 55-115 atm. Flow rate at the outlet of the column adjusted at 1.0 liter/min. (From Ref. 2. Courtesy Dr. D. Bartmann and Dr. G. M. Schneider.)

For compounds of the same molecular weight, operational temperatures in supercritical fluid chromatography are much lower than such temperatures observed in gas chromatography. This is of particular interest for the analysis of high-molecular-weight compounds and thermally labile materials. The analysis of oxygen-containing compounds shown in Figure 10 indicates the feasibility of chromatographing a wide variety of thermally labile compounds at essentially room temperature.

An interesting aspect of the chromatography of compounds of very high molecular weight is that in high-temperature gas-solid chromatography on alumina, peaks are often skewed even when the substrate is specially treated to reduce active sites. In contrast, compounds of still higher molecular weight will yield symmetrical peaks when chromatographed by supercritical fluid chromatography at much lower temperatures, even with untreated alumina as the substrate.

E. Flow Rate

Giddings [12] observed that the minimum in the curve of the HETP-vs-linear mobile phase velocity can be approximated by the following relation:

$$v_{opt} = \frac{D_m}{d_p} \tag{5}$$

where

v_{opt} = flow rate corresponding to the minimum in the HETP-v curve
D_m = diffusivity, or diffusion, coefficient
d_p = average particle diameter of column packing

For a solute with a diffusivity about 10^{-3} cm^2/sec and a column packed with particles with an average diameter of 10^{-2} cm, v_{opt} is calculated to be about 10^{-1} cm/sec. This value is about 10^2 smaller compared with V_{opt} for gas chromatography and about 10^2 larger compared with the value derived for a system with a liquid as the mobile phase.

Since the viscosity of a fluid is about 10^2 smaller than the viscosity of a liquid, pressure drops at the same "reduced" mobile phase velocities [12] and particle diameters are about 10^2 times smaller than pressure drops observed in liquid chromatography. Novotny et al. [30] observed that under comparable conditions of flow, the increase in HETP with flow rate of the test compound chrysene in n-pentane as the mobile phase is much less pronounced in supercritical fluid chromatography compared with the curves shown for liquid chromatography. Experiments with Porasil C (80-100 mesh) and Porasil C (120-150 mesh) show that HETPs in supercritical fluid chromatography remain practically constant about 1.5-2 mm for a range of flow rates up to about 3.0 ml/min. The column diameter in these experiments is 2.0 mm. With liquid pentane at room temperature as the mobile phase, HETPs increase sharply from about 2 mm for very low flow rates to about 10-15 mm at flow rates of 2.0-2.5 ml/min. An interesting observation is that for the smaller-particle-size Corasil I packing (50 μm), the HETPs in both supercritical fluid chromatography and liquid chromatography are almost comparable at about 2-3 mm for flow rates up to 1.0 ml/min.

Figure 8 shows that for a column packed with particles with a diameter of about 100 μm, flow rates should be kept below 2 ml/min for a 2-mm diameter column. Assuming the average densities of liquid and supercritical pentane to be about 0.7 and 0.23 g/ml, respectively, and the void volume about 70%, the computed maximum linear velocity at these conditions is about 4.5 cm/sec, or about 50 times higher than the optimum velocity calculated from Eq. (5). Practical flow rates [18, 35] in supercritical fluid chromatography are actually about a few cm/sec.

Measurements [43] with CO_2 as the mobile phase on the dependence of the HETP on v for various particle diameters show an almost linear increase of the HETP from 2 to 20 mm for linear flow velocities from 0.5 to 15 cm/sec for 50-70 mesh particles. For smaller diameter packings, the slope is less steep. For 120-200 mesh packing, the HETP increase in this range is only from 2 to 4.5 mm.

IV. INSTRUMENTATION

Because of the great degree of similarity between this technique and modern high-pressure liquid chromatography, many instrumental aspects are well covered in the most recent publications on liquid chromatography [7, 16, 25]. A schematic diagram of an assembly for supercritical fluid chromatography is shown in Figure 11. The mobile phase is compressed and brought to the desired pressure in a high-pressure pumping assembly. The output pressure can be adjusted and programmed. A pressure gauge, typically a Bourdon tube gauge, is used to measure and display the column inlet pressure. A preheater or a preconditioning coil or both raise the temperature of the mobile phase to the desired level. Injection is carried out through an injection system, which is usually a septum or a separate bypass assembly. After the sample is fractionated on the chromatographic column, the components are detected in a detector and then collected in a special sample collection assembly. Two options are shown. For eluents that are liquid at ambient conditions, the pressure is reduced to atmospheric levels over a short piece of capillary tubing and a micro-regulating valve. Detection and sample collection are carried out at atmospheric

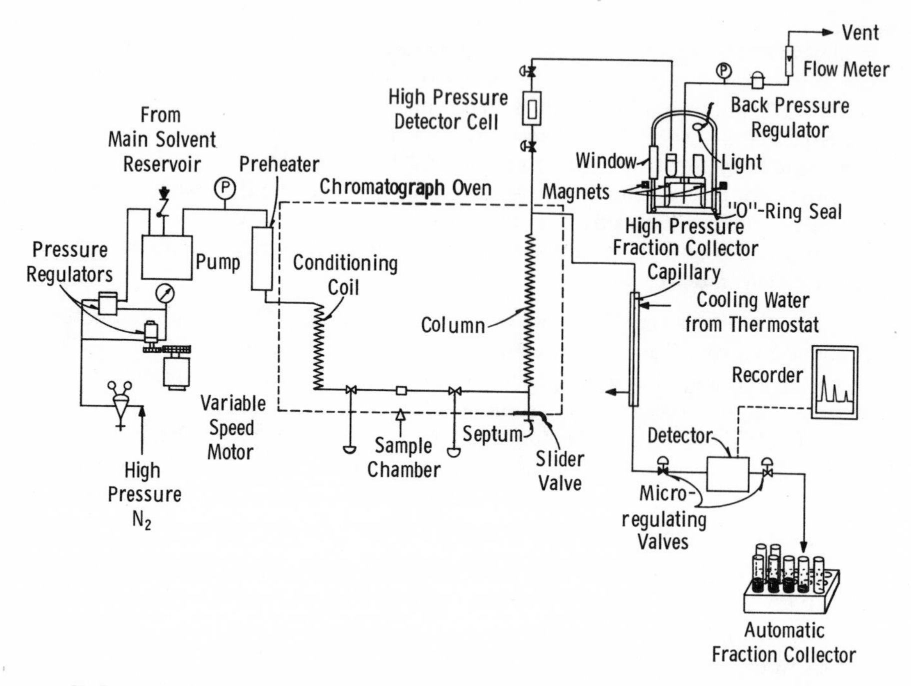

FIG. 11. Diagram for an apparatus for supercritical fluid chromatography.

pressure at room temperatures. For eluents such as CO_2 and N_2O, which are gaseous under these conditions, detection and sample collection are carried out under pressure. The eluents are generally not recycled in this technique. A recorder is used to record the output of the detector as a function of time.

A. Pumping System

Pumps and injectors in modern liquid chromatography have recently been the subject of an excellent survey [7]. In supercritical fluid chromatography, it is necessary to have a pumping system with pressure programming capabilities. In this case, flow becomes a dependent variable and has to be controlled by adjusting the pressure drop over the column, e.g., with a throttling valve at the exit of the column. In addition, it is desirable to have a pump with a pulseless output and a pressure output capability of at least 400 atm.

Three pump types are in general use today, viz., the syringe-type pump, the gas-driven pressure transfer pump, and the reciprocating plunger, or diaphragm, pump. Large, high-pressure syringe-type plunger pumps for liquid chromatography are currently available from many sources. With some modifications, these pumps can be adapted for operation in supercritical fluid chromatography. To ensure uninterrupted flow for these runs, which necessitates more volume than what is available in one unit, two pumps operating synchronously can be used. These pumping systems are quite expensive, however.

Because of the requirements of careful pressure control, the most suitable pumps in supercritical fluid chromatography are those based on transferring pressure from a gas source. A specially constructed pumping system ideally suited for this purpose has been described recently [19]. It consists of two identical stainless steel compound cylinders mounted adjacent to each other and connected by a system of check valves and solenoid valves. Each cylinder acts as the barrel of a large syringe pump and consists of a 7/8-in. ID upper portion connected to a 1.5-in. ID lower portion. A double-headed piston with spring-loaded Teflon piston rings is used to transfer and intensify the pressure from the bottom cylinder to the mobile phase that is present in the upper pump cylinder. Nitrogen, which is introduced in the bottom portion of the cylinders, is used as the primary pressure source. By adjusting the cycling mode of these two cylinders, one can get a continuous, pulseless flow.

Output pressures are controlled by adjusting the nitrogen inlet pressures. This pump has been used to generate output pressures as high as 350 atm. A main advantage of this pumping system is that it can perform pressure-programming operations by simply pressure-programming the nitrogen inlet pressure. This is done conveniently by connecting a variable-speed motor to a system of pressure regulators [18].

Reciprocating piston pumps can be used for mobile phases that are liquid at room temperature, but a damping system is necessary to decrease the pressure and flow fluctuations created by the pump. A simple and practical approach is to use one or more reservoirs or coiled metal tubes, such as are found in Bourdon pressure gauges, and needle valve(s) between the pump and the chromatographic column. A pressure-regulating system is also necessary to adjust the column inlet pressure to the desired levels. Reciprocating pumps are available in a wide variety of output volumes and flow variations. The damping and pressure regulating system has to be designed more or less separately for each type of pump.

If the mobile phase is gaseous at ambient conditions, one can use a mechanical diaphragm compressor to compress the gas and then feed the condensed liquid through a pressure regulator to the chromatographic system [27, 28]. Another convenient approach is to have a bypass line with a regulating valve connecting the output of the pump to a reservoir at the inlet of the pump [2, 4]. An additional pressure drop is taken over a second regulating valve between the pump and the column inlet. Adjusting these two valves during a chromatographic analysis allows the output pressure to the column to be programmed in the desired mode.

B. Columns and Column Packing

Column sizes in supercritical fluid chromatography are similar to those used in high-resolution liquid chromatography. The most popular column diameters are in the range of a few millimeters. Column lengths are generally not longer than 3-4 m because of pressure drop considerations. Most columns are actually much shorter. If an increase in a particular resolution is desired, this can in many cases be accomplished more effectively by modifying the selectivity of the separation instead of increasing the length of the column. This is carried out by changing the mobile phase or the column packing or both. We have also indicated earlier that selsctivity can be influenced by changing the pressure or the temperature or both.

Almost all column packings currently used in liquid chromatography can also be used in supercritical fluid chromatography. In fluid-liquid chromatography, one should take the strong solubilizing power of the mobile phase into account; and unless the stationary liquid phase is bonded to the solid support, stationary liquid phases of very high molecular weight are desirable. From mass transport considerations, the use of these usually very viscous products is not such a deterrent as it is in gas chromatography. In supercritical fluid chromatography, a substantial portion of the mobile phase is dissolved in the stationary substrate. Under supercritical conditions, e.g., approximately 15 wt % CO_2 dissolves in squalane [42]. This phenomenon decreases the viscosity of the stationary phase and increases the solute diffusivity by at least one order of magnitude [2].

In the presence of an appreciable pressure drop over the column, one should consider the danger of rearrangement of the liquid phase by partial solubilization in the front portion and redisposition in the latter sections of the column. This results in an increase in capacity factors during the movement of the solute throughout the column. This is equivalent to the "holding back" phenomenon described earlier under pressure-drop considerations.

Because of these potential problems, one should actually consider only the chemically bonded phases as suitable column packing, where the organic substrate is obviously less apt to dissolve in a moving liquid phase. In addition, these column packings are produced in the form of microspheres in a very narrow mesh range. It is a well-known fact in fluid dynamics that the pressure drop over a column is decreased by reducing the particle size range and by "improving" the shape of the particles into the optimum spherical form. We have already pointed out the necessity of maintaining low pressure drops over the column; the use of these specialty column packings over "regular" packings is therefore preferred.

In the choice of a bonded phase column packing, one should also consider the stability of these packings. Some of the older phases are thermally not very stable and are decomposed by traces of moisture. These are the phases where the bonding is carried out through a Si-O-C bond.

The newer bonded phases are formed through the more stable Si-O-Si-C bond. These packings are available in a wide variety of organic phases bonded to superficially porous microbeads. These beads usually have a fluid-impermeable glass glass core of about 30-40 μm or a porous core of 5-10 μm in diameter.

When a polar substrate is desired, one should also consider carrying out fluid-solid chromatography with a solid substrate such as alumina or Porapak [47]. There is the added advantage that there is much less tendency to form skewed peaks than in gas-solid chromatography. A decrease in entropy is associated with the adsorption of a portion of the mobile phase on the stationary substrate. The presence of this stationary film of the mobile phase in a somewhat different physical form as the bulk eluent will, of course, affect the properties of the stationary phase. Highly active adsorptive sites are absent. In this case, we actually have something like fluid-fluid chromatography, which is totally different in concept from gas-solid chromatography. (See also Ref. 3.)

C. Injector

Injection in supercritical fluid chromatography can be carried out through a septum with a high-pressure syringe. To prevent leakage, it is necessary to use hard Viton septa and specially designed ridged plates or soft septa with a needle guide [7]. Spurious peaks are often observed, either due to the septum material or from previously injected samples. This effect can be reduced by using Teflon-coated septa or by having an additional slider valve to protect and easily change the septum during regular operation.

Instead of a septum, a system of two small O rings has been found to be more reliable [18, 19]. The syringe needle is first inserted through the O rings; a metal nut around these O rings is tightened to increase the pressure on the O rings to prevent leakage, after which the slider bar is positioned to allow the needle to be inserted into the injection chamber proper. After injection, the needle is pulled out halfway and the slider bar is again pushed back to close off the O rings from the mobile phase; the needle can then be pulled out. In this approach, injections against pressures up to 200 atm have been carried out regularly without any problems. Septumless injectors using a Teflon collar (and a sample loop) are now commerically available.

Special sampling valves have also been described that can position the sample in the stream [7, 43]. The advantage of these multiport valve injectors is that they can handle fairly large volumes under moderate pressure with satisfactory reproducibility. Another alternative is to deposit the sample in a separate chamber at ambient pressure; the chamber is then closed off, heated, and raised to the desired inlet pressure before a sequence of high-pressure valves are operated to introduce the vaporized sample into the mobile gas stream [2, 27].

As in all forms of chromatography, the sample should ideally be introduced as a plug dissolved in the mobile phase. In the presence of limited solubility, phase separation takes place. Indeed, many of the very broad bands observed in the chromatograms of higher-molecular-weight compounds can be ascribed to this phenomenon. The general tendency is to overcome this phenomenon by increasing the column temperature in gas chromatography, increasing the solvent power in liquid chromatography, or increasing the pressure in supercritical fluid chromatography. A substantial improvement can actually be obtained by modifying the conditions to increase the solubility of the sample in the <u>injection</u> chamber.

D. Preheater and Column Oven

Temperature control is important in supercritical fluid chromatography because of the variation in selectivity and partition coefficients with a change in temperature. This is especially important in the region very close to the critical point; and if operations are to be carried out in this region, very accurate temperature control will be necessary. The range of temperatures in general use will be from ambient to about 300°C.

For a mobile phase with relatively high critical temperatures, it is necessary to preheat the eluent above its critical temperature before it is allowed to enter the column proper. This can be carried out in a separate preheater or in a conditioning coil mounted in the column oven, or by both means. Both are probably desirable for a fluid with a critical temperature above 80° C. A conditioning coil only is sufficient for fluids such as CO_2 or N_2O.

E. Detectors

The detector in supercritical fluid chromatography is currently the weakest link in the system. The detectors in current use have found prior applications in gas or liquid chromatography. Since these systems generally operate at ambient pressures, the supercritical column effluent pressure has to be reduced to atmospheric conditions before detection can be carried out. Decompression can be carried out in a system of small-diameter capillaries and a pressure-reducing micrometering valve.

If the effluent is gaseous at ambient conditions, however, high-molecular-weight solutes tend to form large clusters or fine droplets during the decompression stage. Detection by the regular gas chromatography detectors becomes erratic and unpredictable. Giddings et al. report, e.g., an unsuccessful attempt to use the flame ionization detector when CO_2 is used as the supercritical mobile phase [27]. Bartmann [2, 3, 4] solved this problem by constructing a modified flame ionization detector in which the expansion associated with the decompression is carried out only a few millimeters below the nozzle of the flame. Under these conditions, the demixing effects because of phase separation of the solute and the mobile phase are negligibly small. A high-pressure stream splitter is used to introduce only a small fraction of the column effluent to the detector. Both the detector and the expansion valve are heated to prevent condensation.

Another way is to subject the effluent to pyrolysis conditions at the outlet of the chromatographic column. In this approach, the macromolecules are reduced to smaller and hence more volatile species, which will form true solutions with the gaseous phase after decompression. Limited data by Myers and Giddings suggest that this approach can be quite useful for many compounds [27].

With a mobile phase that is liquid at ambient conditions, we may have the problem of dissolved gases and part of the mobile phase flashing off during decompression. In this process, bubbles are created that interfere with detection. Some solutes may even precipitate during this stage. A gas removal section, ahead of the detector, solves this problem but introduces additional band spread, especially if additional solvent has to be added to redissolve any precipitates. However, once the effluent is obtained as a homogeneous solution at atmospheric pressure, detection should pose no particular problem.

If the mobile phase has a relatively low vapor pressure at ambient conditions, such as with n-pentane, problems are usually not observed when the pressure is reduced. The use of these mobile phases is, however, limited to solutes that are stable at the high temperature necessary for the eluent to become supercritical. With these mobile phases, almost any detector that has been successfully used in high-resolution liquid chromatography should suffice. A microadsorption detector has, for example, been reported as having been used successfully as the detector in a supercritical fluid chromatographic system [8, 9].

Most of the work by this technique involves an ultraviolet absorption detector. Even though its applicability is limited, this detector can be used in a surprisingly large number of applications. Supercritical fluid chromatography is, as we have noted before, especially applicable to analyzing larger molecules. These compounds very probably possess UV absorbing chromophoric group in their molecular structures.

The absorptivity varies from compound to compound. In the 200-300-nm range, it is as low as 10 for aliphatic olefinic systems and as high as 10^5 for polynuclear aromatic hydrocarbons. The mobile phases should be transparent in the wavelength of interest. Generally, this is not a problem since most eluents are of sufficiently low molecular weight and are nonabsorbing to wavelengths as low as 220 nm. The UV absorption detector is very sensitive and also easy to use. Almost any UV recording spectrophotometer can be modified to perform as a detector, and fairly many inexpensive units are now commercially available [1, 51].

Detection is carried out in tiny flowthrough cells. Detector cells with 10-mm pathlengths and cell volumes as low as 10-20 μl are quite common.

An alternative to detection after decompression is to construct a high-pressure detector cell that is mounted before the decompression section of the chromatographic system. The detector analyzes the effluent with the mobile phase still under high pressure. A UV detector cell that has been used successfully up to 300 atm has been reported [19].

Another advantage of these detectors is that they can be used to identify the compounds eluting out of the column. The flow can be interrupted for a short period while the UV absorption spectrum is scanned in the wavelength region of interest. It is therefore possible with this technique to obtain both qualitative and semiquantitative information on certain components in the sample mixture.

F. Sample Collection and Recovery

The collection and the recovery of solutes can be accomplished in much better yields in supercritical fluid chromatography than in gas chromatography. The problems of trapping high-molecular-weight solutes from a hot gas chromatographic column outlet are well known. Quantitative recoveries are very difficult to achieve because of a tendency for these materials to form a fine fog that cannot be easily condensed or trapped quantitatively. There are at least 100 different papers on trapping solutes from gas chromatographic effluents, which attest to the difficulty of the problem involved.

These problems are obviously absent in liquid chromatography. However, in this technique, higher-molecular-weight solvents are generally used; and complete solvent removal can constitute a somewhat onerous assignment, especially if the solute of interest is somewhat volatile or thermally not stable.

Solvent removal is, of course, also necessary in supercritical fluid chromatography if a mobile phase with a high critical temperature is used, such as pentane or methanol. However, removal of these solvents from high-molecular-weight solutes should not be too difficult. From a practical point of view, one would use low-boiling fluids as the mobile phase if the solute were somewhat volatile or thermally unstable.

Figure 12 shows a high-pressure fraction collector that can be used in those cases in which a low-boiling fluid such as CO_2 or N_2O is used as the mobile phase [19]. It is essentially a fraction collector mounted in a high-pressure container. The inside of the collector vessel can be viewed through the polycarbonate window. A light is mounted in this vessel. The turntable in the unit is magnetically coupled to a movable ring around the base of the unit. Several test vials of varying capacity can be placed in this unit.

The important principle in this sample collector is the possibility of adjusting the pressure in this steel vessel by a back-pressure regulator on the outlet of the system. This regulator is adjusted in such a way that the emerging mobile phase is still liquid at the outlet from the detector. The pressure should not be too low because the mobile phase will be decompressed to a gas and the solute would be dispersed inside the unit. If the bleed through the regulator is adjusted, the mobile phase can be evaporated without "bumping" from the test vials. The solutes are then left behind as residues in the test tubes.

V. CONCLUSIONS

Supercritical fluid chromatography has shown considerable promise as a rapid chromatographic tool for analyzing compounds of medium to high molecular weight. In many cases, exceptional results, which could not have been attained by other chromatographic techniques, have been observed. Although substantial improvements have been reported, the

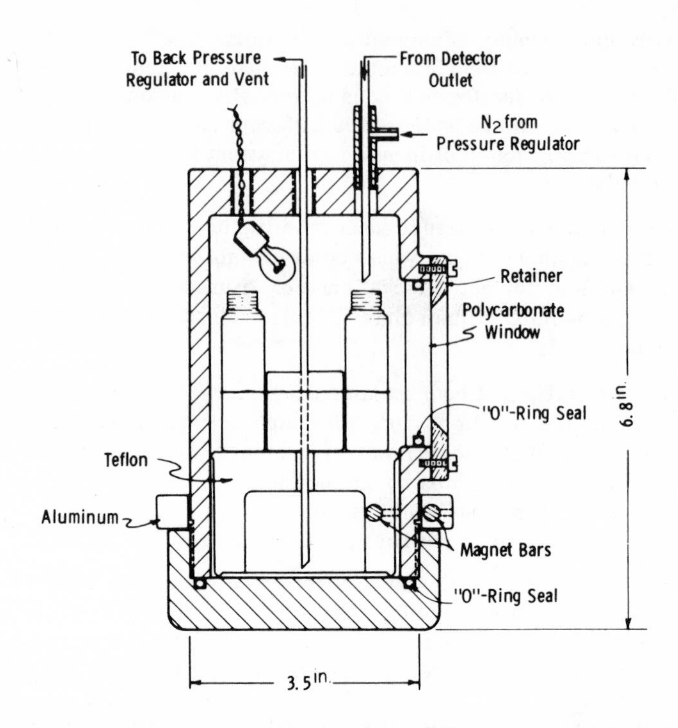

FIG. 12. High-pressure fraction collector for low-boiling eluents such as CO_2 and N_2O. The back-pressure regulator and vent are adjusted to allow eluent to emerge as a liquid from the detector outlet and to allow liquid to evaporate from the vials without bumping. (From Ref. 19. Courtesy American Chemical Society.)

detector section of the chromatographic system is still a problem that has not been solved adequately. The explosive growth of high-resolution liquid chromatography has overshadowed the efforts to develop this technique into a more practical chromatographic tool. On the other hand, many of the important advances in this technique are due to developments in the high-resolution liquid chromatography field. A prime example is the development of the thermally stable bonded stationary phases, which are indispensable for good operation in this technique.

Supercritical fluid chromatography will obviously not displace gas or liquid chromatography, except in some selected fields and in some special applications. Complete chromatographs are currently not yet commercially available. Instrument manufacturers will probably make the first models available as options to their existing lines of liquid chromatographs. In the final analysis, one instrument should be capable of performing both liquid and supercritical fluid chromatography.

Even though some time is necessary before the instrumental problems are solved and the dynamics of the chromatographic process are better understood, supercritical fluid chromatography has a definite place in the future as an important instrumental method of analysis.

ACKNOWLEDGMENTS

We acknowledge the courtesy of the American Association for the Advancement of Science for permission to publish Fig. 1, Verlag Chemie GmbH. for permission to reproduce Fig. 2, Elsevier Publishing Company and the American Chemical Society for permission to include Figs. 4 and 9, respectively, in this paper. We are especially indebted to Dr. D. Bartmann and Dr. G. M. Schneider for valuable information on the current status of their work, for advice, and for furnishing us with Figs. 2, 3, 7, and 10.

REFERENCES

1. S. R. Bakalyar, Amer. Lab. 3, 29 (1971).
2. D. Bartmann, Ph.D. thesis, Ruhruniversität Bochum, Germany, 1972.
3. D. Bartmann, Ber. Bunsen Ges. Physik. Chem. 76, 336 (1972).
4. D. Bartmann and G. M. Schneider, Chem.-Ing. Tech. 42, 702 (1970).
5. D. Bartmann and G. M. Schneider, First Internat. Symp. Column Liquid Chromatog., May 2-4, 1973, Interlaken, Switzerland. J. Chromat. 83, 135 (1973).
6. D. Bartmann and G. M. Schneider, Ber. Bunsen Ges. physik. Chem. 77, 220 (1973).
7. L. Berry and B. L. Karger, Anal. Chem. 45, 819A (1973).
8. J. L. Cashaw, Ph.D. thesis, University of Houston, 1970.
9. J. L. Cashaw, R. Segura, and A. Zlatkis, J. Chromat. Sci. 8, 1363 (1970).
10. J. J. Czubryt, M. N. Myers, and J. C. Giddings, J. Phys. Chem. 74, 4260 (1970).

11. G. A. M. Diepen and F. E. C. Scheffer, J. Am. Chem. Soc. 70, 4081 (1948).

12. J. C. Giddings, "Dynamics of Chromatography," part 1, Marcel Dekker, New York, (1965).

13. J. C. Giddings, M. N. Myers, and J. W. King, J. Chromat. Sci. 7, 276 (1968).

14. J. C. Giddings, M. N. Myers, L. McLaren, and R. A. Keller, Science 162, 67 (1968).

15. T. H. Gouw and R. E. Jentoft, J. Chromatog. 68, 303 (1972).

16. T. H. Gouw and R. E. Jentoft, in "Guide to Modern Methods of Instrumental Analysis," ed. T. H. Gouw, Wiley-Interscience, New York, 1972.

17. J. B. Hannay and J. Hogarth, Proc. Roy. Soc., London, 29, 324 (1879).

18. R. E. Jentoft and T. H. Gouw, J. Chromat. Sci. 8, 138 (1970).

19. R. E. Jentoft and T. H. Gouw, Anal. Chem. 44, 681 (1972).

20. R. E. Jentoft and T. H. Gouw, J. Polym. Sci., part B 7, 821 (1969).

21. N. M. Karayannis and A. H. Corwin, Anal. Biochem. 26, 34 (1968).

22. N. M. Karayannis and A. H. Corwin, J. Chromat. Sci. 8, 251 (1970).

23. N. M. Karayannis and A. H. Corwin, J. Chromatog. 47, 247 (1970).

24. N. M. Karayannis, A. H. Corwin, and E. W. Baker, E. Klesper, and J. A. Walter, Anal. Chem. 40, 1736 (1968).

25. J. J. Kirkland, ed., "Modern Practice of Liquid Chromatography," Wiley-Interscience, New York, 1971.

26. E. Klesper, A. H. Corwin, and D. A. Turner, J. Org. Chem. 27, 700 (1962).

27. M. N. Myers and J. C. Giddings, in "Progress in Separation and Purification," vol. 3, ed. E. S. Perry and C. J. van Oss, Wiley-Interscience, New York, 1970.

28. M. N. Myers and J. C. Giddings, Sepn. Sci. 1, 761 (1966).

29. C. J. van Nieuwenburg and P. M. van Zon, Rec. Trav. Chim. 54, 129 (1935).

30. M. Novotky, W. Bertsch, and A. Zlatkis, J. Chromatogr. 61, 17 (1971).

31. J. R. Partington, "Treatise on Physical Chemistry," vol. 1, Longmans, Green and Co., London, 1949.

32. J. M. Prausnitz, "Molecular Thermodynamics of Fluid-Phase Equilibria," Prentice-Hall, Englewood Cliffs, N. J., 1969.

33. J. M. Prausnitz, Adv. Chem. Eng. 7, 139 (1968).

34. J. E. Ricci, "The Phase Rule and Heterogeneous Equilibrium," Van Nostrand, Toronto, 1951.

35. G. W. A. Rijnders, "5th Intern. Symp. on Separation Methods: Column Chromatography," 1969, Chimia Suppl., p. 192.

36. G. W. A. Rijnders, Chem.-Ing.-Tech. 42, 290 (1970).

37. H. H. Reamer and B. H. Sage, J. Chem. Eng. Data 8, 508 (1963); 10, 49 (1965).

38. J. S. Rowlinson, "Liquid and Liquid Mixtures," 2nd ed., Butterworths, London, 1969.

39. J. S. Rowlinson and M. J. Richardson, "Advances in Chemical Physics," vol. 2, Interscience, New York, 1959.

40. G. M. Schneider, in "Advances in Chemical Physics," ed. J. Prigogine and S. A. Rice, 17, 1 (1970), Interscience, New York.

41. G. M. Schneider, A. Alwani, W. Heim, E. Horvath, and E. U. Franck, Chem.-Ing. Tech. 39, 649 (1967).

42. S. T. Sie, W. van Beersum, and G. W. A. Rijnders, Sepn. Sci. 1, 459 (1966).

43. S. T. Sie and G. W. A. Rijnders, Sepn. Sci. 2, 699 (1967).

44. S. T. Sie and G. W. A. Rijnders, Sepn. Sci. 2, 729 (1967).

45. S. T. Sie and G. W. A. Rijnders, Sepn. Sci. 2, 755 (1967).

46. S. T. Sie and G. W. A. Rijnders, Anal. Chim. Acta 38, 31 (1967).

47. S. T. Sie, J. P. A. Bleumer, and G. W. A. Rijnders, Proc. 7th Int. Symp. Gas Chromatog. and Its Exploitation, ed. C. L. A. Harbourn, Institute of Petroleum, London, 1969, p. 235.

48. Studiengesellschaft Kohle mbH., British Patent 1,057,911, February 8, 1967; British Patent 1,155,872, June 25, 1969.

49. Studiengesellschaft Kohle mbH., Austrian Patent Application 3085 (April 6, 1963), 6005 and 6006 (July 26, 1963), 9310 (November 20, 1963), and 10,203 (December 18, 1963).

50. J. Zernike, "Chemical Phase Theory," Kluwer, Deventer, 1955.

51. G. Zweig, Anal. Chem. 44, 47R (1972).

Chapter 2

GEL PERMEATION CHROMATOGRAPHY: A REVIEW OF AXIAL DISPERSION PHENOMENA, THEIR DETECTION, AND CORRECTION

Nils Friis and Archie Hamielec

Department of Chemical Engineering
McMaster University
Hamilton, Ontario

I. INTRODUCTION

Gel permeation chromatography (GPC) is an analytical or preparative technique in which solute molecules are separated according to their effective volumes in solution. The separation occurs as the solute molecules in the liquid carrier solvent (mobile phase) percolate through a porous bed (stationary phase), and it depends on steric exclusion of solute molecules of a certain size range from the pores of the bed packing, which itself has a pore-size distribution. Smaller solute molecules can permeate a larger fraction of the pores and thus elute later than larger solute molecules.

For several reasons, to be outlined, a chromatogram obtained via the GPC detector can never fully represent the distribution of molecular sizes of the solute in the sample injected. Instrumental spreading, or axial dispersion, of solute molecules of a single species results in elution of a single species occurring over a range of retention times (retention volumes), giving a distribution of retention times for each species of the solute. The chromatogram of the sample as measured by the detector is the superposition of these distributions. With small solute molecules or even oligomers, one would normally obtain a chromatogram involving many obvious but overlapping peaks. With a high polymer, the number of species is very large and the peaks of the individual species are not evident; one usually obtains a unimodal chromatogram, and sometimes a bimodal. Interpretation of a chromatogram must therefore account for this superposition and involves an evaluation of instrumental spreading (peak broadening) and correction of the detector response to obtain the true concentrations of the solute molecules.

Axial dispersion is a serious imperfection in the analysis of high polymers by gel permeation chromatography. GPC has found widespread use in polymer research and production, and today it is the most powerful analytical tool available to polymer chemists and engineers. For these reasons, much research has been concerned with the causes of axial dispersion and methods of GPC data correction for this effect.

Research carried out to date can be classified in three groups:

1. Studies of axial dispersion phenomena in terms of chromatographic theories.

2. Theoretical approach of correcting GPC data with numerical or analytical solutions of Tung's axial dispersion equation.

3. Empirical procedures of correcting GPC data for axial dispersion.

This paper proposes to review and evaluate the various approaches suggested to date. For convenience, the subject matter will be treated in accordance with the above classification. However, we shall first briefly consider the various sources of axial dispersion in the GPC instrument and the influence of axial dispersion on GPC resolution. Our discussion will be limited to the analysis of high polymers.

II. AXIAL DISPERSION PHENOMENA

A. Sources of Dispersion

Axial dispersion can conveniently be separated into two contributions, namely extracolumn dispersion and column dispersion. It is generally agreed that these two contributions are independent and additive.

Extracolumn dispersion is confined to dispersion stemming from a finite pulse input and occurring in the injection valve, tubing, and detector flow cell. Extracolumn dispersion is evaluated using the GPC instrument minus columns.

Column dispersion is composed of two independent factors, namely, interstitial dispersion, i.e., dispersion in the mobile phase, and pore dispersion, i.e., dispersion arising from permeation of the solute molecules into the pores of the gel. Main sources of interstitial dispersion are longitudinal molecular diffusion, eddy diffusion, and flow velocity variations caused by nonuniform packing associated with slow radial diffusion. The latter effect, which produces nonrectangular flow profiles may be important in GPC measurements of relatively viscous, high-molecular-weight polymer solutions, in which radial diffusion is slow due to small diffusion coefficients of the larger-polymer molecules. Eddy diffusion and longitudinal diffusion both usually cause symmetrical peak broadening, whereas flow variations can cause unsymmetrical broadening.

Extracolumn dispersion has been evaluated by several investigators [1-3], and although there is disagreement on the chief cause of extracolumn spreading (tubing or flow detector cell), it is generally concluded that in ordinary GPC measurements involving several columns, the contribution to peak spreading due to extracolumn dispersion is negligible, leaving column dispersion as the main source of axial dispersion.

The relative importance of interstitial-versus-pore dispersion depends on numerous parameters such as molecular weight of the solute, packing efficiency, viscosity of the solution, and pore structure and size. Several of these parameters are difficult to measure and control and it is therefore not possible to draw a general conclusion about which of the two factors is predominant. Although the majority of investigators [3, 4] adhere to the opinion that pore dispersion is the primary cause of peak broadening in GPC measurements with macromolecules that do not exceed the resolution limit, it is not certain whether this is true in general, since the conditions (packing efficiency, packing material) vary from one GPC to another. It may be meaningless to discuss the importance of pore dispersion versus interstitial dispersion without exact specification of the conditions.

The effect of different packing materials on peak broadening has been studied by Kelley and Billmeyer [5]. These investigators observed that peak broadening is significantly greater with Porasil packing than with Styragel. * Kelley and Billmeyer attribute this to the difference in pore structure of the two materials. There is experimental evidence [5, 6] that the pores are much deeper in Porasil than in Styragel. Furthermore, in Porasil the pores are interconnected througout the particle. Thus, once a molecule has entered a pore in Porasil, it may have a much longer diffusion path, with molecules held in the pores for a much longer time than in Styragel. This results in a larger axial dispersion with Porasil. The greater axial dispersion associated with Porasil does not necessarily imply, however, that this packing material gives poorer resolution than Styragel. The deeper pores can give increased peak separation. As shown below, this may compensate for the greater axial dispersion, since the resolution depends on axial dispersion as well as on peak separation with one effect counteracting the other.

In GPC measurements of macromolecular solutes too large to permeate the pores, the principal contribution to axial dispersion is interstitial dispersion. For low-molecular-weight solutes, on the other hand, axial dispersion is due mainly to pore dispersion. A large percentage of the pores are penetrated, while interstitial dispersion is small. Solutions of small molecules have low viscosities. Therefore, velocity profile defects are minor and high diffusion rates of the solute in the radial direction can easily correct these defects. Finally, for partially penetrating species, interstitial and pore dispersion will both contribute significantly to axial dispersion. Thus it appears that the effect of solute molecular weight on axial dispersion will be smaller than might have been expected. Using a reverse flow technique and thereby eliminating apparent axial dispersion due to peak separation, Tung and Runyon [4] measured axial dispersion as a function of molecular weight. In their study, Tung and Runyon observed that the variance of single-species chromatograms obtained with polystyrene standards reaches a maximum at a retention volume corresponding to a

*PORASIL and STYRAGEL are registered trademarks of Waters Associates, Inc., Milford, Mass. 01757.

molecular weight of 400,000. The variance for species too large to penetrate the pores of the packing was approximately the same as the variance for the species of low molecular weight. Hendrickson [3] made similar observations. Tung and Runyon observed furthermore that variances for single species of PVC and polybutadiene were the same as variances for polystyrene at the same retention volume, indicating a universal curve. This observation could be very important if this universal curve can be shown to be of general applicability. The variances of single species for other polymers could then be found using a curve constructed for polystyrene. At present there are too few data to prove its general applicability. However, it should be mentioned that Giddings and Mallik have predicted this universality using chromatography theory.

B. Factors Affecting Resolution

Resolution depends on peak separation and axial dispersion. For a quantitative discussion of resolution, we shall refer to the analytical solution of Tung's axial dispersion equation [8] after Hamielec and Ray [7]. For the special case of a Gaussian instrumental spreading function with constant variance (constant resolution factor) and a linear molecular-weight calibration curve, the analytical solution follows:

$$\frac{M_K(t)}{M_K(\infty)} = \exp\left[\frac{(3 - 2K)\, D_2^2}{4h}\right] \tag{1}$$

where

$K = 1, 2, 3, 4$ corresponds to number-, weight-, z-, and (z+1)-average molecular weights, respectively

$M_K(t)$ = Kth molecular-weight average corrected for axial dispersion

$M_K(\infty)$ = Kth molecular weight average uncorrected for axial dispersion

D_2 = slope of molecular weight calibration curve [$M(v) = D_1 \exp(-D_2 v)$ with D_1 and D_2 constant]

h = dispersion factor (originally called resolution factor by Tung [8]) equal to $1/2\mu$), where μ is the variance about the mean of a single-species chromatogram

$M(v)$ = molecular-weight calibration curve

As Hamielec [9] has shown, Eq. (1) can be the basis for a definition of a specific resolution factor $R_s(K, M_o)$. Perfect resolution is obtained when

$$\frac{M_K(t)}{M_K(\infty)} = 1$$

i.e., when

$$\frac{(3 - 2K)\, D_2^2}{4h} = 0$$

The specific resolution factor is therefore defined:

$$R_s(K, M_0) = \frac{(-1)^{K!}\, 4h}{(2K - 3) D_2^2} \tag{2}$$

The subscripts K, M_0 are used to emphasize the need for specifying the particular molecular-weight average and the molecular weight at the peak retention volume. As shown above, axial dispersion and therefore h vary somewhat with molecular weight. The fact that D_2, the slope of the calibration curve, may in general vary with molecular weight is consistent with the observation that a single column may give good peak separation at an intermediate molecular weight but may be entirely inadequate for higher molecular weights. At the high end and similarly at the low end of the molecular weight spectrum, D_2 increases very rapidly, giving little peak separation and therefore poor resolution even though the variance of single-species chromatograms is small. The appearance of K in the denominator of Eq. (2) emphasizes that imperfect resolution is more serious in correcting higher-molecular-weight averages.

From Eq. (2) it is apparent that to minimize the correction for imperfect resolution, one should choose a single column or column combination for which $D_2^2/4h$ is as small as possible. In other words, D_2, the slope of the molecular-weight calibration curve, should be as small as possible, and h, the dispersion factor, as large as possible (variance of single-species chromatogram as small as possible). This implies that peak broadening due to axial dispersion can be tolerated if peak separation is adequate, i.e., if the slope of the molecular-weight calibration curve is sufficiently small.

Equation (1) permits the resolution to be measured for the polymer in question with a once-through technique. The molecular-weight calibration curve must be available. The polymer sample is injected to find either $M_n(\infty)$ or $M_w(\infty)$. Having knowledge of its true number- or weight-average molecular weight (from osmometry or light scattering) is sufficient to evaluate $D_2^2/4h$ and also h with Eq. (1). Since h can vary somewhat with molecular weight and Eq. (1) is derived on the assumption of a constant h, the polymer sample should not have too broad a molecular-weight distribution, particularly when h is small. It should be mentioned that Bly [10] has derived an empirical expression for the specific resolution

$$R_s = \frac{2(V_2 - V_1)}{(W_1 + W_2)(\ln M_1 - \ln M_2)} \tag{3}$$

where V_1 and V_2 are retention volumes of species 1 and 2, and W_1 and W_2 the corresponding peak widths (width of the baseline of the curve between two tangents drawn at the points of inflexion and extended to the baseline). The symbols M_1 and M_2 represent the corresponding molecular weights.

As Hamielec [9] has shown, Bly's expression is consistent with Eqs. (1) and (2) in the limit of monodisperse standards and for values of K equal to 1 or 2. However, Eqs. (1) and (2) are more general as they apply to any molecular-weight average. Furthermore, since W_1 and W_2 would include the effect of peak separation on axial dispersion if used with standards that are not truly monodisperse, using Bly's expression with existing polymer standards would introduce an error in the calculation of the specific resolution R_s. This limitation does not exist with Eq. (1).

As is well known, it is possible to obtain increased resolution by increasing the number of columns in series or by operating the GPC in a recycle mode. On the basis of Eq. (1), Hamielec [9] has derived the following relation between the ratio of corrected to uncorrected molecular weight averages and the number of recycle passes n,

$$\frac{M_K(t)}{M_K(\infty)} = \exp\left[\frac{1}{n}\,\frac{(3 - 2K)\,D_2^2}{4h}\right] \tag{4}$$

where D and h are values for a single pass. This equation may also be used to investigate the effect of increasing the number of columns of the same kind in series. It thus appears that with increasing pathlength, peak separation grows faster than variance of single species and resolution therefore increases. The most obvious method of increasing resolution is therefore to increase the number of columns in series. Other more subtle effects on resolution will now be discussed.

C. Theories of Axial Dispersion

As with other chromatographic techniques, the height equivalent to a theoretical plate H can be used to characterize column efficiency in gel permeation chromatography of single species. Since the plate height can be related to the variance of the retention-time distribution curve, several workers [11-15] have applied chromatographic theories to derive models that can predict H in GPC. In general, H is expressed in terms of parameters such as column geometry (column length, column radius, and packing particle radius), flow rate, and solute molecular diffusivity.

Although it is often claimed that these models can account for peak broadening in GPC, it cannot be overemphasized, however, that they are of little or no quantitative use in correcting GPC data for axial dispersion. Predictions of shape or variance of single-species chromatograms are of considerable qualitative value at this stage in the development of chromatography theory. Because of these limitations and also because the plate height concept at times is very misleading, we shall refrain from further treatment of these models. It should be mentioned, however, that models that predict plate height are useful in other respects because together with the data on which they are based, they give us qualitative information on the variables affecting axial dispersion. This is very useful information in designing GPC instruments with minimized axial dispersion. (The reader is referred to the excellent review on theory of axial dispersion by Kelley and Billmeyer [16].)

Difficulties in predicting the retention-time distribution curve lie in obtaining a solution of the general chromatographic equation. Ouano and Barker [17] have recently circumvented this problem. With a high-speed digital computer, these investigators obtained numerical solutions of the chromatographic equations and were able to simulate qualitatively the shape of GPC chromatograms. In their model, Ouano and Barker incorporated the contributions to axial dispersion from pore dispersion, eddy diffusion, and longitudinal diffusion. Velocity profile effects were not taken into account, however, since this would have involved a complex multichannel model.

With their model, Ouano and Barker predicted a maximum in the variance-molecular weight relation, a result consistent with the findings of Tung and Runyon [4]. (Another interesting feature of the model is its prediction of skewed chromatograms for single species, showing that skewness increases with increasing flow rate. We shall refer to this observation in the subsequent section.)

Ouano and Barker also used their model to study the effect of pore-size distribution on resolution; they found that an optimum pore-size distribution may exist. Both very narrow and very broad pore-size distributions lead to poorer resolution, whereas optimum resolution is obtained with an intermediate pore-size distribution.

Although the model of Ouano and Barker can predict the shape of chromatograms and certainly is a significant contribution to our knowledge of variables affecting axial dispersion, it cannot, however, be used for quantitative correction of GPC data for axial dispersion. Because axial dispersion is strongly affected by parameters such as pore size and pore structure, which cannot be measured or controlled and hence appropriately incorporated in a chromatographic model, and also because the complex mechanism of velocity profile variations cannot be accounted for, Ouano and Barker's mathematical formulation must be considered a highly idealized, rather than a stringent phenomenological, description of the physical processes causing axial dispersion. The only practical alternative to determining chromatogram shape is to use experimental calibration procedures. (These will be discussed later in the chapter.)

III. CORRECTION OF GPC DATA FOR AXIAL DISPERSION: THEORETICAL APPROACH

To correct GPC data for axial dispersion by mathematical techniques, it is necessary first to establish a relation between the function F(v) representing the experimental chromatogram and the function W(v) representing the chromatogram of the same sample that would have been obtained in absence of axial dispersion.

For single-species chromatograms, the chromatogram height F(v) can be expressed as follows:

$$F(v) = A\ G(v) \tag{5}$$

where A is the area of the chromatogram and G(v) is the normalized instrumental spreading function of that species.

For a polydisperse high polymer with n species, F(v) is the sum of the height contributions of the individual species, i.e.,

$$F(v) = \sum_{i=1}^{n} A_i\ G_i\ (v) \tag{6}$$

where A_i and $G_i(v)$ represent area and normalized spreading function of species i, respectively.

The area A_i is proportional to the mass of species i. When the number of species is very large, A_i can be replaced with the continuous distribution function W(y), where y is the mean retention volume and W(y)dy the weight fraction of species i.

F(v) can thus be expressed:

$$F(v) = \int_0^\infty W(y)G(v, y)\, dy \tag{7}$$

where G(v, y) is the normalized spreading function of the component with mean retention volume y. This equation was first applied by Tung [8], and it is often referred to as Tung's axial dispersion equation.

The solution of Eq. (7) for W(y) involves two problems. First, it is necessary to choose an appropriate instrumental spreading function and to determine numerical values for its parameters. The second problem involves the choice of an appropriate numerical technique to solve Eq. (7).

Often the function G has been approximated by a Gaussian distribution, and in this case Eq. (7) takes the form

$$F(v) = \int_0^\infty W(y) \left(\frac{h}{\pi}\right)^{1/2} \exp\,[-h(v-y)^2]\, dy \tag{8}$$

where it is often assumed that chromatograms of single species all have the same dispersion factor h (uniform spreading function).

In his early work, Tung [8] applied different numerical techniques to solve the equation (8). One method involved the Gaussian quadrature formula and linear programming. The other used a polynomial expansion technique. These techniques and others [21] have been reviewed and comprehensively evaluated by Duerksen and Hamielec [18-20]. Several promising numerical techniques have recently been proposed by Tung [22], Chang and Huang [23], and Ishige and coworkers [24]. (These newer methods will be reviewed later in this section.)

The Gaussian spreading function is inadequate, however, when single-species chromatograms are skewed. Smith [25] proposed using a log-normal spreading function to take skewing effects into account. However, although this spreading function is better than the Gaussian when skewing is present, it is not of sufficient generality to account for variations in skewing with molecular weight. Pickett and coworkers [26] attempted to account for skewing by using a function obtained from the chromatogram shapes of standards with narrow MWD. In contrast to methods using Gaussian shapes, this technique overestimates skewing because the narrow MWD standards used are not truly monodisperse. Hess and Kratz [27] suggested using a spreading function predicted by the plug flow dispersion model. This model does not account for pore dispersion. It predicts skewed chromatograms of single species, but again is not sufficiently general and also leads to computation difficulties [18-20].

Recently, Provder and Rosen [28-29] suggested using a general statistical spreading function to account for skewed single-species chromatograms. This spreading function contains statistical coefficients that describe symmetrical axial dispersion, skewing, and flattening of single-species chromatograms. Earlier methods by Smith [25], Pickett and coworkers [26], and Hess and Kratz [27] have already been extensively reviewed [18-20]. We shall confine our discussions to the general instrumental spreading function Provder and Rosen proposed, since it is the most generally applicable one.

If a uniform instrumental spreading function can be assumed, Eq. (7) can be written

$$F(v) = \int_0^\infty W(y)\, G(v-y)\, dy \tag{9}$$

The variance and higher moments of the spreading function are independent of mean retention volume y. Equation (9) has the form of a convolution integral, and as shown by Hamielec and Ray [7], this form permits an analytical solution for the ratio of corrected to uncorrected moments and molecular-weight averages in terms of GPC parameters D_2 and h. In their solution, these workers used a Gaussian spreading function. However, the analytical procedure is general and can be applied with more complex spreading functions. Thus, Provder and Rosen [28-29] obtained an analytical solution, using the statistical shape function.

It should be emphasized that the method Hamielec and Ray suggested gives a solution for the corrected molecular-weight averages but not for W(y). However, if the GPC measurements are performed with a proper choice of column combinations and if the sample has a broad MWD, the corrections to MWD for axial dispersion are small enough to be negligible. The corrections to molecular-weight averages can be appreciable, however, because small corrections to MWD when integrated over a broad chromatogram can give large corrections to the average molecular weights.

In the subsequent discussions, we shall first consider analytical solutions of Eq. (9) obtained with different instrumental spreading functions, then the various methods of measuring parameters of the instrumental spreading function, and finally the more recent numerical solutions of the integral equation.

A. Analytical Solutions of the Integral Equation

If we assume that chain length is a continuous variable, then the molecular-weight averages can be expressed as follows:

$$\frac{M_K(t)}{M_K(\infty)} = \frac{\int_0^\infty W(v)\, M(v)^{K-1}\, dv \Big/ \int_0^\infty W(v)\, M(v)^{K-2}\, dv}{\int_0^\infty F(v)\, M(v)^{K-1}\, dv \Big/ \int_0^\infty F(v)\, M(v)^{K-2}\, dv} \tag{10}$$

where $K = 1, 2, 3, 4$ corresponds to number-, weight-, Z-, and (Z+1)-average molecular weights respectively. The values $W(v)$ and $F(v)$ are corrected and uncorrected normalized GPC chromatograms respectively and $M(v)$ is the true molecular-weight calibration curve.

For calibration curves that are linear in $\ell n\, M$ over the retention volume range of interest,

$$M(v) = D_1 \exp(-D_2 v) \tag{11}$$

where $D_1, D_2 > 0$ and constant. Combination of Eqs. (10) and (11) gives

$$\frac{M_K(t)}{M_K(\infty)} = \frac{\overline{W}((K-1)\, D_2)/\overline{W}((K-2)D_2)}{\overline{F}((K-1)\, D_2)/\overline{F}((K-2)D_2)} \tag{12}$$

where $\overline{W}$ and $\overline{F}$ denote Laplace transforms of W and F.

By applying the convolution theorem of the Laplace transforms to Eq. (9), we obtain

$$\overline{G}(s) = \frac{\overline{F}(s)}{\overline{W}(s)} \tag{13}$$

where $\overline{G}$ is the Laplace transform of G. Subsituting Eq. (13) into Eq. (12) yields

$$\frac{M_K(t)}{M_K(\infty)} = \frac{\overline{G}((K-2)\, D_2)}{\overline{G}((K-1)\, D_2)} \tag{14}$$

$$\frac{M_1(t)}{M_1(\infty)} = \overline{G}(-D_2) \tag{15}$$

$$\frac{M_2(t)}{M_2(\infty)} = \frac{1}{\overline{G}(D_2)} \tag{16}$$

and so on. This method is generally referred to as the "method of molecular-weight averages." With a Gaussian spreading function, Eq. (14) becomes

$$\frac{M_K(t)}{M_K(\infty)} = \exp\left[(3 - 2K)\left(\frac{D_2^2}{4h}\right)\right] \tag{17}$$

with

$$\frac{M_1(t)}{M_1(\infty)} = \frac{M_n(t)}{M_n(\infty)} = \exp\left(\frac{D_2^2}{4h}\right) \tag{18}$$

$$\frac{M_2(t)}{M_2(\infty)} = \frac{M_w(t)}{M_w(\infty)} = \exp\left(\frac{-D_2^2}{4h}\right) \tag{19}$$

Equation (17) permits the dispersion factor h to be measured with a once-through technique. A standard with known M_n or M_w is chromatographed to find either $M_n(\infty)$ or $M_w(\infty)$. A knowledge of the slope of the calibration curve is then sufficient to calculate h. Once h has been evaluated, the molecular-weight averages can be calculated from Eq. (17).

As mentioned earlier in this paper, Eq. (17) serves as a basis for the definition of a specific resolution factor. Equation (17) is also useful in evaluating the effect of column length or number of recycle passes on GPC resolution.

As Balke and Hamielec [30-31] pointed out, Eqs. (18) and (19) can furnish a criterion for skewing. If peak broadening is symmetrical, then

$$\frac{M_n(t)}{M_n(\infty)} = \frac{M_w(\infty)}{M_w(t)} \tag{20}$$

If skewing is present

$$\frac{M_n(t)}{M_n(\infty)} \neq \frac{M_w(\infty)}{M_w(t)} \tag{21}$$

To correct for skewing, Balke and Hamielec defined an empirical skewing factor SK in accordance with Eqs. (22) and (23):

$$\frac{M_n(t)}{M_n(\infty)} = (1 + \frac{1}{2}\mathrm{SK}) \quad \exp\left(\frac{D_2^2}{4h}\right) \tag{22}$$

$$\frac{M_w(t)}{M_w(\infty)} = (1 + \frac{1}{2}SK) \quad \exp\left(- \frac{D_2^2}{4h}\right) \tag{23}$$

The skewing factor SK and the dispersion factor h can both be evaluated by chromatographing standards of known M_n and M_w and applying Eqs. (22) and (23). Since the skewing factor may vary substantially with MWD breadth, it is recommended that Eqs. (22) and (23) be used only with samples that are similar in breadth to the standard used to evaluate SK. Because of concentration effects, samples with narrow MWD will probably have higher skewing factors than samples with broad MWD.

As Tung and Runyon [4] pointed out, narrow standards of high molecular weight, prepared by anionic polymerization, may contain small amounts of low-molecular-weight material. In measurement of M_n of such standards by osmometry, this low-molecular-weight fraction may not be detected because of the facile passage of low-molecular-weight polymer through the membrane. This will lead to an $M_n(t)$ that is too high and thus, because the GPC detects the low molecular tail, Eqs. (21), (22), and (23) will predict skewing due both to this low-molecular-weight tail and to the instrument. On the other hand, it is indisputable that skewing due to instrumental spreading does occur for high-molecular-weight species. This has been shown by Provder and Rosen [28-29]. These investigators compared skewing factors measured for narrow standards, prepared by anionic polymerization, with the skewing factor measured for an ultra-narrow recycle polystyrene standard, which should have no low-molecular-weight tail at all. The skewing factor of the ultranarrow standard was some 60-70% of the skewing factor observed with standards prepared by anionic polymerization, indicating that the apparent skewing due to low-molecular-weight material may account for 30-40% of the measured skewing. Also, the fact that the model of Ouano and Barker [17] predicts skewed chromatograms of single species indicates that skewing due to instrumental spreading can occur for high-molecular-weight species and is not only an apparent effect caused by the presence of low-molecular-weight material in the standards.

To account for deviation from the Gaussian shape, Provder and Rosen [28-29] proposed using a general statistical shape function to describe instrumental spreading. This function has the form

$$G(v) = \phi(v) + \sum_{n=3}^{\infty} (-1)^n \frac{A_n}{n!} \frac{\phi^n(v)}{(\sqrt{2h})^n} \tag{24}$$

where $\phi(v) = (h/\pi)^{1/2} \exp(-hv^2)$ and $\phi^n(v)$ denotes its nth-order derivatives.

The coefficients A_n are functions of μ_n, the nth-order moments about the mean retention volume μ_1 of the single-species chromatogram. The first two coefficients are of direct statistical significance and are related to the moments as

$$A_3 = \left(\frac{\mu_3}{\mu_2}\right)^{3/2} \tag{25}$$

$$A_4 = \left(\frac{\mu_4}{\mu_2^2}\right) - 3 \tag{26}$$

The variance μ_2 is related to the dispersion factor as follows:

$$\mu_2 = \frac{1}{2h} \tag{27}$$

When applying the method of "molecular-weight averages" and using the general-shape function to describe instrumental spreading, Provder and Rosen obtained the following solutions for corrected molecular-weight averages and intrinsic viscosity:

$$\frac{M_K(t)}{M_K(\infty)} = \exp\left\{\frac{-(2K-3)\ D_2^2}{4h}\right\}\left\{\frac{\left\{1 + \sum_{n=3}^{\infty} (A_n/n!)\ [-(K-2)D_2/\sqrt{2h}]^n\right\}}{\left\{1 + \sum_{n=3}^{\infty} (A_n/n!)\ [-(K-1)D_2/\sqrt{2h}]^n\right\}}\right. \tag{28}$$

$$\frac{[\eta]\ (t)}{[\eta]\ (\infty)} = \exp\left\{\frac{-a^2D_2^2}{4h}\right\}\frac{\{1\}}{\left\{1 + \sum_{n=3}^{\infty} (A_n/n!)\ [(-a\ D_2)/\sqrt{2h}\,]^n\right\}} \tag{29}$$

where a is the Mark-Houwink exponent.

Using the truncated form where $A_5 = 0$, $A_6 = 10A_3^2$, $A_7 = 0$, $A_8 = 0, \ldots$ and introducing the moments, one obtains the following equations for number- and weight-average molecular weights and intrinsic viscosity:

$$\frac{M_n(t)}{M_n(\infty)} = \exp\left\{\frac{D_2^2}{4h}\right\}\left\{1 + \frac{D_2^3\mu_3}{6} + \frac{D_2^4}{24}\left(\mu_4 - \frac{3}{4h^2}\right) + \frac{D_2^6\mu_3^2}{72}\right\} \tag{30}$$

$$\frac{M_w(t)}{M_w(\infty)} = \exp\left\{\frac{-D_2^2}{4h}\right\} \Big/ \left\{1 - \frac{D_2^3\mu_3}{6} + \frac{D_2^4}{24}\left(\mu_4 - \frac{3}{4h^2}\right) + \frac{D_2^6\mu_3^2}{72}\right\} \tag{31}$$

$$\frac{[\eta](t)}{[\eta](\infty)} = \exp\left\{-a^2\frac{D_2^2}{4h}\right\} \Big/ \left\{1 - \frac{a^3D_2^3\mu_3}{6} + \frac{a^4D_2^4}{24}\left(\mu_4 - \frac{3}{4h^2}\right) + \frac{a^6D_2^6\mu_3^2}{72}\right\} \tag{32}$$

Equation (30) and (31) are similar in form to the semiempirical correction equations (Eqs. (22) and (23) proposed by Balke and Hamielec [31]).

The use of calibration standards with known number- and weight-average molecular weights and intrinsic viscosity together with Eqs. (30), (31), and (32) permits the evaluation of h, μ_3, and μ_4. Provder and Rosen [28-29] have done this with polystyrene and poly(vinyl chloride) standards.

The coefficient A_3 or μ_3 provides a measure of skewness and is, as such, equivalent to the skewing factor SK in Eqs. (22) and (23). When A_3 is positive, the chromatogram is skewed towards higher retention volumes with a lowering of the number- and weight-average molecular weights. When A_3 is negative, the opposite is true. Finite values of A_4 give a symmetrical distribution but provide a measure of deviation from the Gaussian shape.

It should be mentioned that the equation of Provder and Rosen, like the equations of Balke and Hamielec (Eqs. (22) and (23), will predict apparent skewing if the number-average molecular weight of the standards is not properly evaluated.

B. Measurement of Parameters of Instrumental Spreading Function

To solve the dispersion equation for W(y), one must evaluate the parameters of the instrumental spreading function G. The number of parameters necessary depends on the complexity of the spreading function. Only one parameter, the variance (or dispersion factor), is necessary to define the Gaussian distribution. However, when skewing is significant, the spreading function cannot be approximated by a Gaussian distribution; in such a case, it may be necessary to evaluate two, three, or more parameters to define the spreading function.

As outlined in the preceding section, using the analytical solution of the dispersion equation permits the parameters of the spreading function to be measured with a once-through technique. Depending on the choice of the spreading function, the parameters can be evaluated from Eq. (17), Eqs. (22) and (23), or Eqs. (30), (31),and (32).

The success of this technique depends greatly on a proper evaluation of the number-average molecular weights of the standards. Narrow standards, prepared by anionic polymerization, should not be used with this technique since such standards may contain low-molecular-weight material that may lead to an erroneous determination of $M_n(t)$. Only ultranarrow standards obtained via GPC fractionation are useful in conjunction with this technique.

Tung, Moore, and Knight [32] suggested a reverse flow technique to evaluate the dispersion factor of a Gaussian spreading function. With this technique, a standard sample is allowed to flow through half of the packing length; the direction of flow is then reversed. Thereby, the size separation due to permeation is also reversed but instrumental spreading continues to broaden the peak. The resulting chromatogram therefore reflects only the instrumental spreading, and the dispersion factor can be determined from the observed chromatogram.

Tung and Runyon [4] used the reverse flow technique to evaluate the dispersion factor for polystyrene, PVC, and polybutadiene, and observed that the dispersion factors of all three polymers yielded data points that fell on a single curve when plotted against retention volume.

The reverse flow technique can be used, however, only when skewing is absent. Skewing is caused by concentration effects and is therefore also reversed when the direction of flow is reversed. Hence, a chromatogram obtained from a reverse flow experiment will always be less skewed than the corresponding "once-through" chromatogram.

C. Numerical Solutions of the Integral Equation

The earlier numerical methods of solving the integral equation [8, 21] are not completely satisfactory in all cases. For instance, when corrections for axial dispersion are appreciable, severe oscillations appear in the corrected molecular-weight distribution. Such oscillations are caused not only by detector noise but also by the mathematical technique involved. This is easily verified by applying the mathematical method in question to synthesized chromatograms in which detector noise is excluded.

There are several recent numerical techniques of solving the integral equation. These techniques are more efficient in terms of computer storage and time requirements, and some of them have been very successful in reducing the oscillation problems encountered in the earlier methods. For mathematical details of the various techniques, the reader is referred to the original papers. We concentrate here on the results of evaluations of the different numerical techniques.

The evaluation of a numerical technique must be based on certain criteria. Among the most important of these are that the method should give

1. good recovery of W(y) with narrow chromatograms

2. good recovery of W(y) with broad and narrow chromatograms when the dispersion factor is small ($h < 0.5$ counts $^{-2}$)

3. good correction of M_z, as well as of M_n and M_w

Additionally, the method should not require large computation time and computer storage.

The method of Chang and Huang [23] is one of the recent methods of solving the integral equation. These workers reformulated the integral equation into an equivalent variational problem of quadratic functional. The method of steepest descent in the function space was then applied to the minimization problem to obtain the true molecular-weight distribution.

As shown by Chang and Huang [23], this method gives excellent recoveries of W(y). Furthermore, it has the advantage of small computation time and storage. Ishige et al. [24] observed, however, in evaluating their method, that it resulted in a relatively poor correction of M_z compared with the iterative method that Ishige et al. [24] proposed.

The original method of Chang and Huang [23] was limited to symmetrical spreading functions. However, these workers later generalized their treatment to remove all restrictions on the shape of the spreading function [33]. At present, the method of Chang and Huang is considered one of the most successful numerical solutions of Tung's axial dispersion equation.

Ishige et al. [24] have proposed two iterative methods of solving the integral equation. In both methods, F(v) is used as the initial guess of W(y) and distributions $F(v)_1$, $F(v)_2, \ldots,$ $F(v)_i$ are consecutively calculated from the integral equation. In Method 1, the iteration procedure uses the difference between F(v) and $F(v)_i$ as the convergence criterion, and Method 2 uses the ratio of F(v) to $F(v)_i$.

Ishige et al. [24] performed a comprehensive evaluation of their methods and compared them with the method of Chang and Huang. Inconsistent oscillations in the tails of the molecular-weight distribution were observed with Method 1. However, with Method 2 excellent recoveries of W(y) were obtained both with narrow and broad chromatograms and with h less than 0.5 count $^{-2}$. Further, in most cases this method yielded a good correction to M_z. Both methods have the advantage of being applicable to any spreading function, and finally, the methods are simple and require small computation time and storage. Method 2 was recently evaluated by Kato and Hashimoto [34]; these investigators have confirmed the excellent recoveries obtained with the method.

It should be mentioned that a method similar to Method 1 of Ishige et al. [24] has been developed by Smit et al. [35]. Like Method 1 of Ishige, the method of Smit also produces inconsistent oscillations in the tails of the distributions. However, the oscillations are small, and when they are omitted in the final calculation, it is possible to obtain good recovery of both broad and narrow molecular-weight distributions. Although the method of Smit seems promising, it has not been evaluated as extensively and critically as the method of Ishige. For instance, this method has been evaluated only with values of h larger than 0.5 $count^{-2}$.

Tung [22] has presented two more methods of solving the integral equation. One method uses a Fourier analysis and can be used with unsymmetrical spreading functions. The other method uses a fourth-degree polynomial and is limited to Gaussian spreading functions. With both methods, the parameters of the instrumental spreading function may vary with retention volume.

Tung [22] evaluated and compared these methods with the method of Pierce and Armonas [21]. In cases with very poor resolution (h = 0.2 $count^{-2}$), both the polynomial method and the method of Pierce and Armonas failed to give good recovery of the molecular-weight distribution. However, the correction by Fourier analysis was very successful. When the limits for the inverse transform were doubled, excellent recovery was obtained with this method, even in cases with poor resolution. In cases with higher resolution (h = 0.4), good recovery was also obtained with the polynomial method, whereas this was not the case with the method of Pierce and Armonas.

A critical evaluation of Tung's Fourier analysis method was performed by Kato and Hashimoto [34]; they confirmed that this method works well with distributions of intermediate breadth. However, with narrow distributions, Tung's method resulted in severe oscillations in the tails and the recovered molecular-weight distribution differed significantly from the true distribution. Method 2 of Ishige [24] was tested under the same conditions and was shown to give excellent recovery, even with very narrow distributions and small dispersion factors.

Rosen and Provder [36, 37] applied a Fourier transform method using the statistical shape function to solve the dispersion equation for W(y). According to the evaluation Rosen and Provder performed, this method seems to work well with distributions of intermediate breadth, when the corrections are relatively small. However, the method has not been tested with narrow distributions and large corrections and therefore cannot be recommended without further evaluation. In comparison with other methods, Rosen and Provder's method involves relatively complicated mathematics and thus has the disadvantage of complex programming and large computation time and storage.

In conclusion, we may say that when the corrections are small and the distribution is not narrow, any method mentioned in this section can satisfactorily solve the integral equation for W(y). However, with large corrections and narrow distributions, Method 2 of Ishige et al. and the method of Chang and Huang, seem to be the most successful. These methods give good recovery of narrow distributions, they require small computation time and storage, and they can be applied with any shape function.

D. Summary of Theoretical Approach

In the preceding sections we have examined the various theoretical approaches for correcting GPC data for axial dispersion. We have not delved into the mathematical details of each method. Rather, we have sought to elucidate the problems involved in using a theoretical approach for the correction of GPC chromatograms for axial dispersion.

To obtain the corrected molecular-weight distribution and averages using any of the aforementioned methods, one must establish the proper instrumental spreading function, determine by calibration numerical values for its parameters, and then solve Tung's axial dispersion equation numerically. In special circumstances an analytical solution can be used.

The mathematical approach of correcting GPC chromatograms for axial dispersion is time-consuming and under certain conditions there may be serious limitations due to skewing of single-species chromatograms. For routine correction of chromatograms, it is recommended that more direct empirical approaches be used.

IV. CORRECTION OF GPC DATA FOR AXIAL DISPERSION: EMPIRICAL APPROACH

There are several empirical approaches to correcting GPC chromatograms. The empirical approach involves finding an effective molecular-weight calibration curve that when used directly with the uncorrected chromatogram F(v) automatically corrects the MWD and molecular weight averages for axial dispersion, as if it were the corrected detector response. When axial dispersion is negligible, the effective and true molecular weight calibration curves are the same.

In contrast to conventional GPC techniques, in which the molecular-weight calibration curve is obtained using a series of narrow standards and in which corrections are either omitted or accomplished by solving the integral equation, the empirical approach uses one or more broad MWD standards to obtain the so-called effective calibration curve, and all necessary corrections are automatically accomplished when this calibration curve is used.

Cantow, Porter, and Johnson [38] were the first to suggest using broad standards for GPC calibration. These investigators obtained a calibration curve for polyisobutene from the known molecular-weight distribution of a single, broad polyisobutene standard. The molecular-weight distribution of the standard was determined by column fractionation [39]. This procedure is rather tedious, and in the subsequent sections we shall show that for many polymers, broad standards with known molecular-weight distributions can be synthesized according to classical kinetic models for chain and condensation polymerization kinetics. Further, we shall outline techniques in which knowledge of two molecular-weight averages is sufficient to construct an effective calibration curve.

The methods used with linear and branched polymers are different and will therefore be treated separately. Corrections for axial dispersion in analyzing copolymers and terpolymers by GPC has not been very well developed and therefore these polymer systems will be treated only briefly at the end of our discussion.

A. Linear Polymers

At least three different techniques are now available to construct an effective molecular-weight calibration curve from broad standards. One technique uses one broad MWD standard with known MWD, from which the calibration curve can be obtained directly by a simple graphical procedure. This technique is considered the most accurate and most easily applicable. A second technique uses the universal calibration curve based on polystyrene together with one broad standard with known M_n and M_w. The effective calibration curve is obtained using a single-variable search. A third technique uses one broad standard with known M_n and M_w together with the assumption that the effective calibration curve is linear. The calibration curve is again obtained using a single-variable search. These methods in order of decreasing generality are as follows.

B. Method Using Broad MWD Standard with Known MWD.

In this method, a broad MWD standard with known MWD is injected into the GPC. The uncorrected detector response is integrated to give the cumulative retention volume distribution, which is plotted together with the true cumulative molecular-weight distribution as shown in Fig. 1. The weight fraction of molecules that have eluted at retention volume V_i equals F_i. The same fraction contains all molecules with molecular weight larger than M_i. Hence, molecules of molecular weight M_i elute at V_i, and taking many pairs V_i, M_i over the entire distribution generates an effective molecular-weight calibration curve.

Abdel-Alim and Hamielec [40, 41] used this method successfully to measure molecular weights of mechanically degraded polyacrylamides. Narrow standards of this polymer are not commercially available. However, it has been shown, by Ishige and Hamielec [42], that broad polyacrylamide standards with known MWD can be obtained by free radical polymerization of acrylamide in aqueous solution. In the polymerization of this monomer, transfer to monomer controls the molecular weight of the polymer, giving a linear polymer with the most probable distribution.

$$W(M) = \frac{M}{M_n^2} \exp\left(-\frac{M}{M_n}\right) \tag{33}$$

where W(M) is weight fraction of polymer with molecular weight M and M_n is the number-average molecular weight. The validity of Eq. (33) was confirmed [42] by measuring the molecular-weight distribution by electron microscopy.* Since the breadth of the molecular-weight distribution increases with decreasing polymerization temperature, a series of well-characterized polyacrylamide standards with different breadths can be obtained by varying the polymerization temperature.

Using a broad MWD polyacrylamide standard, Abdel-Alim and Hamielec [40, 41] obtained the calibration curve shown in Fig. 2. The method was tested by comparing the true MWD of another narrower standard to the MWD obtained by using the calibration curve shown in Fig. 2. From Fig. 3, it is seen that there is excellent agreement between the true and the measured MWDs, thus proving the validity of the calibration curve.

A similar procedure was used by Swartz, Bly, and Edwards [43] to construct an effective molecular-weight calibration curve for Nylon 66.

*In brief, the procedure involves adding a nonsolvent, n-propanol, to a dilute aqueous solution of polyacrylamide (40 wppm) to give a theta solvent (20% water and 80% n-propanol). The solution is then sprayed on to a copper substrate, shadowed with gold-palladium, and protected with carbon. Electron micrographs show individual polyacrylamide molecules as discrete spheres. The sphere size distribution is converted to a molecular-weight distribution.

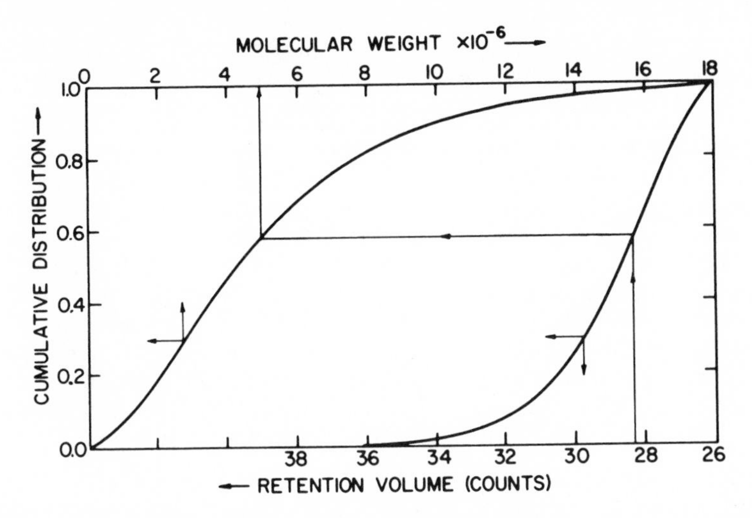

FIG. 1. Cumulative molecular-weight and retention volume distribution of polyacrylamide standard. (From Ref. 41.)

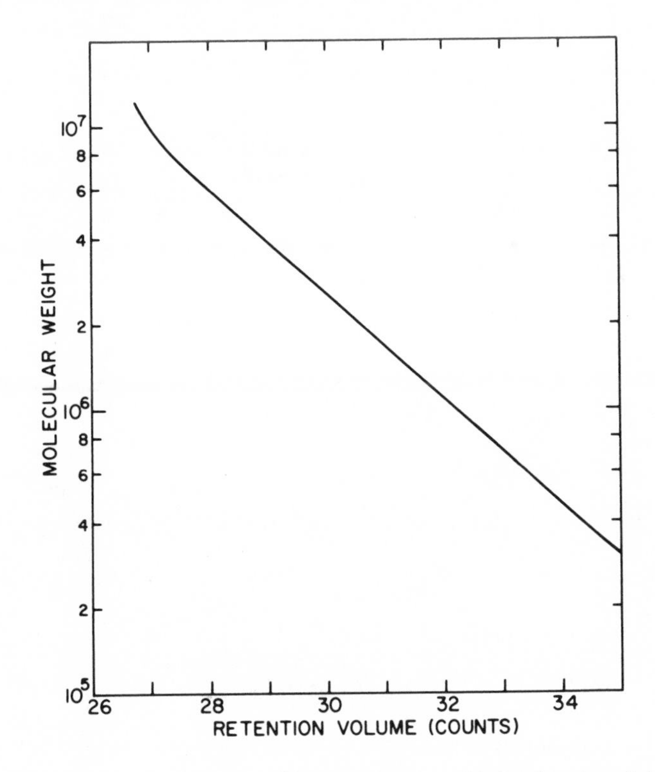

FIG. 2. Molecular-weight calibration curve obtained using data in Fig. 1.

These investigators used a Nylon 66 sample with a most probable distribution for the calibration.

This method, which obviously has great potential, has not found widespread use. Unfortunately broad standards with known MWD are not commercially available at present. However, with a rather complete understanding of the kinetics of polymerization of several monomers, broad MWD standards with known MWD could be manufactured. Thus, polyacrylamides with known MWD have been synthesized. Furthermore, on the basis of kinetic studies on vinyl chloride polymerization, it has been shown that PVC produced in bulk polymerization has the most probable distribution over a wide range of temperatures [44]. Similar predictable synthesis possibilities exist for polystyrene [45] and polymethylmethacrylate [46]. Linear condensation polymers prepared under equilibrium conditions follow a Flory most probable distribution.

A molecular-weight calibration curve obtained using a broad standard cannot be used to evaluate the MWD of narrow MWD polymers. The error involved, of course, vanishes when axial dispersion is negligible. To date, a quantitative assessment of the error involved in using an effective

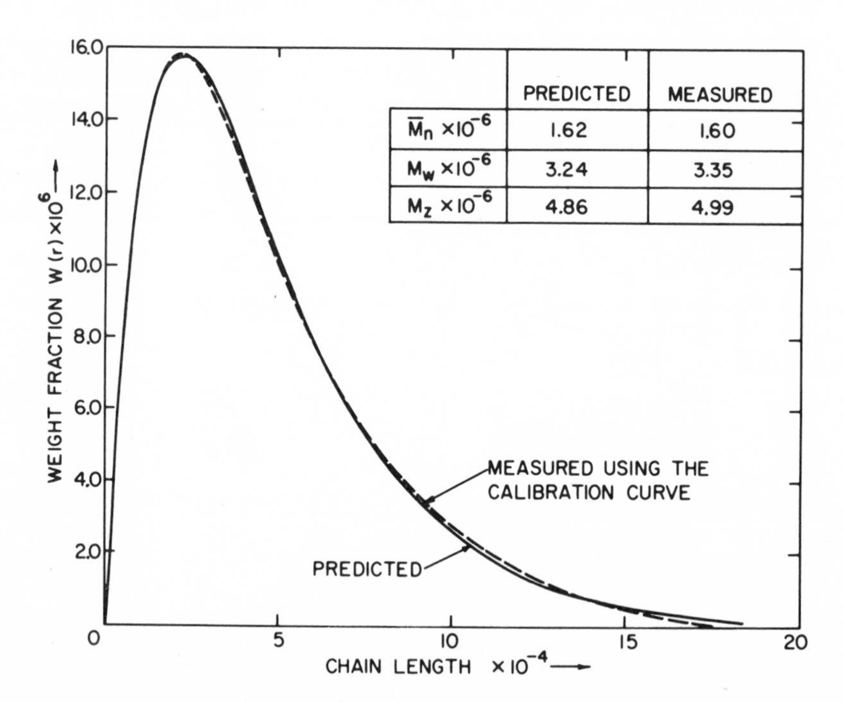

FIG. 3. Comparison between predicted and measured MWD of polyacrylamide. (From Ref. 41.)

molecular-weight calibration curve based on a broad MWD standard with a narrow MWD polymer has not been reported. However, from much evidence [47], we suspect that the errors are small. The authors are now investigating this matter, and it will be the subject of a future publication.

C. Method Based on Universal Calibration Curve and Use of Broad Standard with Known M_n and M_w

Provder, Woodbrey, and Clark [48] were the first to suggest the method based on the universal calibration curve and involving a broad standard with known M_n and M_w. The method is based on the concept of a universal calibration curve that Benoit et al. [49, 50] introduced.

As Benoit et al. have shown, narrow fractions of several polymer types, e.g., polystyrene, PMMA, and PVC, have a common calibration curve, the so-called universal calibration curve, when a plot of the product of intrinsic viscosity $[\eta]$ and molecular weight M versus retention volume is made. The intrinsic viscosity of linear polymers can be determined with the Mark-Houwink equation and it is therefore possible, using the universal calibration curve based on polystyrene, to construct molecular-weight calibration curves (ℓn M versus v) for other linear polymers. The method assumes knowledge of the Mark-Houwink constants K and a. Effective values of K and a can be obtained using a single-variable search and the following equations:

$$\frac{M_w(t)}{M_n(t)} = \left\{ \int_0^{\infty} F(v) \; \phi^{-\alpha}(v) \; dv \right\} \left\{ \int_0^{\infty} F(v) \; \phi^{\alpha}(v) \; dv \right\} \tag{34}$$

$$M_w(t) = K^{-\alpha} \int_0^{\infty} F(v) \phi^{\alpha}(v) \; dv \tag{35}$$

where $\phi(v) = M\,[\eta] = KM^{1+a}$ and $\alpha = 1/(1 + a)$.

The effective molecular-weight calibration curve $M(v) = K^{-\alpha}\phi^{\alpha}(v)$ can thus be obtained with one broad standard of known number- and weight-average molecular weights through the use of the universal calibration curve. In general, two pieces of information are required for the broad MWD standard. In some circumstances it may be desirable to use the intrinsic viscosity $[\eta]$ (t) in place of the weight-average molecular weight $M_w(t)$.

The Mark-Houwink constants obtained in this manner are not true values; their magnitudes depend on the correction for axial dispersion. If the correction is negligible, these constants are then the true Mark-Houwink constants.

Provder [48] used this method to construct effective molecular-weight calibration curves for PMMA, PVAc, and certain polyamides. This method has also been applied by several other workers. Thus, Abdel-Alim and Hamielec [44] obtained effective molecular-weight calibration curves for PVC by this method.

It should be mentioned that a calibration curve obtained in this way, using a broad standard as in the previous method, should not be used with samples having a narrow MWD. This method is less general than the one involving a knowledge of the entire molecular-weight distribution curve. Corrections to M_n and M_w should be valid. However, corrections to M_z and higher averages may be less than adequate.

D. Effective Linear Molecular Weight Calibration Curve

The method first proposed by Balke et al. [51] using an effective linear molecular-weight calibration curve assumes a curve of the form

$$M(v) = D_1 \exp(-D_2 v) \tag{36}$$

A broad MWD standard with known M_n and M_w is injected into the GPC to generate an uncorrected chromatogram, F(v). The constants D_1 and D_2 are calculated by solving Eqs. (37) and (38) using a single-variable search:

$$\frac{M_w(t)}{M_n(t)} = \left\{\int_0^\infty F(v) \exp(-D_2 v)\, dv\right\} \left\{\int_0^\infty F(v) \exp(D_2 v)\, dv\right\} \tag{37}$$

$$M_w(t) = D_1 \int_0^\infty F(v) \exp(-D_2 v)\, dv \tag{38}$$

This method has recently been evaluated by Swartz, Bly, and Edwards [43]. These investigators compared the effective molecular-weight calibration curve obtained using a broad MWD standard (Nylon 66) with a known MWD (most probable distribution) with the calibration curve obtained by the present method. The two calibration curves were in excellent agreement over almost the entire molecular-weight range, with small deviations occurring at the extremes. We agree with their comments that the success of the method using an effective linear molecular-weight calibration curve depends on the shape of the true molecular-weight calibration curve. If the true calibration curve is linear and the correction for axial dispersion is not too great, then the effective linear calibration curve should adequately correct M_z and higher moments as well as M_n and M_w. Again, the effective linear molecular-weight calibration curve obtained using a broad MWD standard should not be used with narrow MWD polymers. The intrinsic viscosity $[\eta]$ (t) can be used in place of $M_w(t)$.

E. Branched Polymers

Ram and Miltz [52] have suggested a method for the dispersion correction of GPC chromatograms for branched polymers. This method uses the universal calibration. For linear polymers, the intrinsic viscosity is related to molecular weight by the Mark-Houwink equation.

$$[\eta] = KM^a \quad \text{(linear polymer)} \tag{39}$$

It is assumed that for branched polymers this relation holds up to a threshhold value M_0:

$$[\eta] = KM^a \quad \text{for } M < M_0 \quad \text{(branched polymer)} \tag{40}$$

To make use of the calibration curve over the entire molecular-weight range, Ram and Miltz expressed $[\eta]$ as follows:

$$\ell n\,[\eta] = \ell n\,K + a\,\ell n\,M + b(\ell n\,M)^2 + c(\ell n\,M)^3 \tag{41}$$

The constants K and a are the same as for the linear polymer and are assumed known. The constants b and c are obtained by trial-and-error search, using the known intrinsic viscosity of the whole branched polymer. The intrinsic viscosity of the whole polymer can be expressed

$$[\eta] = \sum_i^{\infty} W_i\,[\eta]_i = \int_0^{\infty} F(v)\,[\eta]\,(v)\,dv \tag{42}$$

An effective molecular-weight calibration curve can now be constructed using assumed (guessed) values of b and c. This calibration curve gives a relation between molecular weight and retention volume and hence between intrinsic viscosity and retention volume. One then evaluates the integral in Eq. (42) to calculate the intrinsic viscosity of the whole polymer. This computational procedure is repeated until the intrinsic viscosities of the whole polymer match within a specified tolerance. The constants b and c are effective since their values depend on the magnitude of the correction for axial dispersion.

This method was used to determine the MWD of low-density polyethylene. Goosney [53] recently used very similar method to determine the MWD of highly branched PVAc.

The method has the advantage of not requiring the use of a specific branching model when calculating the MWD for a branched polymer.

F. Copolymers

We can find no reference to proposed methods for correcting for axial dispersion with copolymers or terpolymers. When dealing with these polymers, one usually neglects axial dispersion corrections, since the means of finding a suitable relation between molecular weight and retention volume is not clear.

If the proposal of a universal relation for the dispersion factor versus retention volume by Tung and Runyon [4] proves to have general applicability, chromatograms for branched polymers or copolymers could be corrected by solving Tung's axial dispersion equation. In this instance, it would be necessary for the chromatograms of single species to be Gaussian.

REFERENCES

1. F. W. Billmeyer, Jr., and R. N. Kelley, F. Chromatogr. 34, 322 (1968).
2. A. C. Ouano and F. A. Biesenberger, J. Appl. Polym. Sci. 14, 483 (1970).
3. F. G. Hendrickson, J. Polym. Sci. A-2, 6 1903 (1968).
4. L. H. Tung and F. R. Runyon, J. Appl. Polym. Sci. 13 2397 (1969).
5. R. N. Kelley and F. W. Billmeyer, Jr., Anal. Chem. 42 399 (1970).
6. A. J. de Vries, M. LePage, R. Beau, and C. L. Guillemin, Anal. Chem. 39, 935 (1967).
7. A. E. Hamielec and W. H. Ray, J. Appl. Polym. Sci. 13, 1319 (1969).
8. L. H. Tung, J. Appl. Polym. Sci. 10, 375 (1966).
9. A. E. Hamielec, J. Appl. Polym. Sci. 14, 1519 (1970).
10. D. D. Bly, J. Polym. Sci. C 21, 13 (1968).
11. J. C. Giddings and K. L. Mallik, Anal. Chem. 38, 997 (1966).
12. S. T. Sie and G. W. A. Rinjders, Anal. Chim. Acta 38, 3 (1967).
13. F. W. Billmeyer, Jr., G. W. Johnson, and R. N. Kelley, J. Chromatogr. 34, 316 (1968).
14. R. N. Kelley and F. W. Billmeyer, Jr., Anal. Chem. 41, 874 (1969).

15. J. A. Biesenberger and A. C. Ouano, J. Appl. Polym. Sci. 14, 471 (1970).

16. R. N. Kelley and F. W. Billmeyer, Jr., Separ. Sci. 5, 437 (1970).

17. A. C. Ouano and J. A. Barker, "A Computer Simulation of Linear Gel Permeation Chromatography," IBM Report R J 1188, March 29, 1973.

18. J. H. Duerksen and A. E. Hamielec, J. Polym. Sci. C 21, 83 (1968).

19. J. H. Duerksen and A. E. Hamielec, J. Appl. Polym. Sci. 12, 2225 (1968).

20. J. H. Duerksen, Comparison of different techniques of correcting for band broadening in GPC, page 81 in "Gel Permeation Chromatography," ed. K. H. Altgelt and L. Segal, Marcel Dekker, New York, 1971.

21. P. E. Pierce and J. E. Armonas, J. Polym. Sci. C 21, 23 (1968).

22. L. H. Tung, J. Appl. Polym. Sci. 13, 775 (1969).

23. K. S. Chang and Y. M. Huang, J. Appl. Polym. Sci. 13, 1459 (1969).

24. T. Ishige, S.-I. Lee, and A. E. Hamielec, J. Appl. Polym. Sci. 15, 1607 (1971).

25. W. N. Smith, J. Appl. Polym. Sci. 11, 639 (1967).

26. H. E. Pickett, J. R. Cantow, and J. F. Johnson, J. Polym. Sci. C 21, 67 (1968).

27. M. Hess and R. F. Kratz, J. Polym. Sci. A-2 4, 731 (1966).

28. T. Provder and E. M. Rosen, Instrument spreading correction in GPC. 1., p. 243, in "Gel Permeation Chromatography, ed., K. H. Altgelt and L. Segal, Marcel Dekker, New York, 1971.

29. T. Provder and E. M. Rosen, Sepn. Sci. 5, 437 (1970).

30. S. T. Balke and A. E. Hamielec, Paper presented at the 6th International GPC Seminar, Miami Beach, October 1968.

31. S. T. Balke and A. E. Hamielec, J. Appl. Polym. Sci. 13, 1381 (1969).

32. L. H. Tung, F. C. Moore and G. W. Knight, J. Appl. Polym. Sci. 10, 1261 (1966).

33. K. S. Chang and R. Y. Huang, J. Appl. Polym. Sci. 16, 329 (1972).

34. Y. Kato and T. Hashimoto, Kobunshi Kagaku 30, 409 (1973).

35. J. A. M. Smit, C. J. P. Hoogervorst, and J. A. Staverman, J. Appl. Polym. Sci. 15, 1479 (1971).

36. E. M. Rosen and T. Provder, Instrument spreading correction in GPC. II, p. 291, in "Gel Permeation Chromatography," ed., K. H. Altgelt and L. Segal, Marcel Dekker, New York, 1971.

37. E. M. Rosen and T. Provder, Sepn. Sci. 5, 485 (1970).

38. M. J. R. Cantow, R. S. Porter, and J. F. Johnson, J. Polym. Sci. A-1 5, 1391 (1967).

39. M. J. R. Cantow, R. S. Porter, and J. F. Johnson, J. Polym. Sci. C 1, 187 (1963).

40. A. H. Abdel-Alim and A. E. Hamielec, J. Appl. Polym. Sci. 17, 3769 (1973).

41. A. H. Abdel-Alim and A. E. Hamielec, "GPC Calibration for Water-Soluble Polymers", J. Appl. Polym. Sci. 18, 297 (1974).

42. T. Ishige and A. E. Hamielec, J. Appl. Polym. Sci. 17, 1479 (1973).

43. T. D. Swartz, D. D. Bly and A. S. Edwards, J. Appl. Polym. Sci. 16, 3353 (1972).

44. A. H. Abdel-Alim and A. E. Hamielec, J. Appl. Polym. Sci. 16, 783 (1972).

45. A. W. Hui and A. E. Hamielec, J. Appl. Polym. Sci. 16, 749 (1972).

46. S. T. Balke and A. E. Hamielec, J. Appl. Polym. Sci. 17, 905 (1973).

47. A. H. Abdel-Alim and A. E. Hamielec, J. Appl. Polym. Sci. 17, 3033 (1973).

48. T. Provder, J. C. Woodbrey, and J. H. Clark, Gel permeation chromatography calibration. I., p. 493, in "Gel Permeation Chromaotgraphy," ed. K. H. Altgelt and L. Segal, Marcel Dekker, New York, 1971.

49. H. Benoit. Z. Grubisic, P. Rempp, D. Dekker, and J. G. Zilliox, J. Chem. Phys. 63, 1507 (1966).

50. H. Benoit. Z. Grubisic, and P. Rempp, J. Polym. Sci. B 5, 753 (1967).

51. S. T. Balke, A. E. Hamielec, B. P. LeClair, and S. L. Pearce, Ind. Eng. Chem., Prod. Res. Develop. 8, 54 (1969).

52. A. Ram and J. Miltz, J. Appl. Polym. Sci. 15, 2639 (1971).

53. D. Goosney, M. Eng. thesis, McMaster University, 1973.

54. D. D. Novikov, N. G. Taganov, G. V. Korovina, and S. G. Entelis, J. Chromatography 53, 117 (1970).

Chapter 3

CHROMATOGRAPHY OF HEAVY PETROLEUM FRACTIONS

Klaus H. Altgelt and T. H. Gouw

Chevron Research Company
Richmond, California

I. INTRODUCTION

A. Special Problems Encountered with Heavy Petroleum Components

If the title of this article appears to be somewhat vague, this is so partly because it reflects the vagueness in the nature of petroleum itself. When we try to define crude oil, or petroleum, we immediately observe the enormous diversity of different crude oils, which extend from very light ones to the very heavy types found in so-called asphalt lakes. Aside from the differences in viscosity, there are tremendous variations in composition as reflected, e.g., in the content and length of the paraffinic chains, in the number of aromatic carbon atoms, in the degree of ring fusion, and in the kind and amount of hetero atoms.

The large variety in the types of compounds in petroleum leads to complex schemes of classification, even for the lightest oils. This is illustrated in Fig. 1, which shows a graph designed by Snyder [1] for the purpose of categorizing crude oil components and for facilitating the selection of methods for their separation. In addition to the three depicted variables, viz., the alkyl carbon number, the aromatic ring number, and the cycloalkyl ring number, petroleum molecules may show differences in the number and position of substituents and in the kind, the number, and

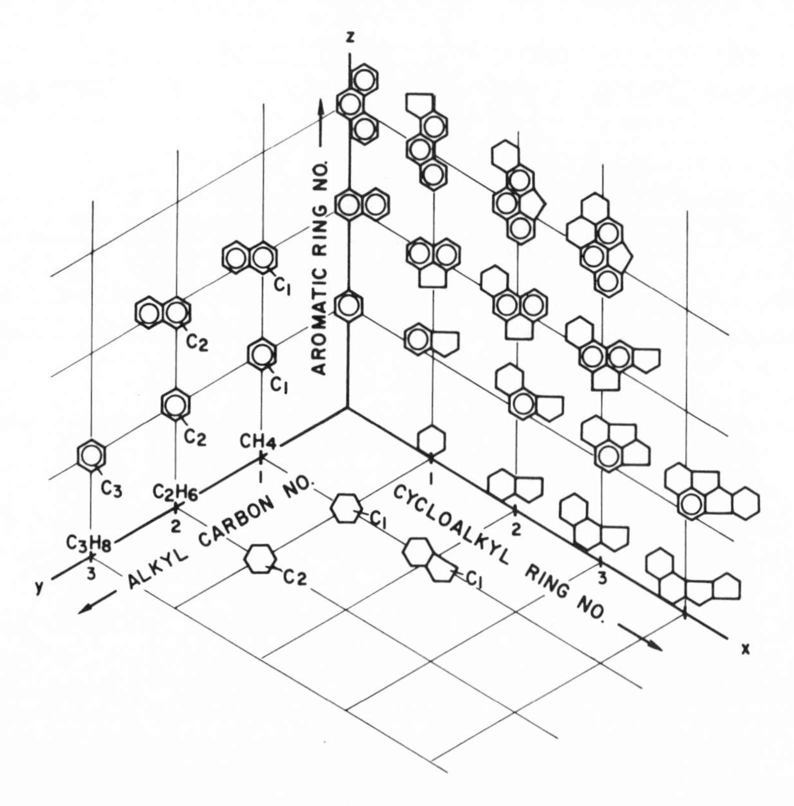

FIG. 1. Petroleum hydrocarbons: Their structural range. (From Ref. 1. Courtesy Acc. Chem. Res.)

the functionality of hetero atoms present - the latter being mainly S, N, O, and metals. Considering all these variabilities, we begin to comprehend the complexity of even the lighter parts of petroleum. However, matters become increasingly complicated with components of higher molecular weight. In the "heavy ends," we find molecules of such complexity that we can no longer categorize them unequivocally. These molecules generally contain several functional groups of different type and more than one substituent. They are therefore often classified only by such general designations as asphaltenes, heavy and light resins, and oils. For many practical purposes, e.g., in most aspects of asphalt research, this crude distinction is sufficient. It is then not necessary to separate the components of a residuum into narrower classes. On the other hand, more refined separations and detailed characterizations are required for other purposes such as characterizing a crude oil, deciding on specific refining processes, manufacturing special high-quality asphalts, and for fundamental investigations of the origin of petroleum.

Since there is a place for both degrees of sophistication, we shall describe separately the gross fractionations of heavy ends into compound classes and the detailed separation schemes for subsequent identification of specific compounds or narrow compound types.

B. Scope of This Review

Another vagueness in our title resides with the definition of heavy fractions. This term encompasses distillation residua and heavy distillates, usually those that boil higher than 400° C (700° F). In some cases we have included much lighter cuts, down to a boiling point of about 250°C (400°F). In our selection, we were guided more by the type of separation, by whether and how it could be applied to truly heavy fractions, than by an arbitrary choice of a distillation temperature.

In this article, we describe primarily useful chromatographic methods and their combinations in different separation schemes in such a way that they can be immediately applied by the reader who is familiar with general chromatographic techniques. A number of papers will be mentioned in different sections of this review, first under "Separation Schemes" and again, in more detail, under one or several subsections of "Various Types of Chromatography." In most instances, we have refrained from cross references in the text for the sake of better readability. Apparatus, technique, and fundamental theory of chromatography are not discussed but only touched on in a few cases where they serve the understanding of specific situations. These topics have been more than adequately covered in the chromatographic literature. We only wish to cite the excellent book

"Modern Practice of Liquid Chromatography," edited by J. J. Kirkland [2]. A useful compilation of chromatographic supplies and also experimental hints are found in the "Handbook of Chromatography," vol 2 [3].

II. SEPARATION SCHEMES

Chromatographic separations are usually performed for the purpose of determining the composition of a sample. In preparative separations, the fractions may subsequently be analyzed for their chemical composition and structure, and their molecular weight, boiling point, density, viscosity, and other properties. In analytical fractions, composition and molecular weight are determined indirectly from R_f values or from calibration curves. Even with such complex samples as heavy ends of petroleum, much information about the chemical structure of a fraction can be gained from the separation itself.

Haines and Snyder [4] describe the goal succintly as follows: "The best separation scheme provides the desired detail of information in the least fractions." In certain cases, a single chromatographic run on silica, alumina, or a GPC column may be sufficient to give the necessary detail. In others, complex schemes involving several methods or repetitive separations or both are required. About very detailed separations, Snyder and Buell [5] offer some fundamental thoughts:

> Each of these separation methods (acid and base extraction, ion exchange, adsorption chromatography, paper chromatography, gas chromatography, complex formation, liquid thermal diffusion, and other techniques) has potential advantages and limitations in the analysis of petroleum heterocompounds; but in the past little attention has been given to the evaluation and optimization of such procedures in terms of actual petroleum separations. Similarly, little work has been reported on their optimum integration into an overall separation scheme.
>
> Ideally an integrated separation scheme for the analysis of petroleum heterocompounds should meet several requirements. First, the various heterocompound types should be concentrated into a reasonably small number of distinct fractions, each of which contains only a few heterocompound types. In particular it is necessary that most of the heterocompounds be separated from the hydrocarbons and sulfur compounds which constitute the bulk of petroleum distillates. Second, the separation should be both repeatable and predictable. By this we mean that the approximate distribution of various compound types among the separated fractions should be known in advance of the separation, regardless of the

particular petroleum sample involved. This can greatly simplify the identification of different compound types within individual fractions (see, e.g., Ref. 6). A standard separation scheme of this type can also facilitate comparisons of the findings of different workers in studies of the nature of petroleum heterocompounds. Third, the separation scheme should be applicable to high-boiling distillates and residual fractions because these contain the bulk of the heterocompounds in a crude oil. Fourth, the separation procedures should be experimentally convenient and involve a minimum of work. Finally, the overall separation should give near quantitative recoveries of the various heterocompound types present in the original sample, with no significant chemical alteration or contamination of these compounds.

In our evaluation, none of the separation schemes published so far completely meets all of these requirements, although a few come reasonably close.

A. Gross Separations into Compound Classes

In practically all fractionations of petroleum, the first step is the distillation of the sample. For the subsequent separations, we can distinguish three main schemes. The first one is based on adsorption chromatography, often in conjunction with nonchromatographic separation methods such as precipitation with pentane or other light hydrocarbons. Another one incorporates GPC; the third one includes ion-exchange chromatography. Details of the different methods are give in Sect. III of this review. Most of the chromatographic work was performed on the pentane- or hexane-soluble part of the sample.

1. Adsorption Chromatography

In a rather sophisticated separation scheme (Fig. 2) applied to extracts of cores taken from petroleum wells, Oudin [7, 8] isolated seven main fractions. The extracts were first precipitated with n-hexane, yielding the insoluble asphaltenes and the soluble maltenes. The latter were divided by liquid chromatography on activated alumina with hexane, chloroform, and ether as solvents into hydrocarbon and resins. The hydrocarbons were further separated on a silica gel column into a saturated (extraction with heptane) and an aromatic fraction (extraction with benzene). The latter was finally subdivided by alumina chromatography into mono-, di-, and polyaromatics. With the taking of appropriate cuts and monitoring of the UV absorption at 254 nm, the saturates fraction could be obtained with only 1% aromatic contamination and the aromatic fraction with only 7% saturated compounds as determined by mass spectrometry. The asphaltenes were subfractionated into two cuts by carbon tetrachloride precipitation.

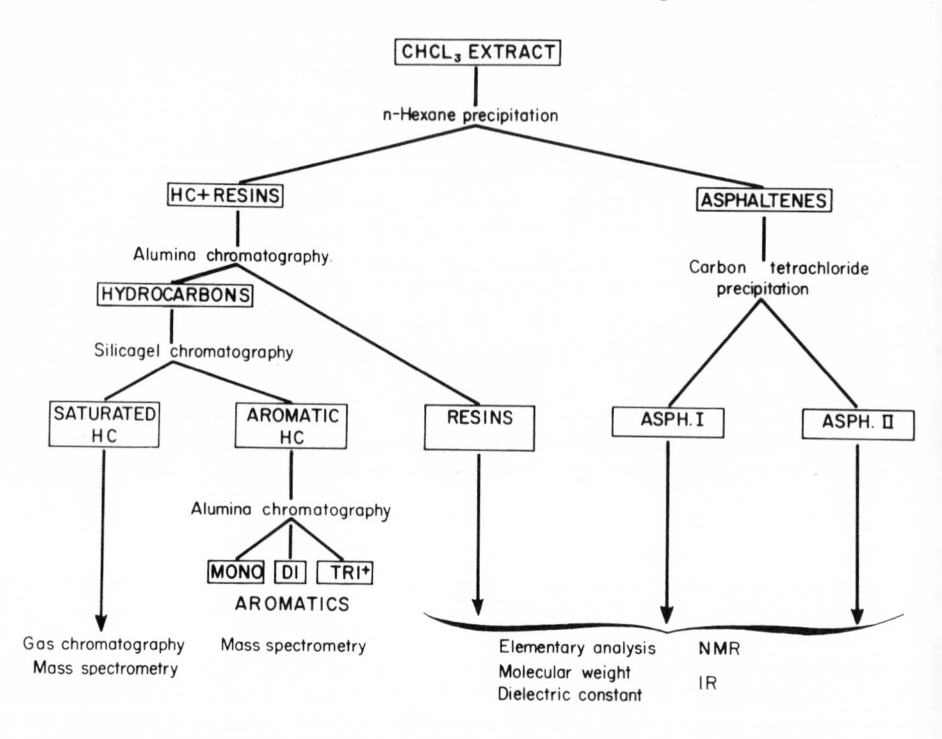

FIG. 2. Oudin's separation scheme. (From Ref. 7. Courtesy Inst. Français Pétrole Rev.)

Oudin started out with 1 g extract. All separation steps are fairly fast. The whole fractionation could probably be finished in two or three days.

Anders et al. used a similar approach for analyzing cycloalkane [9] and aromatic hydrocarbons [10] in a benzene-soluble bitumen from Green River shale. Poulet and Roucaché [11] fractionated crude oils by distillation and chromatography of the distillation cuts and residua on silica into saturated and aromatic fractions.

Middleton [12, 13] used gradient elution chromatography over alumina to separate petroleum residua into six fractions: saturates, monoaromatic and noncondensed diaromatic oil, polynuclear aromatics, soft resins, hard resins, and eluted asphaltenes. A small portion of the asphaltenes remained on the columns and was neglected. Corbett [14] first separated the asphaltenes from the maltenes by precipitation with n-pentane or other light hydrocarbons [15]. The maltenes were then further segregated by adsorption chromatography on activated alumina. Stepwise elution with heptane, benzene, and a mixture of methanol, benzene, and trichlorethylene yielded three fractions: saturates, naphthene-aromatics, and polar aromatics. O'Donnell [16] also started out by precipitating the asphaltenes. The maltenes were further separated by distillation and freed from sulfides and nitrogen compounds by complex formation with $HgCl_2$. The fractions were then chromatographed either over silica or over a combination of alumina and silica. Elution with isopentane, benzene, and finally alcohol yielded saturates, aromatics, and resins, respectively. Griffin et al. [17] later used the same scheme with modifications in the nonchromatographic steps. Dinneen et al. [18] and Robinson et al. [19, 20] also devised combinations of separation techniques, which included adsorption chromatography of their shale oil distillates on Florisil, alumina, and silica. Bestougeff [21] isolated the asphaltenes from a number of crude oils by precipitation and extraction with heptane. After separating them into two portions based on their solubility in ether or other polar solvents, he fractionated both parts separately on silica to obtain a total of 20-30 fractions of different composition and molecular weight.

Kleinschmidt [22] chromatographed asphalt samples on Fuller's earth after removal of the asphaltenes with pentane. Elution with n-pentane, methylene chloride, methyl ethyl ketone, and acetone-chloroform mixture (actually acetone + 10% water followed by a chloroform rinse) yielded, respectively, water-white oils, dark oils, asphaltic resins, and a small fraction called acetone-chloroform desorbed.

An interesting version of a single-operation fractionation of residua was recently developed by Duffy [23, 24]. The sample is deposited on Teflon powder. A column packed with this coated powder is placed ahead of a silica column. Successive elution with hexane and ether-ethanol has two effects. The sample is first selectively extracted from the Teflon, and the extract is then chromatographed on the silica. This combination gives a very efficient and reproducible separation into the three components, oils, resins, and asphaltenes. The asphaltenes are removed from the Teflon column with chloroform and methanol and collected directly, i.e., without being passed through the silica gel column.

The great advantage of this method seems to be the efficient and reproducible separation of the asphaltenes from the maltenes. However, without modification the method does not lend itself to further subdivision of the maltenes into a greater number of distinct fractions. For this purpose, the columns would have to be separated or arranged in such a way as to allow the complete chromatographic development on the silica column of each separate fraction obtained from the Teflon column. The reason for this contention is that several different solvents would be used for the chromatography on silica. These solvents would extract the sample components from the Teflon in a sequence different from that in which they would migrate through the silica column. Two examples may suffice to support this view. Higher-MW members of a homologous series of saturated hydrocarbons will ordinarily be more strongly retained in a film because of their higher degree of entanglement with other molecules and their higher diffusion coefficient. Because of steric exclusion, on the other hand, these large molecules will be eluted faster on a column packed with silica, which, in contrast to the Teflon, is porous. In addition to the size effects, there are differences between solubility and adsorption properties. For instance, tributyl amine is extracted off the inert Teflon more readily than a high-MW polyisobutylene or a phenol. On the acidic silica column, however, the amine moves much more slowly than the hydrocarbon.

2. Gel Permeation Chromatography

The second type of separation scheme makes use of gel permeation chromatography (GPC) either by itself or combined with other methods. GPC is unique in that it discriminates (in principle) only on the basis of one property, molecular size. All other methods discriminate by properties which are affected by molecular size and chemical structure (adsorption, solubility, etc.). Combination of GPC with another separation technique, therefore, allows the fractionation of a sample separately by molecular weight and by chemical structure. This is particularly advantageous for the characterization of heavy meterials, especially in cases where their catalytic cracking behavior or their rheological properties are of interest [27]. Altgelt was the first to apply this technique to the fractionation of asphaltenes [25, 26], maltenes [26], and whole asphalts [25, 27, 29]. He further chromatographed the maltene fractions over deactivated alumina into an oil and three resin fractions. In this way he obtained a matrix of fractions differing in molecular weight and in chemical structure [28, 29]. He used preparative GPC columns of capacities between 100 ml and 70 liters. Sample sizes of 300 g could be run on the large columns. Most of the fractions were large enough to allow rheological tests in addition to structural analyses.

Other early applications of GPC to asphaltic materials were reported by Richman [30], Minshull [31], Branthaver and Sugihara [32], Rosscup and Pohlmann [33], Edstrom and Petro [34], and Dickie and Yen [35]. Hayes et al. [36] also combined adsorption chromatography with GPC. The latter first chromatographed their samples, 40-g portions of maltenes, on alumina, eluting them successively with n-heptane, n-heptane plus 10% benzene, benzene, pyridine, and a 1:1 mixture of benzene and acetone. The resulting five fractions were then further separated by GPC on two different column combinations, which were chosen to give maximum resolution in the high- and the low-MW ranges, respectively.

Coleman et al. [37] followed the same course with distillation cuts, boiling in narrow ranges between 400° C and 525° C (700-900°F), in that they first fractionated their pentane-deasphalted samples by adsorption chromatography on silica columns. Their center fractions, the aromatic concentrates, were then further separated by GPC. The distillation residue was directly subjected to GPC without prior segregation by functionality.

One of the questions arising at this point concerns the sequence of the separation. Does one first carry out GPC and then adsorption chromatography, or vice versa? To answer this question, we have to look at the two methods more closely. At least in theory, GPC separates only by molecular size [38, 39]. All of the other chromatographic methods separate by functionality and by MW.

Fractionating first by GPC avoids the overlap of functionality otherwise caused by MW differences in the subsequent separation by other chromatographic methods.* Furthermore, because of the limited peak capacity of GPC, the number of GPC fractions after consolidation will usually be smaller than the number of fractions obtained by alternate chromatography. This means fewer separation steps if GPC is performed first. In special cases, such as with unsubstituted polynuclear aromatics, the reverse order may be preferable.

Before going on to the next scheme, we want to mention the papers by Haley [40], Bynum et al [41], and Dougan [42]. These papers deal with the separation by GPC and the characterization of asphalts before and after oxidation, by airblowing, by exposure to air in pavements, or by aging in the Rolling Thin-Film Oven. Also, we want to point out the use of GPC in the analytical characterization of mg amounts of petroleum and petroleum products by establisheing so-called elution patterns or fingerprints [43-45]. This method permits very fast characterization of a crude oil, an oil spill,

*An example for the interactions of MW and structural effects is the different behavior of an amine with a long alkyl chain and of one with a a very short alkyl group. With n-pentane as the solvent, polybutene amine elutes more rapidly from an alumina column than monobutylamine because its polar moiety is much smaller in proportion to the nonpolar remainder. Here the elution order is the same as in GPC. In reversed-phase partition chromatography between pentane and water, the polybutene amine is strongly retained in the pentane phase and therefore elutes after the more highly water-soluble butylamine. The situations would be reversed with a multifunctional polyamine in place of the monofunctional polybutene amine.

or other samples of unknown origin. Because of its speed and distinctive patterns, it is a very valuable screening technique. Figure 3 shows four examples by Done and Reid [43]. Albaugh and Talarico [44] expanded this method by taking fingerprints with three detectors (refractive index, flame ionization, and UV absorption) in series. Figure 4 illustrates the different responses obtained with these detectors.

Oelert [45] suggested the use of only RI detection for routine fingerprinting of high boiling petroleum distillates. For hydrocarbons of C_{30}, the refractive index is so close to its asymptotic value that, for practical purposes, it can be considered constant as long as the average chemical structure of the (MW) fractions is the same.

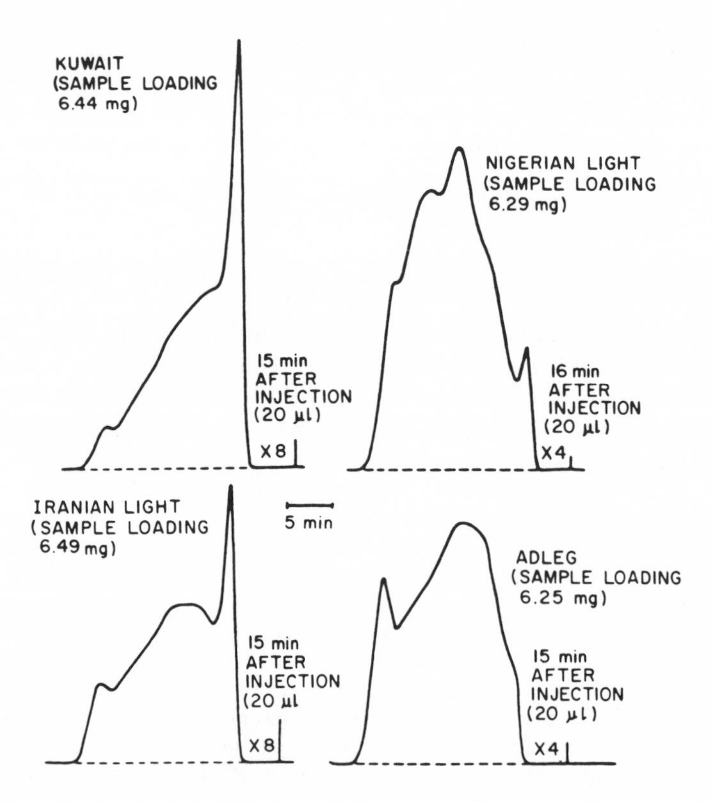

FIG. 3. Examples of GPC fingerprint chromatograms of crude oils. Column: 2 ft by 3/8 in. Stainless Steel, 60 Å Styragel; solvent, tetrahydrofuran; flow rate, 70 ml/hr average. (From Ref. 43. Courtesy Separation Sci.)

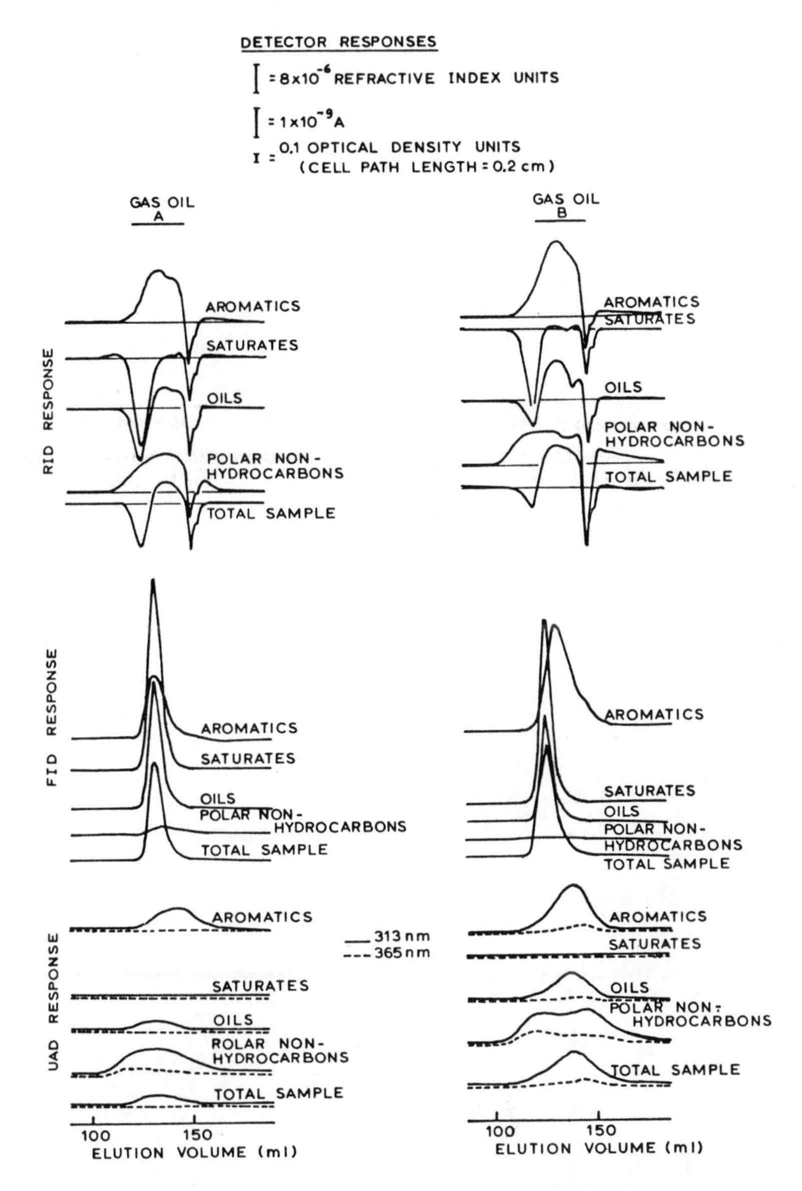

FIG. 4. Fingerprint elution patterns of gas oils and gas oil fractions obtained by Albaugh and Talarico. RID, refractive index detector; FID, flame ionization detector; and UAD, ultraviolet absorption detector. Columns 1 x 10^4 Å, 1 x 10^3 Å, 250 Å; solvent, benzene; temperature, 50°C; flow rate, 1.7 ml/min.; sample size, 5 mg. (From Ref. 44. Courtesy J. Chromatog.)

3. Schemes Involving Ion Exchange

The American Petroleum Institute Research Project 60 group developed a separation scheme [4, 46-51] that gives considerably cleaner separations of pentane-soluble heavy fractions into compound classes than the ones previously described, though at the expense of rapidity. Its outline is shown in Fig. 5. First, anion exchange is used to remove the acids, followed by cation exchange to isolate the bases. Next, the compounds with neutral nitrogen groups are complexed by $FeCl_3$ deposited on a clay column [48]. The remainder is finally separated on silica or on a dual silica-alumina column into saturates and aromatics. Large-sample sizes can be processed by this method since the ratio of ion-exchange resin to sample is only 2:1. With the $FeCl_3$-clay column, the ratio of support to sample is still a low 10:1. Only the final silica separation was carried out with a high adsorbent to sample ratio of 200:1. More technical details are given in Sects. IIIC and IIIE1. Here it may suffice to point out that several fractions can be obtained from each ion-exchange column by desorption with solvents of increasing solvent power and polarity.

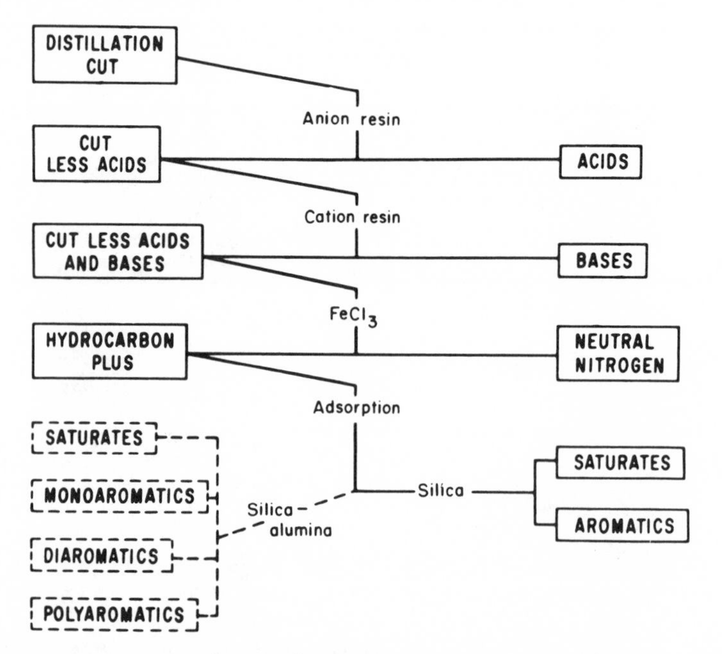

FIG. 5. Separation scheme of API Research Project 60 Grous. (From Ref. 49. Courtesy API Div. Refining Preprint.)

This scheme gives such a high resolution that it actually should be classified in Sect. IIB. We have included it here because in an abridged version it should give outstanding separations into relatively few fractions in a tolerable time. As reported, the amount of time required for a complete separation is considerably more than ordinarily expended for gross separations. Forty-two hours were spent only on depositing the sample on the anion-exchange column and on washing unreacted sample off it. The same length of time was needed for the cation-exchange step. It can be assumed that at least one more day was required for the subsequent two separations. In addition to this time, three 24-hour periods were expended on each of the ion-exchange columns for successive extraction of the reacted species with solvents of increasing polarity. This adds up to at least one week of separation time with up to three columns simultaneously operated, not even counting all the time needed to condition the resins and to pack and clean the columns.

Recently, however, Bunger reported that the deposition of the sample onto the ion exchange column could be speeded up from 42 to only 16 hours without loosing significant resolution [52].

Jewell et al. [53] improved this technique by designing an integrated apparatus that contains two columns and three detectors. The sample enters first a composite column, which is packed in four segments with cation-exchange resin, anion-exchange resin, $FeCl_3$-clay, and again anion-exchange resin. The other column is filled with highly activated silica gel. A schematic diagram of the design is shown in Fig. 6. According to the authors, rapid analyses of residual products into saturates, aromatics, resins, and asphaltenes (SARA) are possible with this apparatus. Apparently, they can still charge reasonably large amounts of sample because they mention the possibility of applying preparative GPC to their fractions. This combination of their functional separation method with GPC should be the most powerful general fractionation method for petroleum residua available today. Oelert [53] adopted this principle but packed the components of the composite column into segments which, after a run, can be separately extracted in a Soxhlet.

A recent paper by Jewell et al. [55] gives a detailed account of their SARA method and its application to four residuals. In this paper the authors also discuss the use of GPC, GLC, and spectroscopic measurements for the characterization of their fractions.

The aromatic fraction obtained by this technique was subfractionated by Jewell, Ruberto, and Davis [56] into mono-, di-, and polyaromatics on

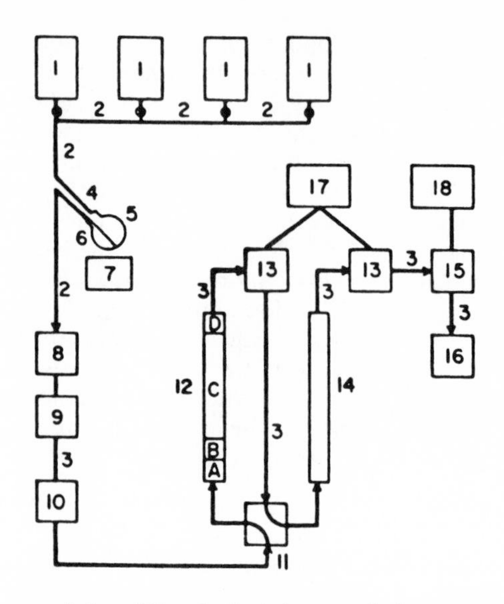

FIG. 6. Diagram of Jewell's et al. integrated chromatographic system. 1. Solvent reservoirs, 1ℓ. 2. Teflon tubing, 1/8-in. O.D. 3. Teflon tubing, 1/16-in. O.D. 4. Adapter with two 1/8-in. holes and ₮ 24/40. 5. 100 ml round bottom flask. 6. 1/2-in. magnetic bar. 7. Magnetic stirrer. 8. Milton Roy minipump. 9. Flow-through pressure gauge. 10. Injection valve-sample loop. 11. Bypass valve. 12. Composite column. 15-in. x 1/2-in. A. H^+ ion-exchange resin, 2 in. B. OH^- ion-exchange resin, 2 in. C. $FeCl_3$-Kaolin, 10 in. D. OH^- ion-exchange resin, 1 in. 13. UV detectors, 254 nm. 14. Silica-gel column, 25 in. x 1/2-in. 15. Differential R. I. detector. 16. Fraction collector. 17. UV dual channel recorder. 18. R. I. recorder. (From Ref. 56. Courtesy ACS Div. Petrol. Chem.)

activated alumina using solvent gradients of cyclohexane in n-hexane, chloroform in cyclohexane, and methanol in chloroform, respectively. Ultraviolet detection at three wavelengths distinguished the different fractions. The reliability of the separation method was tested with several mixtures of known aromatic compounds.

Snyder developed a somewhat different scheme [57, 58]. Figure 7 depicts only the part that gives the appropriate gross separation into compound classes. The sample (nondeasphalted) is separated on deactivated silica into two fractions, S_0 and S_{14}.* S_0 was extracted off the column with n-pentane and contains all the neutral components. These are further divided on a

*For a description of Snyder's nomenclature, see Sect. IIB3.

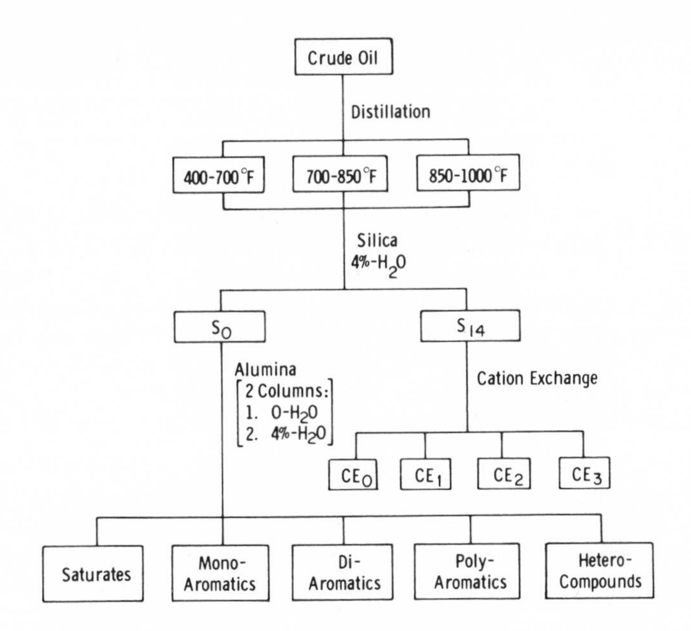

FIG. 7. Snyder's separation scheme. (From Ref. 57. Courtesy Anal. Chem.)

two-column system into saturates, A, monoaromatics, B, diaromatics, C, polyaromatics, D, and neutral hetero compounds, E. The columns are packed with deactivated and activated alumina, respectively, and operated in series. A three-way valve between them allows the more polar fractions to be withdrawn directly from Column 1 and thus to bypass Column 2. By appropriate choice of solvents and of running modes, viz., the use of only Column 1 or both columns in series, and by regular monitoring of the effluent, it seems to be possible to effect the separation into the mentioned five compound classes repeatably and with overlaps of 3-15%. Fraction S_{14} is the other fraction off the silica column. It is obtained by elution with a benzene-methanol mixture and consists of the polar N-, S-, and O-containing compounds, which can be further separated by ion-exchange chromatography.

This scheme has so far been employed only for the separation of medium and heavy distillates and not yet, at least to our knowledge, for residua. It has the advantage of simplicity and standard conditions as far as column packings and solvents are concerned. However, this advantage is partly offset by the rather complicated elution procedure with the two alumina columns. It is also necessary to observe constantly, or at least frequently, the UV absorption of the effluent at two wavelengths in order to decide when to change solvents.

B. Detailed Separations for Identifying Specific Compounds or Narrow Compound Types

1. Hydrocarbons and Other Constituents of Low Polarity

Extension of the method that the API Research Project 60 group developed leads to separations that go far beyond the scope of the gross separation discussed in the previous section. In the same paper [53] in which Jewell et al. describe the improvement of the technique (as discussed in the last section), they point to further subfractionation by additional methods. Molecular sieves allow segregating saturates into n-paraffins and non-n-paraffins. Chromatography over alumina renders the aromatics in separate cuts of mono-, di-, and tri- plus polyaromatics. The complete schematic diagram is presented in Fig. 8.

McKay and Latham [59] of the API 60 group, in another extension of their regular separation scheme, submitted their acid concentrate (obtained by ion-exchange chromatography from a 335-550°C = 640-1020°F distillate) to GPC. Each of the later fractions (No. 44-50) was subfractionated by TLC (thin-layer chromatography) on a 2:1 mixture of aluminum oxide and acetylated cellulose. Eleven polyaromatic hydrocarbon and two heterocyclic ring systems could be identified in the fractions by fluorescence spectroscopy. Their chemical structures and their distribution in the GPC fractions are shown in Fig. 9.

The fact that hydrocarbons were isolated from the acid concentrate points to limitations in the first separation. This is another reminder of the difficulties of separating such complex mixtures; however, the amount of hydrocarbons in the acid concentration is rather small (5%) and does not detract from the basic value of the separation scheme.

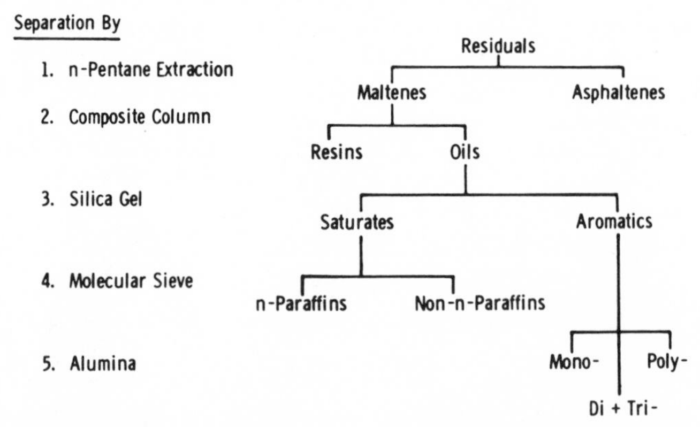

FIG. 8. Jewell's et al. sequence for studying residuals. (From Ref. 56. Courtesy ACS Div. Petrol. Chem.)

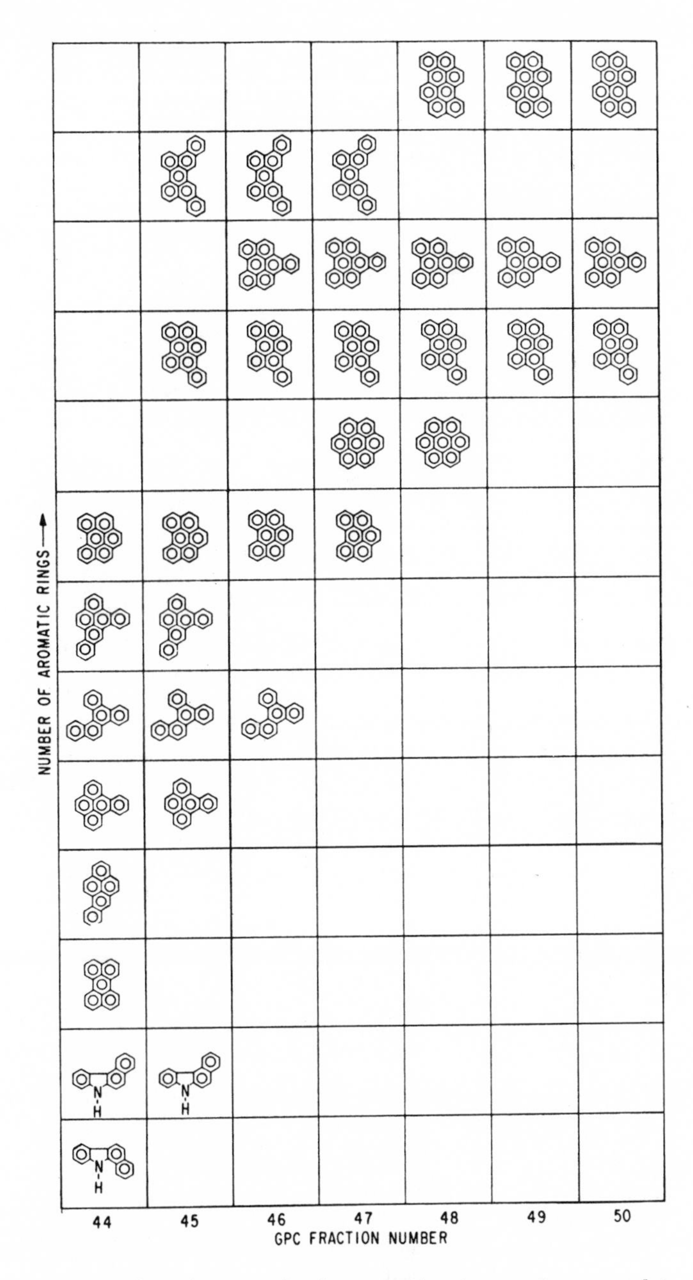

FIG. 9. GPC distribution of polyaromatic ring systems as determined by McKay and Latham. (From Ref. 59. Courtesy Anal. Chem.)

Coleman et al. [51] combined the elaborate version of the API Project 60 method with GPC and a number of modern analytical techniques such as high- and low-voltage mass spectrometry, GLC, and NMR. Their results obtained on the hydrocarbon composition of a 370-535°C (700-1000° F) distillate are an example of the wealth and the detail of information that can be generated today. Table 12 lists the composition of the saturated and aromatic hydrocarbon fractions in terms of mass spectral Z numbers, which relate to the number of rings in a molecule and their degree of fusion. A similar study is the characterization of two crude oil residua by Oelert et al. [60] in terms of structural and MW distribution. These authors used a somewhat simpler separation scheme than the one of the API 60 group. The saturated hydrocarbon fraction was treated with urea to obtain a concentrate of naphthenes, in which 14 sterane prototypes were postulated on the basis of GPC and mass spectrometry measurements.

Snyder's work in this area has been described (Sect. IIA3). Also, some of Seifert's separations should be mentioned here, since he converted his acids to hydrocarbons before final chromatography. His general separation schemes are discussed in Sect. IIB2, and details of his hydrocarbon chromatography are shown in Tables 10 and 11. Anders et al. [9] were able to separate and identify 52 cycloalkanes, many of them related to steroids or steroid precursors, by combining LC on alumina and silica with preparative GC and mass spectroscopy. They started out with 890 g of a shale extract, which they fractionated in 50-g batches on washed alumina. The pentane fraction was further separated on silica into alkanes and aromatics. After removal of the n-paraffins with molecular sieves, the branched alkanes were subjected to distillation and further chromatography over alumina. The final step in the separation scheme was preparative gas-liquid chromatography. Distillation of the aromatic fraction and preparative gas chromatography combined with mass spectral, IR, and dehydrogenation studies led to at least approximate identification of numerous compounds [10].

Oudin's general separation scheme [7] was described in Sect. IIA1. First, he separated the saturated hydrocarbons of his well core extract from the aromatic hydrocarbons on activated silica gel. Then, applying a method of Stewart [61], he subdivided the aromatics on activated alumina into mono-, di-, and polyaromatics. The polyaromatics contained also some nitrogen compounds. Heptane and a 1:1 mixture of heptane and ethyl ether were used as solvents. Increasing the eluent flow (pressure) allowed Oudin to compress the separation into a reasonable time span. See Fig. 10 for more details. The eluate was monitored by UV adsorption at 280 nm. Oudin studied his fractions with gas chromatography and mass spectrometry. His results were part of a comprehensive study of the origin and evolution of hydrocarbons in a certain shale oil [7].

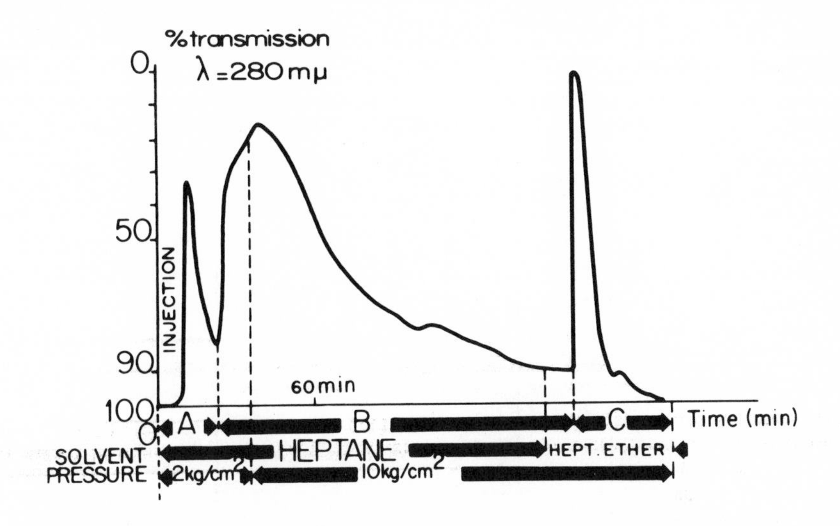

FIG. 10. Oudin's separation of mono-, di-, and polyaromatics on activated alumina. (From Ref. 7. Courtesy Inst. Français Pétrole Rev.)

Whitehead and coworkers [52] have used liquid chromatography on silica and on alumina and preparative gas liquid chromatography in combination with several other nonchromatographic separation methods for enrichment and identification of steranes and other biogenic hydrocarbons in various crude oils.

2. Acids

McKay, Jewell, and Latham [63] applied a different extension of the API 60 method to the study of acids in heavy petroleum distillates. They segregated the acid concentrate further by a number of steps involving $FeCl_3$-kaolin and two types of anion-exchange chromatography. Carbazoles, amides, quinolines, phenols, and carboxylic acids were concentrated in different subfractions. In another approach, the same group applied GPC to their original acid concentrate [63, 64] and obtained fractions that had not only different MWs but also different chemical structures. They combined their 50 GPC cuts into a total of only four main fractions. The composition of the first fraction, the one with the highest MWs, was too complex for identification. The next one consisted mainly of carboxylic acids in the form of dimers. The third contained nitrogen compounds and phenols; and the fourth, representing only 1% of the acid concentrate, consisted mainly of aromatic hydrocarbons. Table 1 compares certain aspects

TABLE 1

Comparison of Two Separation Methods for Acids in Petroleum Fractions by McKay, Jewell, and Latham[a]

Category	Method I Ferric chloride and anion resin	Method II GPC
Time required for separation	Days	Hours
Compound types isolated	Nitrogen compounds, weak phenols, entrained materials	Carboxylic acids, heavy MW materials, fluorescent aromatic materials
Percentage of material lost during separation	15-20	Less than 2
Relative acidity of compound types	Can be determined	Cannot be determined

[a]From Ref. 59.

of the two methods that are complementary, as the authors point out. We tend to prefer the GPC method because of its rapidity and low losses. The lack in discrimination between phenols and nitrogen compounds in Fraction 3 could likely be overcome by an added separation step, e.g., on $FeCl_3$-kaolin (following McKay's et al. preference) or on acidic or basic alumina.

Seifert and coworkers [65-69] used a different approach for separating and identifying acids in crude oils. Their goals were to extract quantitatively an entire virgin crude to obtain all components substituted by anionic functional groups and then identify compound classes of maximum interfacial activity at alkaline pH. Phenols turned out to be inactive [66] and carboxylic acids most active [65], so that Seifert's et al. subsequent chromatographic work deals mainly with carboxylic acids of various polarity. They started out with extracting the entire virgin crude oil with ethanolic NaOH [65]. The extract was further separated by ion-exchange chromatography on strongly and weakly basic resins followed by adsorption chromatography on silica, which effected quantitative separation of acids from phenols (Fig. 11). Central to their schemes is reducing pure carboxylic acid fractions to the corresponding alcohols and then to the corresponding hydrocarbons [67] or the deuterated hydrocarbons [70]. The hydrocarbons are further separated by silica chromatography and GPC (Fig. 12) or by chromatography over neutral and then over acidic alumina (Fig. 13).

High-resolution mass spectrometry combined with UV, IR, NMR, fluorescence spectroscopy, and GC made possible the unambiguous identification of numerous phenolic and carboxylic components [68, 69] in the fractions, including several individual steroids [70, 71]. The carboxylic acid classes identified range from C_{15} to C_{31}. Therefore, Seifert's techniques are of general enough interest to be included in this review. The starting material for his extensive operations consisted of 36-g NaOH extract. During the final separations, he had only mg quantities at his disposal. For this reason, he also used preparative thin-layer chromatography in one of his early schemes [72].

3. Nitrogen and Oxygen Compounds

Some of L. R. Snyder's publications [4, 5, 57, 73-74]* show how an analytical procedure is developed by first applying it to relatively simple distillate cuts and then progressing to very heavy ones. The first paper [5]

*L. R. Snyder is one of the pioneers of liquid chromatography in general and of this technique as applied to petroleum fractions in particular. He has published so proficiently in these fields that we will not attempt to quote all his relevant papers. We will only cite the more recent ones or others that we feel are especially pertinent for our review, even though some of the methods described in them may have been improved or superseded.

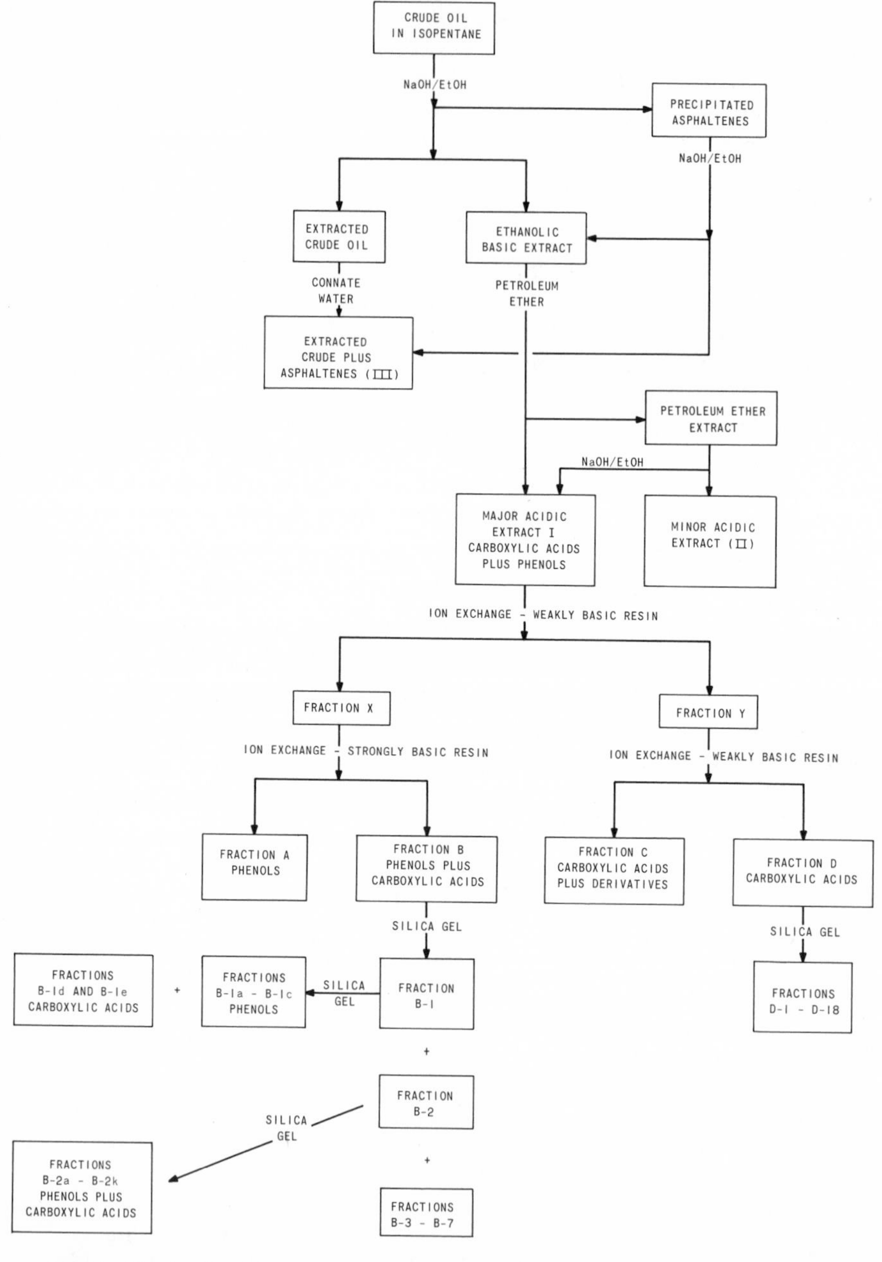

FIG. 11. Seifter's et al. extraction and separation scheme for acids in crude oil. (From Ref. 65. Courtesy Anal. Chem.)

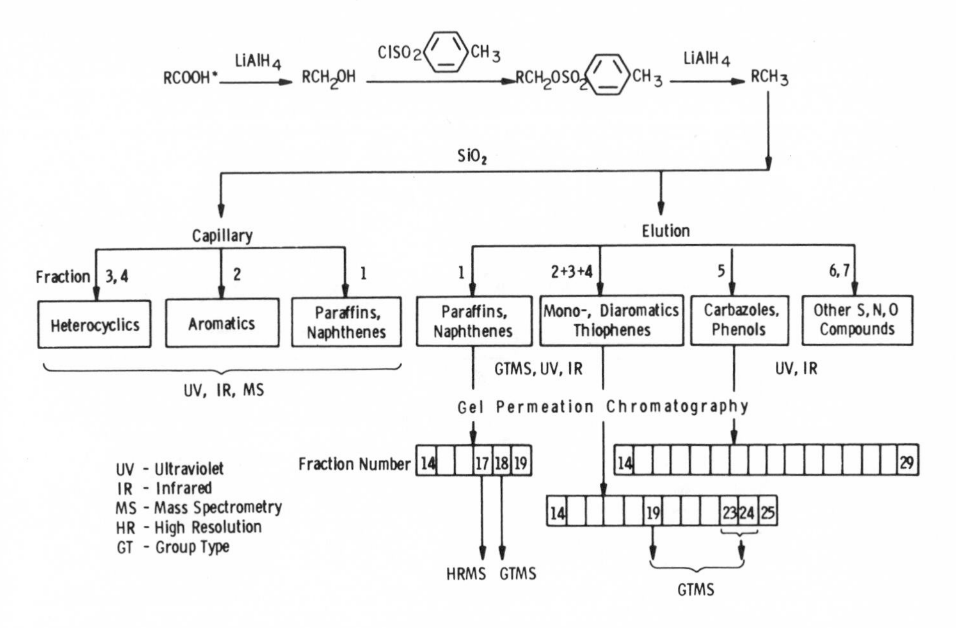

FIG. 12. Seifert's et al. Scheme I of transformation of carboxylic acids and separation of derived hydrocarbons. (From Ref. 67. Courtesy Anal. Chem.)

describes the separation scheme, experimental details, and tests to prove that the technique indeed meets Snyder's requirements, as cited at the beginning of Sect. II. The results obtained by the application to a 700-800°F (370-430° C) crude distillate are presented in his next paper [73].

A subsequent communication [74] reports some improvements of the method and results obtained with a 400-700° F (200-370° C) cut of the same crude oil. The improved method was finally applied, with only minor modifications [57], to an 850-1000° F (450-550° C) distillate cut. The results obtained on the different distillates checked very well, confirming again the repeatability and the general applicability of the separation scheme. Remarkable also is the large amount of sample that can be processed by this technique. As much as 250 g can be used in the initial stages.

Figure 14 illustrates Snyder's improved scheme. It will be necessary at this point briefly to describe Snyder's fraction nomenclature. The capital letters refer to the adsorbent: S for silica, A for alumina, CE for cation exchange. The subscripts indicate which one of six standard solvents (0 through 5) was used. The solvents are listed in Ref. 5 and also in Tables 4 and 5 of this chapter. (For more detail, see Sect. IIIA3.) A

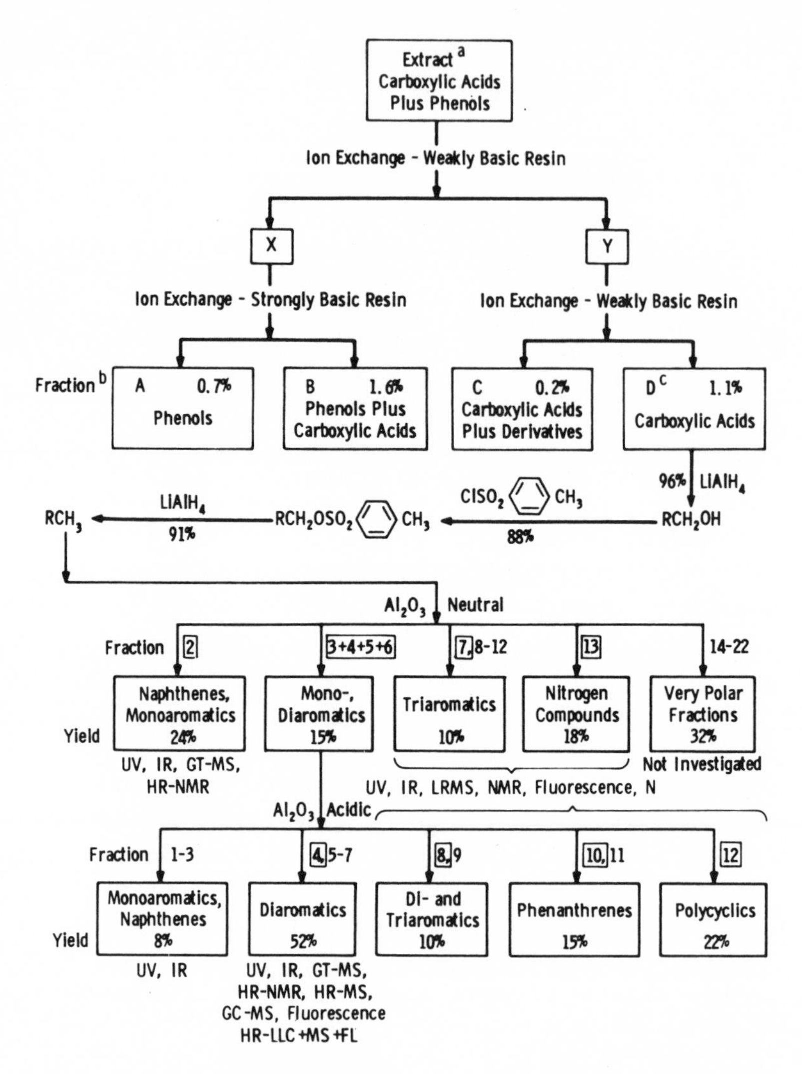

FIG. 13. Seifert's et al. Scheme II of transformation of carboxylic acids and separations of derived hydrocarbons. [a]This extract represents 3.54% of the total crude oil; 2.5% based on crude oil is RCOOH. [b]Percentages are based on total crude oil. [c]Represents 40% of all carboxylic acids present in crude oil. (From Ref. 69. Courtesy Anal. Chem.)

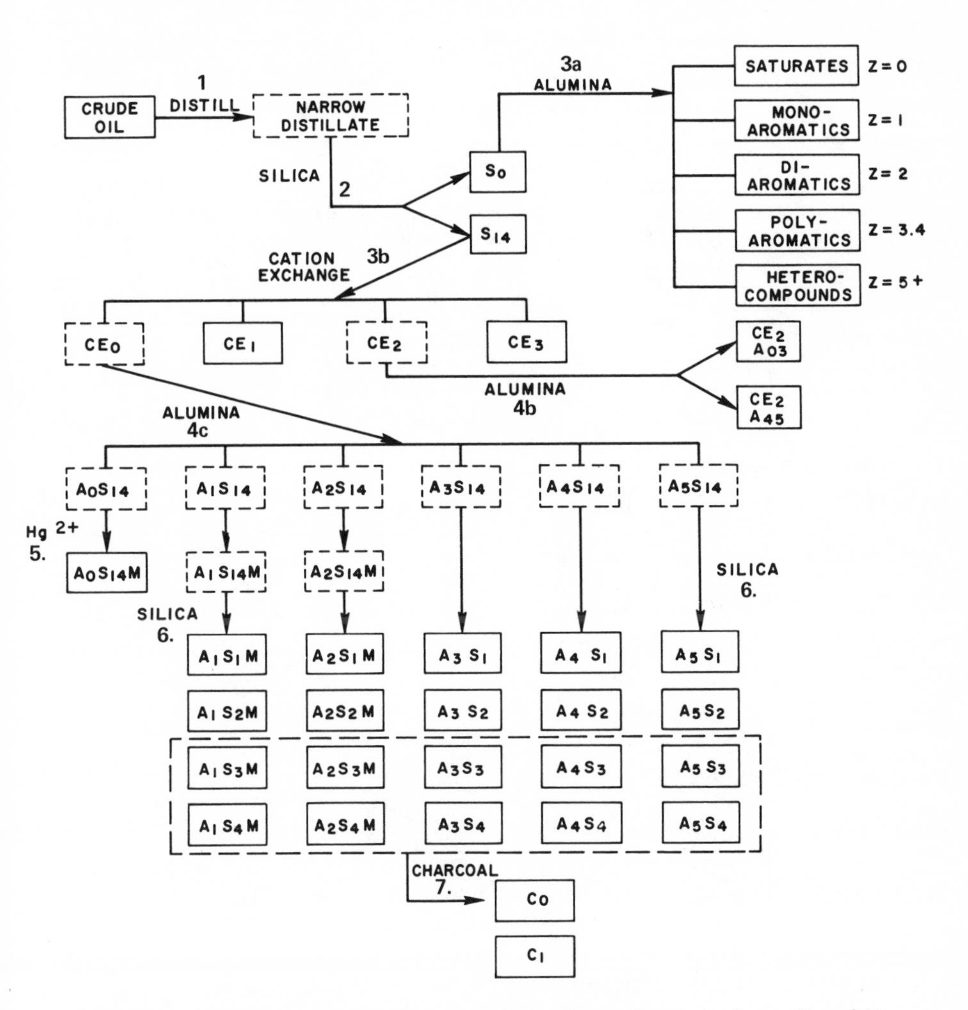

FIG. 14. Snyder's standard separation scheme for petroleum N and O compounds. (From Ref. 1. Courtesy Acc. Chem. Res.)

double subscript such as in A_{14} indicates that the solvent denoted by the second subscript has been used and that all the solvents described by lower subscripts down to and including the first one have been skipped. In Example A_{14}, Solvent 4 was used to elute the sample left on the column by Solvent 0, skipping Solvents 1-3. Multiple designations indicate several separation steps. Fraction A_0S_{14} thus had been eluted off an alumina column with Solvent 0, put onto a silica column, freed from unwanted components by extraction there by Solvent 0, and finally eluted off the silica column with Solvent 4.

Now we can understand Figure 14. Chromatography on silica first yielded two fractions: S_0, containing all nonpolar compounds, and S_{14}, containing all polar N, S, and O compounds. S_0 can be separated on deactivated alumina [57] into the five components shown in the right by the procedure described at the end of Sect. IIA. S_{14} is chromatographed on cation-exchange resin into four fractions, two of which are further separated on deactivated alumina. Sulfides are isolated from the fractions A_0S_{14}, A_1S_{14}, and A_2S_{14} by Hg^{2+} complexation, which is indicated by an M (for mercury) in Figure 14. Subsequent chromatography of two of the Hg^{2+}-treated samples and the untreated fractions A_3S_{14} to A_5S_{14} on deactivated silica separates the different kinds of compounds containing O, O and N, and N. Fractionation on charcoal finally isolates the aliphatic hetero compounds from the aromatic ones. Reference 57 lists in great detail the results obtained with this separation scheme from the 850-1000°F (450-540°C) distillate cut. Reference 1 gives a rather comprehensive discussion of the hetero compounds found in all of the fractions boiling higher than 200°F. Table 2 summarizes in greatly abbreviated form the results presented there and gives an impression of the degree of discrimination afforded by Snyder's separation scheme.

4. Sulfur Compounds

For determining S compounds, Snyder worked out a different approach [73]. Following a similar procedure as described above, he isolated alkyl thiophenes along with the saturates and monoaromatics and determined their concentration in this fraction by UV absorption at 245 nm and 270 nm. Sulfides are separated from the sample on a deactivated (3.8% H_2O) alumina column. Isooctane removes the alkyl monosulfides along with other saturates. Subsequent elution with 50% CCl_4-isooctane yields a fraction containing all the aromatic and polysulfide sulfur, which is measured quantitatively by reaction with iodine in isooctane and UV absorption.

The isolation of sulfides from petroleum by Hg^{2+} complexation [57] has already been mentioned (Sect. IIB3). There the sulfides had been regarded more as contaminants that had to be removed to facilitate isolating N and O compounds.

Drushel and Sommers [75] were actually interested in the S compounds themselves and designed a scheme for isolating and characterizing these compounds. They chromatographed a narrow distillate (425-455°C = 800-850°F) over activated silica, as shown in Fig. 15. Oxidation of the benzene eluate converted the S compounds (even the thiophenes) to sulfones. These are now sufficiently polar to allow their separation from the aromatic hydrocarbons present in the same fraction by another chromatographic run over silica.

A second method, described in the same paper [75] and outlined in Fig. 16, discriminates more between different types of S compounds. The first separation on silica is simplified by eluting immediately with benzene. The resulting fraction is gently oxidized to convert only the aliphatic sulfides to sulfoxides, which are then isolated from the rest by chromatography over silica. The unreacted fraction is then oxidized under more severe conditions to convert the aromatic S compounds. After separation by silica chromatography, the oxidation is repeated several times to ensure complete reaction. Reduction of the sulfoxides back to the sulfides makes them nonpolar again and thus easily separable from the polar non-S compounds by another chromatographic separation over silica.

Although this method is applicable to fairly heavy distillates, as mentioned above, apparently it did not work satisfactorily with the much more complex residua. When Drushel investigated S compounds in a deasphaltened residuum [76], he resorted to simple chromatography over active alumina and determination of S in the fractions by X-ray fluorescence. The failure of Drushel's more intricate methods with residua is easy to understand in view of the much higher molecular weight and particularly the presence of mixed compounds containing other hetero atoms together with the S.

Giraud and Bestougeff [77] published an interesting method of characterizing high-molecular-weight sulfur compounds in petroleum by pyrolysis and gas chromatography. However, since the heavy molecules are broken down into small fragments before the chromatographic steps, their method is only mentioned here but not further discussed.

III. VARIOUS TYPES OF CHROMATOGRAPHY

A. Adsorption Chromatography

1. General Considerations

In this section, we depend extensively on L. R. Snyder's work, especially on his book "Principles in Adsorption Chromatography" [78]. Since 1960 [79], Snyder has elucidated in numerous papers the mechanism and the different variables involved in liquid adsorption chromatography to the point where almost any separation can be effected predictably and

TABLE 2

N- and O-Compound Types Isolated by Snyder's Standard Separation Scheme from Several Distillate Cuts of a Crude Oil[a]

Compound type	% (Wt) in each distillate[b]			
	No. 2 (200–370° F)	No. 3 (370–450°F)	No. 4 (450–540°F)	No. 5 (540–700° F)
Aromatic compounds[c]				
Pyridines	0.6	2.7	5.0	8.2
Indoles	0.3	4.5	6.1	15.2
Amides	0.2	1.2	2.1	9.3
Benzofurans	0.5	1.1	1.8	3
Dihydrobenzofurans	0.1	0.1	0.1	
Phenols	0.3	1.0	1.3	1.2
Polyfunctional compounds[d]	0.003	0.1	0.4	8.1
Aliphatic compounds				
Carboxylic acids	0.5	1.7	1.3	4
Neutral oxygen compounds				
Monofunctional	0.5	0.6	1.7	2.2
Polyfunctional	0.0	0.1	0.6	1.1
Total N, O compounds	(3.0)	(13.2)	(21.0)	(52)
Distillate yields (wt % of petroleum)	25.7	16.5	13.7	17.9

[a]From Ref. 1.

[b]Fractions 1 (200° F minus) and 6 (700° F plus) were not analyzed in detail; Fraction 6 contains 95% N, O compounds.

[c]Plus higher benzologs (e.g., pyridines include quinolines, benzoquinolines, etc.).

[d]Does not include sulfur groups (N, S, or O, S compounds).

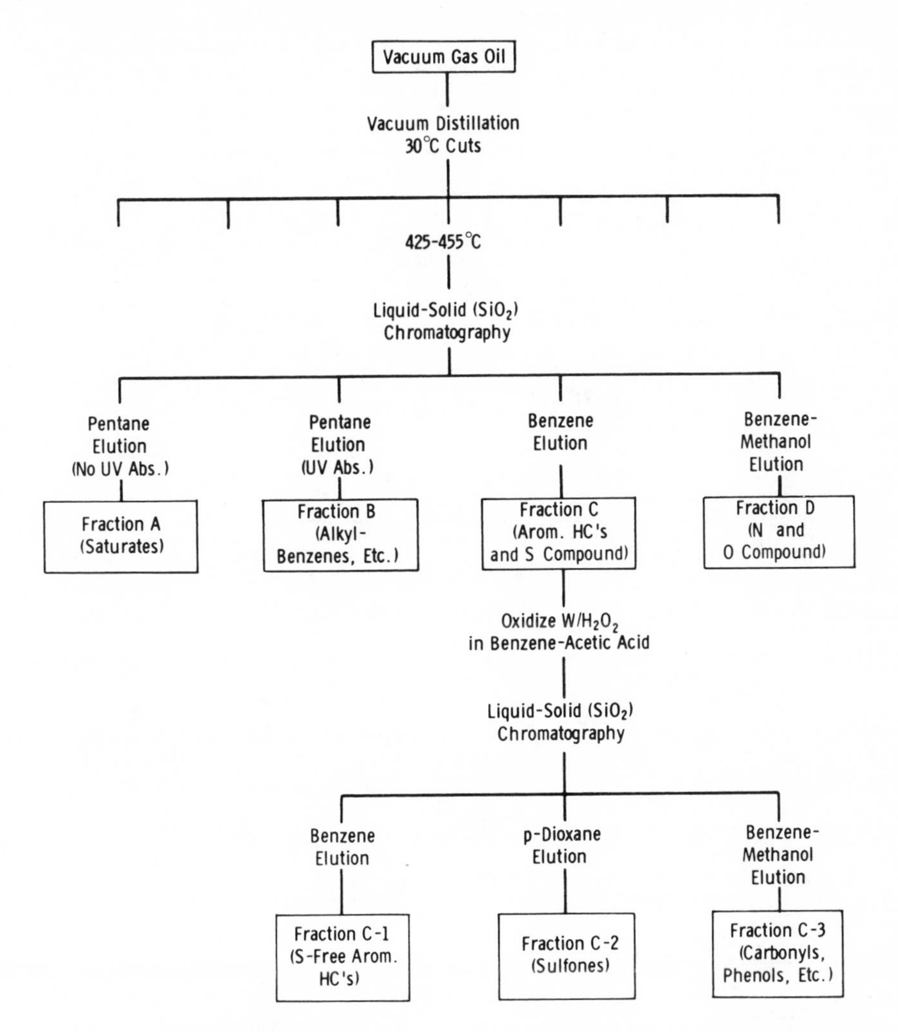

FIG. 15. Drushel and Sommer's sulfide separation Scheme I. (From Ref. 75. Courtesy Anal. Chem.)

reliably with proper choice of adsorbent and solvent. From the beginning, he has applied his insights to the separation and analysis of petroleum [80-83]. We can discuss here only the most important aspects of this topic, but it seems appropriate to summarize his discussions on adsorbents and on solvents and also some of his ideas on linear and gradient elution.

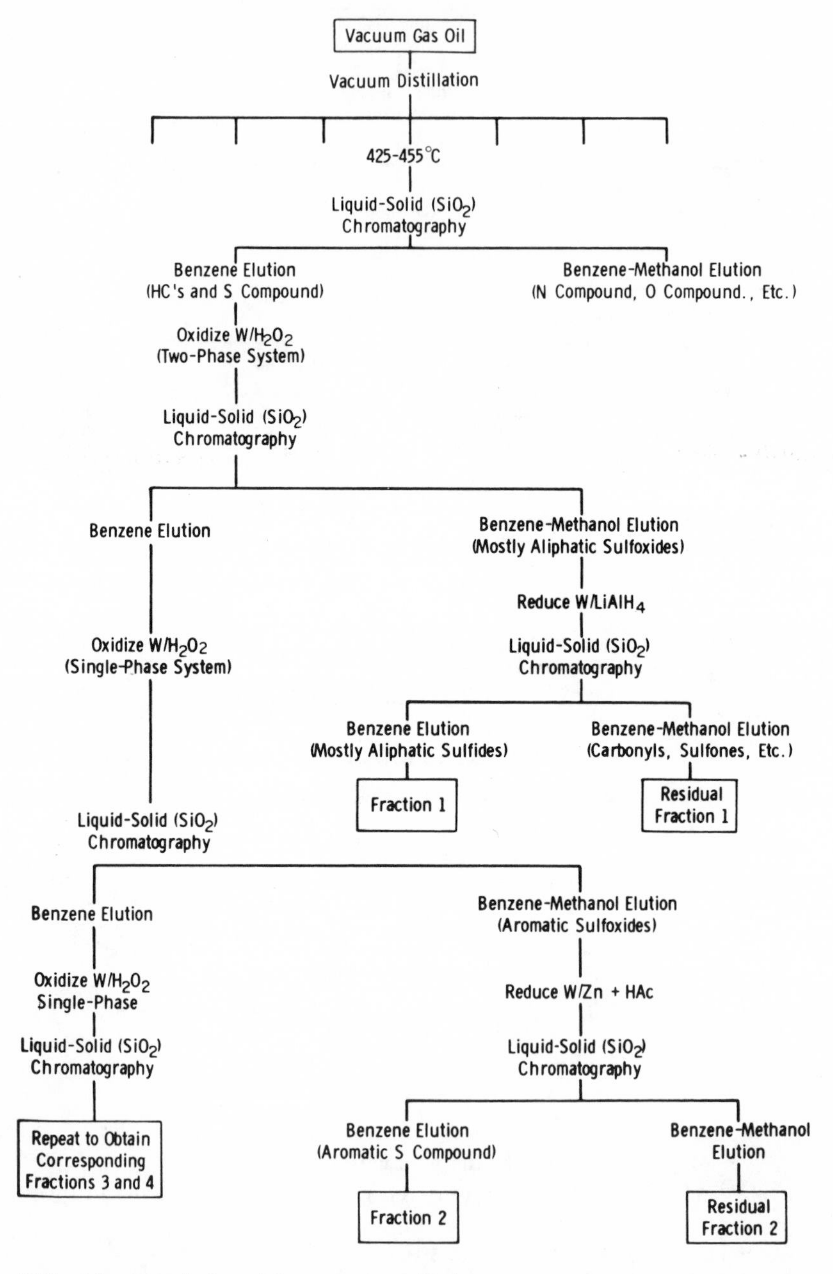

FIG. 16. Drushel and Sommer's sulfide separation Scheme II. (From Ref. 75. Courtesy Anal. Chem.)

Even though his rules are strictly valid only for the simpler low-MW petroleum molecules that fit into his categories of families and groups of compounds, they can be applied within certain limits also to the heavier, more complex molecules. However, while adsorption chromatography may afford rather good separations even with the complex heavy molecules, it cannot guarantee the reliability expected with the simple lighter ones. To reduce this complexity to a more manageable degree, Snyder resorted to ion-exchange chromatography in his routine separation scheme for heavy petroleum fractions. (See Sect. IIB2.) After removal of the basic constituents, adsorption chromatography can better resolve the remaining multitude of different species.

In the following three sections, we shall first try to summarize the important principles of liquid adsorption chromatography in general and then examine their application to heavy petroleum components.

We shall not discuss column efficiency in this article. This subject has been dealt with authoritatively elsewhere [84-86]. It should suffice to mention the importance of well-packed columns and adsorbents suitable in particle size and shape. For preparative separations of the kind we are most interested in, viz., heavy petroleum fractions, granular packings of 50-200 mesh are mostly used. Smaller particles increase column efficiency but also the back pressure. With modern high-pressure chromatographic equipment, back pressure is no longer a deterrent; and much smaller particles can be used. Adsorbents of 5 - 10 m diameter are now readily available. Spherical beads are preferred over irregularly shaped granules because of higher column efficiency as a consequence of better packing.

While in general the complexity of petroleum fractions is to our disadvantage, the contrary is the case as far as linear column capacity is concerned [78]. This capacity increases with the amount of adsorbent contacted by the sample, i.e., for a given column, with the sample bandwidth. This means first, that the capacity can be increased at the expense of column efficiency and vice versa, and secondly, that the capacity is greater for a mixture of compound classes with broad bands than for a mixture of pure compounds exhibiting narrow bands. According to Snyder [78] optimum efficiency for separating compound classes is achieved if the ratio of bandwidth (in terms of Δ elution volume) to elution volume W/R is between 0.2 and 0.3.

2. Adsorbents

a. Deactivated Adsorbents. A very important contribution to liquid chromatography was the use of deactivated adsorbents. Snyder systematized this approach in his concept of linear elution adsorption chromatography (LEAC) [78, 79, 87]. Adding small amounts of water to silica or alumina deactivates the relatively few very active adsorption sites, but does not effect the great majority of sites with a lower and much more uniform activity. In this way, two important improvements in the chromatographic behavior are realized: (1) the capacity of the adsorbent is increased, typically from 5-100-fold; and (2) the adsorption isotherms become linear and independent of compound classes at reasonable sample concentrations, which makes the process amenable to theoretical treatment.

Maximum linear capacity usually occurs at 30-60% surface coverage, which is achieved by adding 1-2% water per 100 m^2/g of adsorbent surface. For instance, an alumina sample with 200 m^2/g surface area would be optimally deactivated with 2-4 g water per gram adsorbent. Any strongly adsorbed compound can be used as deactivator as long as it is more strongly adsorbed than the sample to be separated. Water is the most common deactivator for silica, alumina, and other strongly polar adsorbents because of its convenience and stability (in contrast, e.g., to amines). It does not work, of course, on nonpolar adsorbents such as charcoal. Here high-molecular-weight organic compounds, e.g., cetyl alcohol or stearic acid, have been used [88, 89].

In working with deactivated adsorbents, care must be taken not to desorb the deactivator. In most practical cases, elution with deactivator-saturated solvent will be sufficient. A simple and reliable way to ensure continued water saturation is to put a water-soaked piece of filter paper into the presaturated solvent in the reservoir. The optimum procedure is to calculate the amount of deactivator required in the solvent and to use the corresponding mixture. Special cases where additional deactivator in the solvent may change the column characteristics do not usually occur in the systems we are here concerned with.

Sometimes, it may be preferable to forgo the advantages of deactivated adsorbents for the sake of better separation. Thus, Jewell et al. use freshly activated silica [48, 53] or alumina [55] for complete separation of monoaromatic, diaromatic, and polyaromatic hydrocarbons. Hexane elutes all monoaromatic hydrocarbons from active alumina and leaves all diaromatics adsorbed, irrespective of the structure and MW of their saturated moieties. Gradient elution with cyclohexane speeds desorption of monoaromatics without affecting the diaromatics. The diaromatics are eluted with a cyclohexane-chloroform gradient, the polyaromatics with a chloroform-methanol gradient. Separation is monitored by UV absorption at 254 nm for the monoaromatics, at 313 nm for the diaromatics, and at 365 nm for the polyaromatics.

Application of their method to model compounds reveals some important rules as shown also in Fig. 17. Substituted mononuclear aromatics elute in the sequence: short multibranched chains < straight chains < condensed cycloparaffins. Two or more benzene rings separated by several saturated

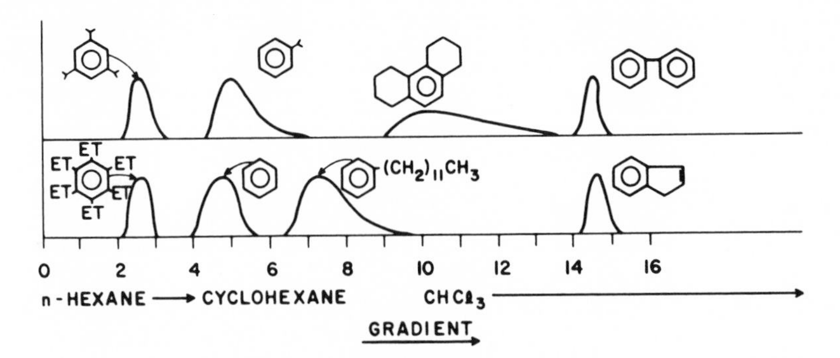

FIG. 17. Separation of model aromatics by Jewell's et al. technique. (From Ref. 56. Courtesy Anal. Chem.)

carbons are isolated as diaromatics. Benzenes containing more than three condensed cycloparaffin rings may appear in the "diaromatic" fraction. "Monoaromatics" are hence defined as molecules as isolated by this technique, containing one benzene ring and three or fewer cycloparaffin rings. The contribution of substituents to the adsorption or di-, tri-, and polyaromatic compounds is relatively small, judged from the limited number of model compounds tested. However, alkylindenes are shifted into the monoaromatics region and alkylphenanthrenes into the diaromatic region; alkylanthracenes are found in the polyaromatic region [56].

The effect of deactivation on the separability of weakly and strongly adsorbed species can be quantitatively assessed. Snyder [78] relates the linear corrected retention volume R° (ml of eluent per gram of adsorbent)* of a sample species with adsorbent properties on the one hand and sample and solvent properties on the other hand by the equation

$$\log R^\circ = \log V_a + \alpha \cdot f(X, S) \tag{1}$$

Here V_a is the volume of adsorbed phase (unimolecular layer of solvent) per unit of adsorbent; α is a parameter describing the activity of the adsorbent; and f(X, S) is a function of the sample adsorptivity X and solvent adsorptivity S. The water content affects only the adsorbent properties V_a and α, not f(X, S). This theoretically derived function is quite closely obeyed in many instances. A graphical representation is shown in Fig. 18. The shaded areas represent regions of impractical chromatographic conditions.

*$R^\circ = (R' - V_0)/W$, where R' = retention volume in linear range; V_0 = column void volume; W = weight of adsorbent.

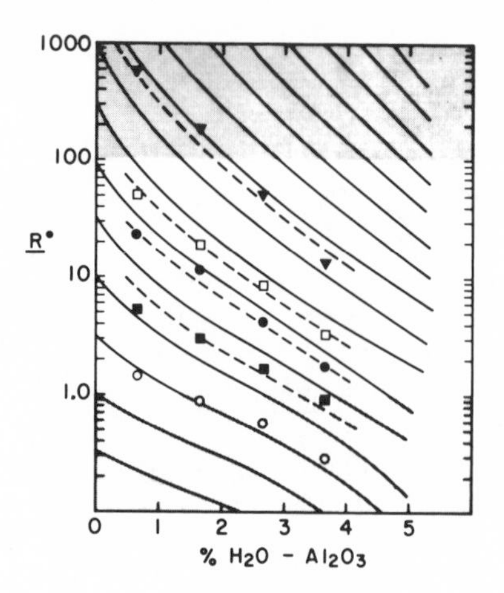

FIG. 18. Linear corrected retention volume R° as a function of adsorbent water content and sample adsorptivity (different curves). Solid and dashed lines calculated from Eq. (1) and tabulated adsorbent parameters. Points signify experimental values: benzene, quinoline, naphthalene, anisole, and phenanthrene, all eluted with n-pentane. (From Ref. 78. Courtesy Marcel Dekker.)

The values of R below 1 indicate insufficient adsorption with resultant poor separation. Values of R above 100 signify too strong adsorption and excessive running time.

Snyder has tabulated V_a and α values as a function of the water content for some of the most common adsorbents. Further, he has listed some representative f(X, S) values for pentane and other light paraffinic hydrocarbons [78]. These tabulations enable us to choose optimum conditions for specific separations, or at least to assess the separation limits for a set of given conditions.

Equation (1) represents only an approximation based on the assumption that the interactions of sample and solvent molecules are negligible or cancelled out. This assumption is generally quite good for nonpolar solutes and weak solvents such as pentane, benzene, and ethyl ether. It becomes less valid with increasing polarity of either solute or solvent, and it may lead to large errors in the presence of hydrogen bonding. With water-miscible solvents, Eq. (1) is no longer applicable.

Recently, columns containing as much as 0.6 g water or other polar liquid per 1 g of wide-pore silica (350-500 m^2/g) were recommended for high-yield routine chromatography [90]. Such heavy loading of adsorbent shifts the mechanism from adsorption to partition chromatography. No application to petroleum fractions was reported.

A potentially very useful suggestion comes from a Dupont sales brochure for chromatographic equipment, which claims that small amounts of methanol added to the solvent can be used instead of water with good results [91]. Thus, 0.1% of methanol should be equivalent to 60% surface coverage of silica by water. Unfortunately, no reference or supportive evidence is given.

b. Standardization of Adsorbents. For separations where reproducibility is desired, the same type of adsorbent should be used all the time; and its activity must be adjusted to the desired level. The adsorbent is first fully activated by heating it in air at some specified temperature for at least four hours or overnight. Then, the desired amount of water is added in liquid form to the adsorbent in a jar or other closed glass container. The jar is shaken until the lumps disappear and then stored at least 24 hours for final equilibration, preferably at some higher temperature. This method is quite convenient and gives good results. The adsorbent will keep its water content indefinitely when stored in a closed container. For best results, the adsorbent is tested with one or several standard samples. If the R° values are off, the mixture is readjusted by adding some more water or freshly activated adsorbent. More details on the standardization of the adsorbents are found in Snyder's book [78].

c. Specific Adsorbents. By far the most widely used substrates in modern adsorption chromatography of petroleum fractions are silica and alumina. Charcoal is sometimes used for separating aromatic (strongly adsorbed) from nonaromatic (weakly adsorbed) hydrocarbons [57]. Florisil (a magnesium silicate) was in vogue 15 years ago but has largely been displaced by the more versatile silicas and aluminas.

Silica is a good general adsorbent that comes in a variety of forms having different properties. Compared with alumina, it has the advantage of a higher linear capacity by 5-15 times for optimally deactivated samples. Figure 19 illustrates adsorbent linear capacity as a function of adsorbent type and water content. Silica comes in several pore size ranges between 40 Å and 150 Å. The narrow-pore silicas may exclude the larger sample molecules, whereby their effective capacity is greatly reduced. On the other hand, they contain the highest proportion of so-called reactive, i.e., hydrogen-bonded, hydroxyls, which adsorb more strongly than the "free" widely spaced, and therefore not hydrogen-bonded, hydroxyls. The wide-pore silicas have sufficient capacity but show much lower adsorption energy. Consequently, silica is often not the best choice for separating heavy oil components. For special applications, e.g., the relatively small molecules and amounts of petroleum acids Seifert et al. [68, 69] have studied, wide-pore silicas do give optimum performance.

The activation temperature for silicas should not exceed 200°C. Most workers prefer the range 150-200°C.

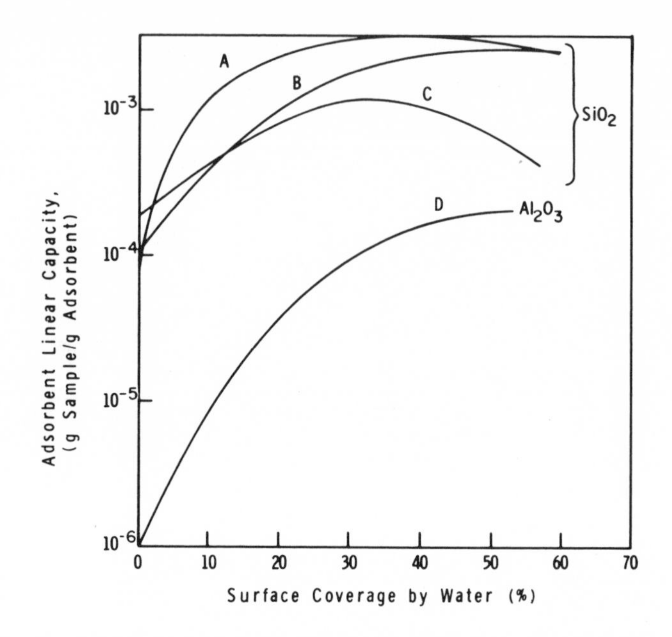

FIG. 19. Adsorbent linear capacity of silica and of alumina as a function of deactivation by water. (a) Narrow-pore silica (Davison Code 12, 190° C preactivation); (b) medium-pore silica (Davison Grade MS, 190° C preactivation); (c) wide-pore silica (Davison Code 62, 190° C preactivation); (d) medium-pore alumina (Alcoa F-20, 400° C preactivation). (From Ref. 78. Courtesy Marcel Dekker.)

Alumina is currently the best adsorbent for separating heavy petroleum components. Just like silica, it is commercially available in several different grades of highly standardized quality. Its surface usually has an area of 100-200 m^2/g and contains three different types of adsorption sites: acidic (positive field), basic (proton acceptor), and charge transfer (electron acceptor). Each of these sites is important for certain compound groups. Polycyclic aromatic hydrocarbons and other easily polarized molecules are preferentially adsorbed on the charge transfer sites, and excellent separations are achieved. Especially activated alumina is highly sensitive to aromatics of different size and shape, which is the reason for its wide use in this field. An interesting peculiarity of alumina is its tendency to adsorb more strongly the more linearly condensed polynuclear aromatics, e.g., anthracene over phenanthrene. The separation order of petroleum-related compounds on alumina as Snyder and Buell [80] have reported is discussed in Sect. IIIA4.

As shown in Fig. 19, the capacity of alumina can be increased 100-fold by deactivation with water, although at the expense of separation efficiency. For the application to heavy petroleum fractions, 4% water on alumina is a good general value. Deactivation also decreases somewhat the chemisorption of strongly acidic compounds on alumina and its tendency to alter certain samples. Before adjustment with water, the alumina is heated for 8-16 hours to temperatures between 400° C and 600° C.

Alumina is sold in basic, neutral, and acidic versions. Basic alumina has a surface pH of about 12. The neutral variety has been extensively washed with distilled water; the acidic one has been acid-treated. In acidic and, to some extent, neutral aluminas, the chemisorption of carboxylic and other strong acids ($pK_a < 5$), which takes place with basic alumina, is avoided. Basic alumina is preferred for separating the stronger bases. The separation order may vary with the different aluminas. Acetone should be avoided in preparative separations on alumina since it polymerizes by catalysis on the basic surface, yielding high-boiling products that contaminate the sample fractions.

Charcoal is now predominantly used for separating aromatic from nonaromatic compounds and to a much smaller extent for fractionating different aromatic molecules. Olefinic double bonds do not increase the adsorption of a species. Charcoal can be used as received, although adding filter aid may sometimes be required. Some workers prefer cellulose to Celite as filter aid because it is free of silica, which tends to interfere in certain cases.

Florisil is a strongly acidic coprecipitate of silica and magnesia. After it is deactivated with water, it is intermediate between silica and alumina in behavior. Its moderate tendency to alter samples is comparable to that of acid-washed alumina. It has the great disadvantage of strongly chemisorbing many aromatic hydrocarbons, basic nitrogen compounds, esters, and other polar compounds. Although water deactivation tempers the chemisorption to some extent, there is no compensating advantage that would recommend Florisil over silica or alumina.

Magnesia has not been used to any great extent in the chromatography of petroleum compounds although it seems to give better separations of polynuclear hydrocarbons than on alumina, especially when it is adequately deactivated [93].

Sephadex LH-20 is potentially interesting but did not give consistent results with petroleum samples. Applying it to oil fractions, Mair et al. [94] separated paraffins from cycloparaffins and in two instances alkylbenzenes from cyclanobenzenes. Wilk et al. [95] have observed increased retention of aromatics with increasing number of fused rings on Sephadex LH-20. Oelert [96, 97] tested Sephadex LH-20 with crude oil fractions and model compounds using different solvents. Chloroform allowed primarily discrimination by molecular weight while isopropanol stressed sorption effects. The results were rather complex and do not appear to lend themselves to clear-cut separations of compound classes. Oelert feels, on the basis of another study [54], that Mair's results cannot be generalized and alicyclics cannot be separated on Sephadex LH-20 by ring number and size.

3. Solvents

Ordinarily, choice of solvent is very important in adsorption chromatography; and elaborate lists of solvent-adsorbent systems have been compiled. In the chromatography of petroleum fractions, only relatively few solvents and their mixtures have been used. The reason is once again the complexity of the samples, which defies the isolation of single compounds and makes one think it unnecessary to select specific separation systems. More recently the use of a greater range of solvents was suggested [98, 99]. We will go into this later. But first it seems appropriate to lay a foundation by briefly delineating the role of the solvent in general adsorption chromatography.

When a sample molecule is adsorbed it displaces one or several solvent molecules on the adsorbent surface. The net adsorption (displacement) energy is the difference between the adsorption energies of sample and solvent if we neglect the much smaller solution interaction energy terms:

$$\Delta E_a = E_a\,(\text{sample}) - E_a\,(\text{solvent}).$$

This net energy translates directly into retention values. From the concepts in Sect. IIIA2c, we rewrite Eq. (1) in a slightly different form:

$$\log R^\circ = \log V_a + \alpha\,(S^\circ - A_s \epsilon^\circ) + \Delta \qquad (2)$$

R°, V_a, and α have the same meaning as before, viz., the linear corrected retention volume, the volume of adsorbed solvent per gram adsorbent, and the adsorbent activity. The previous term f(X, S) has been replaced by $S^\circ - A_s \epsilon^\circ$. The symbol S° designates the <u>sample</u> adsorption energy on the standard activity ($\alpha = 1$) adsorbent; A_s, the area occupied by the <u>sample</u> on the adsorbent; and ϵ°, the <u>solvent</u> adsorption energy per cm^2 on the standard activity surface. The symbol Δ takes into account secondary effects observed with strong solvents. It is negligible with weak and moderate solvents. With strong solvents, it is usually negative and largest in systems with hydrogen bonding. For our present discussion, the important term in Eq. (2) is $S^\circ - A_s \epsilon^\circ$, i.e., the difference between the sample adsorption energy and the adjusted solvent adsorption energy.

The fact that R° depends on the product of A_s and ϵ° means that the elution sequence of two samples of different size can be reversed by going from a weak solvent (small ϵ°) to a strong one (large ϵ°).

Table 3 shows a part of Snyder's [78] eluotropic series, i.e., a number of solvents ranked by increasing solvent strength, ϵ°. Also listed are the solvent viscosities, which should always be low to ensure low column back pressure.

TABLE 3

Snyder's Eluotropic Series[a]

Solvent	ϵ° (Al_2O_3)[b]	Viscosity (cP, 20°)
Fluoroalkanes	-0.25	
n-Pentane	0.00	0.23
Isooctane	0.01	
Cyclohexane	0.04	1.00
Carbon tetrachloride	0.18	0.97
Xylene	0.26	0.62-0.81
i-Propyl chloride	0.29	0.33
Toluene	0.29	0.59
Benzene	0.32	0.65
Ethyl ether	0.38	0.23
Chloroform	0.40	0.57
Methylene chloride	0.42	0.44
Tetrahydrofuran	0.45	
Ethylene dichloride	0.49	0.79
Methylethylketone	0.51	
p-Nitropropane	0.53	
Acetone	0.56	0.32
Ethyl acetate	0.58	0.45
Methyl acetate	0.60	0.37
Dimethyl sulfoxide	0.62	2.24
Nitromethane	0.64	0.67
Acetonitrile	0.65	0.67
Pyridine	0.71	0.94
i-Propanol, n-Propanol	0.82	2.3
Ethanol	0.88	1.20
Methanol	0.95	0.60

[a]From Ref. 67.
[b]Solvent strength parameter for aluminum oxide as adsorbent.

In this treatment, the solvent strength, ϵ°, depends to some extent on the adsorbent. Different adsorbents give different absolute values of ϵ°; however, the sequence of solvents remains the same, unless the character of the major adsorbent sites is different.

Another way of rating solvents is by Hildebrand-Scott's solubility parameter [100], δ, especially in the differentiated version of Hansen [101] and others [102]. δ is the square root of the cohesive energy density which in turn is the energy of evaporation per molar volume,

$$\delta = (E/V)^{1/2}$$

Thus δ is only a function of the solvent and independent of the adsorbent. Three main forces contribute to the energy of evaporation, dispersion (London) forces, polar (dipole-dipole) forces, and hydrogen-bonding forces. Hansen [101, 103, 104] and others [102, 105-107] devised ways of measuring or estimating the contributions from these different forces and thus to divide the total solubility parameter up into its main parts:

$$\delta^2 = \delta_{\alpha}^2 + \delta_p^2 + \delta_H^2$$

δ_H itself can be subdivided into its proton donor and acceptor shares, $\delta_{H'}$ and δ_a, leading to

$$\delta^2 = \delta_{\alpha}^2 + \delta_p^2 + \delta_{H'}^2 + \delta_a^2$$

This concept appears very satisfying theoretically since it describes pure solvent properties unaffected by interactions with solutes or adsorbents. However, in practical situations these interactions are very real; and so far it has been easier to deal with them in terms of Snyder's elution strength parameter, ϵ°.

Recently, Snyder published another approach to the classification of solvents for liquid chromatography [108, 109]. It too is based on chromatographic data, viz., the Rohrschneider constants [110, 111]. A polarity index, P', is defined by the ability of a solvent to interact with three somewhat arbitrarily, though carefully, selected polar test solutes: ethanol, dioxane, and nitromethane:

$$P' = \log (K'')_{\text{ethanol}} + \log (K'')_{\text{dioxane}} + \log (K'')_{\text{nitromethane}}$$

K'' is the excess retention of a sample over that of an alkane of equivalent molar volume. Division of each K'' term by P' gives the selectivity parameters

$$x_e = \log (K'')_{ethanol}/P'$$

$$x_d = \log (K'')_{dioxane}/P'$$

$$x_n = \log (K'')_{nitromethane}/P'.$$

These turn out to be very useful since they permit the classification of solvents into the eight groups listed in Table 4.

TABLE 4

Classification of Solvents by Selectivity Parameters[a]

Group	Solvents	
I	Aliphatic ethers, trialkyl amines, tetramethylguanidine (strong proton acceptors, strong proton donors, intermediate dipole moments)	
II	Aliphatic alcohols (strong proton acceptors and donors)	
III	Pyridines, tetrahydrofuran, amides (except the more acidic formamide) (intermediate proton acceptors, weak donors, intermediate dipole moments)	
IV	Glycols, glycol ethers, benzyl alcohol, formamide, acetic acid (stronger proton donors, weaker dipole moments than III)	
V	Methylene chloride, ethylene chloride, tricresyl phosphate	(increasingly stronger proton donors and weaker dipoles, weak proton acceptors)
VIa	Alkyl halides, ketones, esters, nitriles, sulfoxides, sulfones, aniline, and dioxane	
VIb	Nitrocompounds, propylene carbonate, phenyl alkyl ethers, aromatic hydrocarbons	
VII	Halobenzenes, diphenyl ether	
VIII	Fluoroalkanols, m-cresol, chloroform, water	

[a]From Ref. 108.

A triangular plot of x_e, x_a, and x_n very nicely illustrates this classification [108]. We believe Snyder's new rating system could be particularly valuable for selecting solvents for separations of the more polar petroleum fractions.

For our discussion of basic considerations in adsorption chromatography, we shall now return to Snyder's elution strength parameter.

Very often, particularly with petroleum fractions, mixed solvents are used. Their elution strength ϵ°_{AB} can be calculated from their component properties by Eq. (3):

$$\epsilon^\circ_{AB} = \epsilon^\circ_A + \frac{\log (N_B 10^{\alpha n_B(\epsilon^\circ_A - \epsilon^\circ_B)} + 1 - N_B)}{\alpha n_B} \quad (3)$$

Here ϵ°_A is the solvent strength of the weaker solvent A; ϵ°_B, the solvent strength of the stronger solvent B. The symbol N_B is the mole fraction of B in the solution phase; n_B, the molecular size of B on the adsorbent surface. The plot of ϵ°_{AB} versus volume percentage of B in the solvent mixture in Fig. 20 shows graphically the relation of Eq. (3). It demonstrates quantitatively the empirically well-known fact that even small amounts of a strong solvent can have a very large effect on a weaker one.

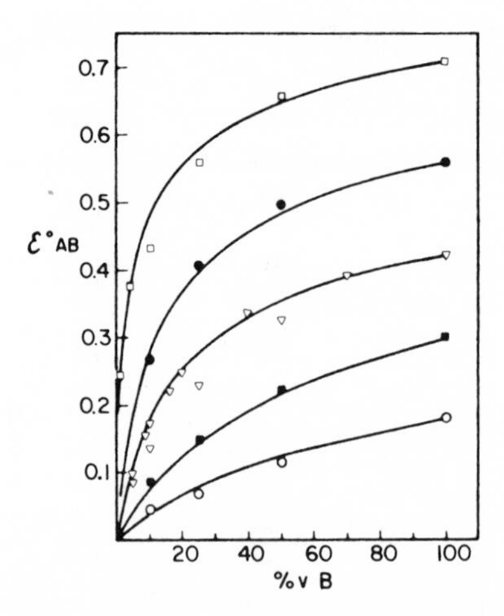

FIG. 20. Solvent strength ϵ° of binary solvents vs solvent composition (percentage of Component B in an A + B Mixture) on 3.8% H_2O-Al_2O_3. O Pentane (A) - CCl_4(B); ▪ pentane (A)-n-propyl chloride (B); F ▽ pentane (A)-CH_2Cl_2(B); pentane (A)-acetone (B); ● pentane (A)-pyridine (B); F □ solid lines calculated from Eq. (3). (From Ref. 78. Courtesy Marcel Dekker.)

It should be noted that in contrast to pure solvents, which are independent of α, the solvent strength ϵ°_{AB} of a binary mixture increases with increasing adsorbent activity α. For example, the experimental ϵ° value of 10 vol % benzene-pentane is 0.066 on 16% H_2O-SiO_2 (small-pore), but 0.098 for 1.6% H_2O-SiO_2 [112]. For solvent mixtures of components of greatly different strength, Eq. (3) can be simplified to

$$\epsilon^{\circ}_{AB} = \epsilon^{\circ}_{B} + \frac{\log N_B}{\alpha n_B} \tag{4}$$

Equation (4) is reliable if $\alpha(\epsilon^{\circ}_B - \epsilon^{\circ}_A) > 0.2$ and $N_b > 0.2$.

Table 5 lists a number of binary solvent mixtures that Snyder and Buell [5] recommended for separating petroleum compounds on alumina and silica. These solvents have the advantage of ready availability and low viscosity. They cover the whole required range of solvent strength. Snyder and Buell demonstrated their adequacy by determining the distribution of a number of petroleum compound classes in fractions obtained with these solvents. Table 6 shows their specificity. Only 3 compound classes out of 15 were distributed over more than 2 fractions. This spread was probably caused more by the diversity of the members of those classes than by an inadequacy of the separation method.

4. Sample Structure

Snyder and Buell have published an extensive study on the effect of sample structure on the retention (R°) by 3.8% H_2O-alumina [92]. Table 7 gives an example of the type of data that were compiled. The first column lists the aromatics are distinguished by their ring number, the second column distinguishes them by their ring arrangement. The symbols S° and A_S represent the quantities used in Eq. (2); $_1$ is a composite measure of adsorptivity defined by Eq. (5):

$$\epsilon_1 = \frac{\alpha S^{\circ} + \log V_a}{\alpha A_s} \tag{5}$$

Thus, ϵ_1 is the solvent-independent part of R°. The last three columns in Table 6 represent the ϵ_1 values of the mixtures of molecules that are related to the parent compounds in Column 2 by different substituents and that are found in petroleum distillate cuts of the temperatures indicated on the column heads.

TABLE 5

Selected Binary Solvents for Separations on Alumina and Silica[a]

Alumina		Silica	
ϵ°	Solvent	ϵ°	Solvent
0.00	100% P	0.00	100% P
0.06	3.5% B-P	0.04	4% B-P
0.12	9.2% B-P	0.08	10% B-P
0.18	20% B-P	0.12	19% B-P
0.24	41% B-P	0.16	32% B-P
0.30	70% B-P	0.20	52% B-P
0.36	21% E-B	0.24	90% B-P
0.42	60% E-B	0.28	6% E-B
0.48	10% A-B	0.32	18% E-B
0.54	28% A-B	0.36	40% E-B
0.60	60% A-B	0.36+	50% M-B
0.66	12% I-B		
0.72	25% I-B		
0.78	57% I-B		
0.78+	50% M-B		

Note: Solvent compositions refer to vol % of indicated solvent pairs; A, acetonitrile; B, benzene; E, ethyl ether; I, isopropanol; M, methanol; P, pentane; ϵ° is eluent strength parameter [6].

[a]From Ref. 5.

The effect of alkyl substitution on the adsorption strength S° of different aromatic ring systems is illustrated in Fig. 21. The symbol ΔS° indicates the increase in S° of the substituted molecule over the unsubstituted parent molecule. The adsorptivity of a ring system by an alkyl chain is increased about proportionally to its number of aromatic carbon atoms.

Figure 22 ranks a number of oxygen and nitrogen compounds found in petroleum in order of increasing ϵ_1. Comparing these data with Table 5 makes very clear the need for separating by ion exchange (or some other means) the hetero compounds from the hydrocarbons before applying adsorption chromatography.

TABLE 6

Distribution of Certain Petroleum Compound Types among the Standard Fractions from Alumina Chromatography[a]

Compound type	ϵ_1[b]	% of compound type in indicated fraction					
		A_0	A_1	A_2	A_3	A_4	A_5
Benzofuranes	0.03-0.04	100					
Dibenzofuranes	0.10-0.13	38	49	13			
Dibenzothiophenes	0.15	51	49				
N-alkyl indoles	0.16-0.17		42	58			
Triaromatic hydrocarbons[c]	0.16-0.19		37	63			
Naphthobenzothiophenes	0.22-0.23			100			
Naphthobenzofuranes	0.22-0.23			98	2		
Tetraaromatic hydrocarbons[d]	0.20-0.26			76	24		
Indoles[e]	0.28-0.33			67	33		
Sulfoxides	0.34-0.38			4	2	22	72
Carbazoles[e]	0.34-0.41				37	63	
Benzcarbazoles[e]	0.41					100	
N-alkyl quinolones	0.44-0.58					100	
Phenols	0.45-0.62				1	15	84
Quinolones[e]	0.6-0.8					6	94
ϵ_1 Range[b]		<0.12	0.12-0.17	0.17-0.24	0.24-0.30	0.30-0.45	0.45+

[a]From Ref. 5.
[b]Compound type adsorptivity.
[c]Fluorenes and phenanthrenes; principally C_nH_{2n-16} through C_nH_{2n-20}.
[d]Benzofluorenes, chrysenes, etc.; C_nH_{2n-22} through C_nH_{2n-23}.
[e]N-H derivatives (not alkyl-substituted on the nitrogen atom).

TABLE 7

Adsorption Parameters of Aromatic Hydrocarbons Found in Petroleum Fractions on Deactivated (3.8% H_2O) Alumina[a]

		Parent compound data			ϵ_1 for petroleum fractions, °F			Previously found	
Compound type	Parent compound[b]	S^{o}[c]	A_s[d]	ϵ_1[e]	600	800	1000	Petr.	Other[f]
Monoaromatic hydrocarbons	Benzene	1.86	6.0	-0.16	-0.09	-0.07	-0.06	(21)	(27)
Diaromatic hydrocarbons	Naphthalene	3.10	8.0	0.04	0.06	0.06	0.07	(21)	(27)
	Biphenyl	3.41	9.7	0.06	0.07	0.07	0.08		
	ϕ—CH_2—ϕ	3.2	11	0.06	0.07	0.07	0.08		
	ϕ—$(CH_2)_{10}$—ϕ	3.7	19	0.04	0.04	0.05	0.05		
Triaromatic hydrocarbons	Fluorene	4.3	9.5	0.15	0.17	0.16	0.16	(21)	(27)
	Phenanthrene	4.5	10.0	0.17	0.17	0.18	0.18		
	Anthracene	4.6	10.0	0.18	0.18	0.19	0.19		
Tetraaromatic hydrocarbons	3,4-Benzphenanthrene	4.9	12.0	0.17	0.17	0.17	0.19	(21)	(27)
	Pyrene	5.0	11.0	0.20	0.20	0.22	0.20		
	Benzfluorene	5.7	11.5	0.25	0.25	0.25	0.25		
	Triphenylene	5.7	12.0	0.24	0.24	0.22	0.24		

	Benzanthracene, chrysene	5.9	12.0	0.26	0.26	0.24	0.26
	Naphthacene	7.5	12.0	0.39	0.39	0.38	0.35
Pentaaromatic hydrocarbons	Benzpyrenes (i)	6.4	13.0	0.28	0.28	0.28	0.29
	Dibenzfluorenes (i)	7.0	13.5	0.31	0.31	0.31	0.32
	Dibenzanthracenes (i)	7.2	14.0	0.31	0.31	0.31	0.32
	Pentacene	10.2	14.0	0.53	0.53	0.53	0.48

[a] From Ref. 1.
[b] ϕ refers to a phenyl ring; (i) includes all unsubstituted isomers.
[c] S° = sample adsorption energy.
[d] A_s = area occupied by sample on the adsorbent.
[e] ϵ_1 = composite adsorptivity. (See Eq. (5).)

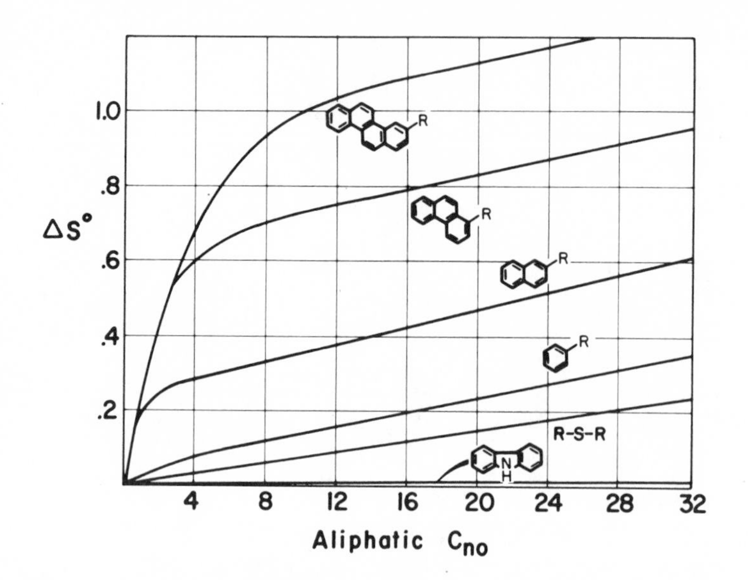

FIG. 21. Effect of aliphatic substitution (chain length n) on the adsorption energy S° of different aromatics. $\Delta S^\circ = S^\circ$ (substituted aromatic) - S° (parent aromatic). (From Ref. 92. Courtesy J. Chem. Eng.)

5. Stepwise and Gradient Elution

Because of the complexity of petroleum samples, it would be impossible to separate them into their components with only one solvent in a reasonable length of time. Stepwise elution with different solvents, starting with n-pentane, progressing to stronger solvents, and ending with methanol usually alleviates this problem. Suitable solvent sequences for alumina and silica are given in Table 5. Better in many cases are "logarithmic" solvent sequences [98]. With them the retention volume changes linearly (though strictly speaking, as a step function) with the logarithm of k, the partition coefficient of a sample species:

$$\log k = a - bR^\circ \qquad (6)$$

for small R°. It also changes linearly with solvent strength (going from solvent 1 to solvent 2):

$$R_1^\circ = R_2^\circ \; C' \; (\epsilon_2^\circ - \epsilon_1^\circ) \qquad (7)$$

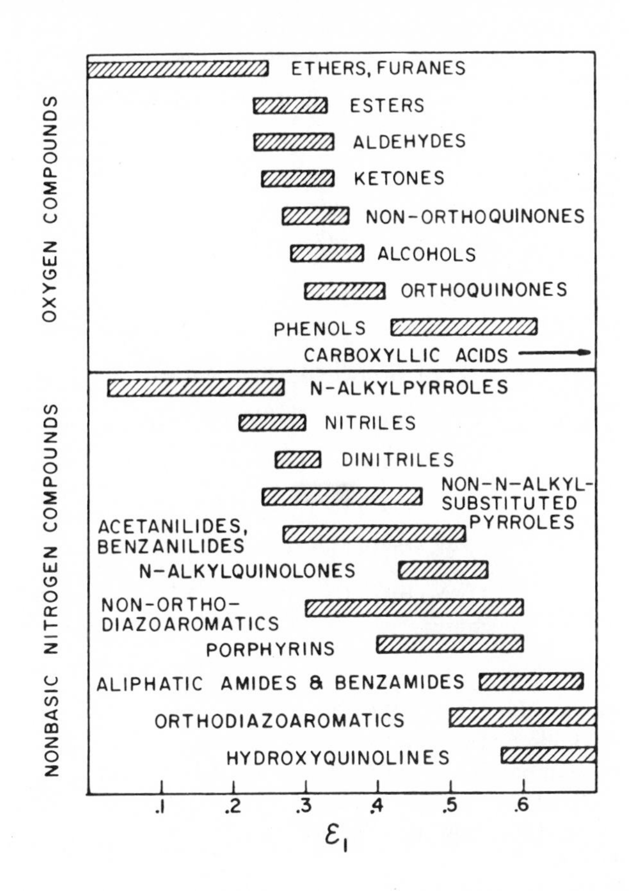

FIG. 22. Adsorptivity ranges of various O- and N-compound types on 3.8% H_2O-alumina. (From Ref. 92. Courtesy J. Chem. Eng.)

with $C' = \alpha A_S/b$. Thus a logarithmic sequence can be obtained by a series of solvents whose ϵ° change by the same amount from one pair to the next as shown in Table 8. For strong solvents we must add a solvent-solute-adsorbent interaction term, Δ

$$R_1^\circ = R_2^\circ \; C' \; (\epsilon_2^\circ - \epsilon_1^\circ) + \Delta \tag{8}$$

where Δ is the product of a specific solute contribution, Δ°, and a contribution from the two solvents, $(m_2 - m_1)$. These terms reflect the competition of strongly adsorbing solute and solvent molecules for the same adsorption sites and have been determined for numerous solvents and model

TABLE 8

Optimum Solvent Series for Solvent Programmed Separation (Logarithmic) on Silica[a]

Silica Solvent	Composition	ϵ_o[b]
1	70%v pentane/FC-78[c]	-.05
2	pentane	0.00
3	4.2%v i-propylchloride/pentane	0.05
4	10%v i-propylchloride/pentane	0.10
5[d]	21%v i-propylchloride/pentane	0.15
6	4%v ethyl ether/pentane	0.20
7	11%v ethyl ether/pentane	0.25
8	23%v ethyl ether/pentane	0.30
9	56%v ethyl ether/pentane	0.35

[a]From Ref. 98.
[b]Values calculated as in Eq. (3), using a measured value of ϵ° for i-propylchloride (0.28).
[c]FC-78 is a perfluorocyclic ether from the 3M Co.
[d]0.8%v ethyl ether added to this solvent to minimize demixing of the following ethyl ether/pentane solvent.

compounds [112]. Although for heavy petroleum fractions Δ has little predictive value, the general concept will help the understanding of some of the problems encountered.

In their detailed discussion of solvent programs and their optimization, Snyder and Saunders [98] pointed to the advantages of a logarithmic sequence. These include constant resolution and constant peak width throughout the separation, regardless of sample composition, and the possibility to calculate compound retention volumes. Thus, the nature of material eluting from a well-defined column can be more or less well identified.

Snyder and Saunders [98] recommend the use of solvents with rather close ϵ° values for mixtures in stepwise or gradient elution programs. Otherwise solvent demixing may take place on the column and severely disturb the separation. Solvent demixing, i.e., a change in mixing ratio due to preferential adsorption of the stronger solvent, has been observed with mixtures of pentane and methanol, but also with ethyl ether at low percentages. Pentane and i-propylchloride are close enough for all practical mixing ratios. Table 9 gives a selection of solvents for optimum solvent programs.

For best results, ϵ° should change by about 0.04 unit per column volume of eluate [98]. Snyder and Saunders also show the importance of starting with a solvent that is weak enough to avoid rapid elution of the least strongly adsorbed sample components. With heavy petroleum fractions it is advisable always to start an elution with pentane rather than benzene or chloroform-containing mixtures.

With modern equipment it is often more convenient and at least as good to use gradient elution, where the solvent strength is changed continuously. Again, several different solvent pairs can be used, e.g., pentane-isopropylchloride, isopropylchloride-acetone, acetone-methanol. Benzene can take the place of isopropylchloride if its high UV absorption can be tolerated (no UV detector). By the proper choice of solvents, gradient function, and gradient rate, it is possible to combine excellent resolution with relatively

TABLE 9

Solvents of Low Viscosity for Optimum Solvent Programs[a]

Solvent	ϵ^{o}[b]	Viscosity (cP. at 20°)
n-pentane	0.00	0.23
n-hexane	0.00	0.32
CS_2	0.15	0.37
i-propyl ether	0.28	0.37
i-propylchloride	0.30	0.33
ethyl ether	0.38	0.23
acetone	0.56	0.32
methyl acetate	0.60	0.37
diethylamine	0.63	0.38
acetonitrile	0.65	0.37

[a]From Ref. 98.
[b]values for alumina; ϵ° values for silica are roughly 3/4 as large.

short running time. A number of gradient generators are commercially available today. A simple, inexpensive, but very useful, apparatus for the generation of various gradients was described by Snuder and Warren [114]. An exponential gradient will usually give the desired linearity in $\alpha\epsilon^{\circ}$. For very small differences between the ϵ° of the two solvent components, the mathematical approximation (Eq. 4) becomes less satisfactory and ϵ° follows a more complex function, viz., that of Eq. (3), where N_B is calculated by:

$$N_B = \frac{V_B}{(1 - V_B)(d_A/M_A)(M_B/d_B) + V_B} \tag{7}$$

V_B is the volume fraction of the strong solvent B in the mixture; d_A and d_B are the solvent densities; and M_A and M_B, their molecular weights [114]. The functionality of the gradient (linear, exponential, etc.) usually refers to the change in the volume fraction with time, not to the change in the eluent strength $\alpha\epsilon^{\circ}$. The gradient rate, which is inversely proportional to the total amount of solvents A and B used in a separation, will depend on the type and complexity of the sample and on the degree of resolution desired. An initial estimate can be made with Eq. (3). A trial run will show how good the estimate was.

Just as in linear elution, mixed solvents containing a water-miscible component (acetone, methanol) will displace the water from a deactivated column. The consequence is a change in the adsorption behavior, which will result in a distorted elution pattern. Adding 1-2% water to the strong solvent will diminish this effect.

Adding a stronger solvent in gradient elution has two effects. First, it increases the solvent strength. In addition, however, it tends to deactivate the adsorbent further. Very small amounts of the stronger solvent in the mixture can cause this deactivation. The effect can therefore take place in the very beginning of a gradient. The stronger solvent can now displace several sample components rather than just the next stronger one from the adsorbent and thus cause a band of unresolved components to move down the column. Scott and Kucera [84] call this the displacement effect. They suggest, for mitigating this phenomenon, using several closely spaced gradients of solvents with only moderate increases in solvent strength rather than using only two or three gradients with very different solvents, as is widely practiced. They established a series of solvents that increase in solvent strength by about the same amount, each giving a retention volume about 2.5 times smaller than the previous one. According to Snyder [5, 115], this small step size is unnecessary besides being time-consuming. Our experience with heavy petroleum fractions bears this out [116]. In the choice of binary solvent combinations, consideration will be given not only to resolving power but also to such fundamental properties as solvency, viscosity (should be low to keep resolution and flow rate high), detector compatibility (e.g., UV absorption in case of UV detectors), and boiling temperature (in preparative separations).

6. Selected Examples

The number of solvent and adsorbent variations that actually have been applied to the adsorption chromatography of heavy petroleum fractions has been rather limited. In Tables 10 through 12, we have accumulated some typical examples that yielded good separations of compound classes and of mono-, di-, tri-, and polyaromatics.

The extent of separation afforded by adsorption chromatography on an alumina/silica dual column was recently illustrated by Coleman et al. [51] and is briefly summarized in Table 13. By mass spectrometry, the saturate fraction was shown to consist of 27.7% normal paraffins and isoparrafins, 22.3% mononaphthenes, 15.8% dinaphthenes, etc. The small amount of monoaromatics in the fraction (0.7%) consists mainly of alkylbenzenes. The second fraction contains the bulk of the monoaromatics. A large proportion of these are fused with up to seven naphthene rings. The diaromatic fraction consists predominantly of alkylnaphthalenes with and without fused naphthene rings. The last fraction was very complex, containing numerous polyaromatic and mixed aromatic-naphthenic hydrocarbons as well as sulfur and nitrogen compounds. Gel permeation chromatography further subdivided the mixtures to the point at which mass spectroscopy and NMR gave sufficiently simple spectra for a detailed analysis. The GPC fractions are further discussed in Sect. IIIB4.

B. Gel Permeation Chromatography (GPC)

1. Advantages and Methodology

Gel permeation chromatography (GPC), or exclusion chromatography, has been extensively employed in analyzing petroleum heavy ends. Besides being now a rapid and simple routine method, it has the unique advantage of separating a mixture by molecular size only. (This facet was mentioned in Sect. IIA2.) There are exceptions to this rule; they are treated in some detail in Sect. IIIB2. Nevertheless, a rapid analytical GPC run without any further analysis or with only a few MW determinations will reveal the approximate molecular weight (MW) distribution of a sample. Changes in the MW distribution due to aging, weathering, oxidation, etc. are easily and rapidly detected by GPC of samples taken before and after the change or at different stages. Examples of these applications are the elution patterns of petroleum pitches and coal tars that Edstrom and Petro [28] established, the fingerprinting of crude oils that Done and Reid [43] and later albaugh and Talarico [44] described, the aging studies of asphalts by

TABLE 10

Selected Examples of Separations by Adsorption Chromatography

Sample, adsorbent, authors	Solvent	Eluted species
1. Hydrocarbon fraction from distillates, > 400° F, on O-H_2O alumina. D. M. Jewell, R. G. Ruberto, and B. E. Davis [56].	n-Hexane/cyclohexane gradient Cyclohexane/chloroform gradient Chloroform/methanol gradient	Monoaromatics Diaromatics
2. Hydrocarbon fraction from distillation cut on O-H_2O silica. D. M. Jewell, J. H. Weber, J. W. Bunger, H. Plancher, and D. R. Latham [48].	n-Pentane Chloroform followed by methanol	Saturates
3. Hydrocarbon fraction of 375-535° C petroleum distillate on O-H_2O alumina-silica column. H. J. Coleman, J. E. Dooley, D. E. Hirsch, and C. J. Thompson [51].		Saturates Monoaromatics Diaromatics Polyaromatics and polar compounds
4. Hydrocarbon fraction from distillates, 650-1000° F on O-H_2O. alumina-silica column. D. E. Hirsch, R. L. Hopkins, H. J. Coleman, F. O. Cotton, and C. J. Thompson [47].	n-Pentane Pentane + 5% benzene Pentane + 15% benzene 60% methanol + 20% diethyl ether + 20% benzene followed by methanol	Saturates Monoaromatics Diaromatics Polyaromatics + polar compounds

5. Petroleum acids converted to hydrocarbons on 3% H_2O neutral alumina. (Activity II). W. K. Seifert and R. M. Teeter (64). More detail is given in Tables 10 and 11. See also Fig. 11.	Cyclohexane Cyclohexane/diethylether gradient	1. Naphthenes, monoaromatics 2. Diaromatics 1. Fluorenes, phenanthrenes, naphthobenzthiophenes 2. Phenanthrenes, carbazoles 3. Unidentified 4. Phenols, S, N, O compounds
6. Hydrocarbon fraction from distillates, 400-1000° F on alumina + 4% H_2O. L. R. Snyder [58].	n-Pentane n-Pentane n-Pentane + 50% ethylether	Saturates Monoaromatics Diaromatics
7. Hydrocarbon fraction from distillates, 400-1000° F on O-H_2O alumina. L. R. Snyder [58].[a]	Pentane + 25% benzene 50% benzene-methanol	Polyaromatics Neutral heterocompounds
8. Hydrocarbon fraction from distillates, $>400°F$, on O-H_2O alumina. Snyder and Roth [117].	n-Pentane	Total saturates (fast routine analysis; relation between elution volume and C number)
9. Hydrocarbon fraction of a petroleum well core extract on O-H_2O silica. J. L. Oudin [7].	n-Heptane Benzene	Saturates Aromatics

TABLE 10 (cont)

Sample, adsorbent, authors	Solvent	Eluted species
10. Aromatic fraction from above (Oudin) on O-H_2O alumina.	Heptane, 2kg/cm^2	Monoaromatics
	Heptane, 10 kg/cm^2	Diaromatics
	Heptane/ether 1:1, 10 kg/cm^2	Polyaromatics
11. Deasphaltened residuum on O-H_2O alumina. L. W. Corbett [14].	n-Heptane	Saturates
	Benzene	Saturates
	Benzene	Naphthene-aromatic
	Methanol-benzene	Naphthene-aromatic
	Trichloroethylene	Polar aromatics
12. Residua, 250°C on 4.5% H_2O alumina. W. R. Middleton [13].	n-Hexane	Saturates
	Gradient in layers of n-Hexane + dichloromethane + THE + methanol	Monoaromatics
		Diaromatics
		Polyaromatics
		Soft resins
		Hard resins
		Asphaltenes
13. Deasphaltene asphalt fractions on 4%-H_2O alumina. K. H. Altgelt [28, 29]. For the composition of the fractions see Ref. [29].	n-Pentane	1. Oil-mostly alkyl naphthenic-(low)aromatic hydrocarbons
		2. Resin 1. Similar to oil except aromatic and some O, N, and S.
	Cyclohexane	Resin 2. Mainly prophyrins
	Benzene + 10% methanol	Resin 3. More aromatic, much more O, N, and S than Resin 1.

14. Distillate 425-455°C on O-H_2O silica. H. V. Drushel and A. L. Sommers [75].	n-Pentane	1. Saturates 2. (Alkylbenzenes, alkylthiophenes)[b]
	Benzene	Aromatic hydrocarbons and S compounds
	Benzene + 10% methanol	N and O compounds
15. Benzene fraction from above [Drushel and Sommers] oxidized on silica (O-H_2O).	Benzene	Aromatic hydrocarbons
	p-Dioxane	Sulfones
	Benzene + 10% methanol	Carbonyls, pehnols, etc.
16. Mixture of aromatic hydrocarbons, sulfides, and sulfoxides on silica.	Benzene	Aromatic hydrocarbons and sulfides
Drushel and Somers [75].	Benzene + 10% methanol	Sulfoxides

[a]Similar earlier separations by Snyder are found in Refs. 81, 92, and 117.
[b]Not identified.

TABLE 11

Seifert's Chromatography of Hydrocarbons on Neutral Alumina[a]

Fraction	Eluent, %				Yield wt.-%	Ultraviolet bands and shoulders, nm[b]											Infrared absorptions, cm^{-1}[c]				Assignments (Friedel-Orchin)
	C[d]	E[d]	M[d]	ml		230	250-260	270	280	290	300	310	320	330	340	360	1700-1740	3020-3070	3483	3610	
2	100	0	0	100	24.1	x	x	x	x								0	x	0	0	Naphthenes and monoaromatics: benzenes, indanes tetralins, octa-hydrophenanthrenes.
3-6	100	0	0	1500	15.4	xxxx	xx	xx	xx	xx	x	x	x (s)	x	x		x	xx	0	x	Diaromatics: acenaphthenes, naphthalenes, benzthiophenes.
7-10	95-99	1-5	0	1400	6.1	x	xxx	xx	x	(s)	(s)	x	x	s	s		x	x	0	x	Fluorenes, phenanthrenes, naphthobenz-thiophenes.
11-12	90-93	7-10	0	1000	4.2	xx	xxx (d)	x	x	xxx	xx		s	x	s	x	x	x	x	0	Phenanthrenes, carbazoles
13	75	25	0	500	3.5	xx	xxx	x	x	xxx	xx		s	s	s	s	x	x	xx	0	

TABLE 11 (cont)

Fraction	Eluent, %				Yield wt.-%	Ultraviolet bands and shoulders, nm[b]											Infrared absorptions, cm^{-1}[c]				Assignments (Friedel-Orchin)
	C[d]	E[d]	M[d]	ml		230	250–260	270	280	290	300	310	320	330	340	360	1700–1740	3020–3070	3483	3610	
14–18	0–60	40–100	0	2600	14.8						← Similar to Cut 13 →						x	x	Hydrogen bonding	xx	Phenols, S, N, O compounds ↓
19	0	99	1	300	20.3		xxx ←				Nondescript →						x	x	Hydrogen bonding	xx	
20–22	0	0	100	1300	11.5						Nondescript						x	0	Hydrogen bonding	x	
Total					99.9																

[a] From Ref. 69.
[b] x = weak; xx = medium; xxx = strong; xxxx = very strong; (s) = shoulder; (d) = doublet.
[c] In carbon tetrachloride solvent.
[d] c = cyclohexane; E = diethylether; M = methanol.

TABLE 12

Seifert's Chromatography of Hydrocarbons on Acidic Alumina[a]

Fraction	Eluent, %				Yield wt. %	Ultraviolet bands and shoulders nm[b]									Infrared absorptions, cm^{-1} [c]			Assignments
	C[d]	E[d]	M[d]	ml		228	232	255	273	285	290	300	305	328	1710–1740	3020–3070	3610	
1-3	100[f]	....	...	1100	7.8	← Inactive →									xx	0	0	Naphthenes, carbonyl compounds
4	99.5	0.5	...	20	32.7		xxxx	xxx			(s)		(s)	(s)	x	xx	0	Trisubstituted naphthalenes, acenaphthenes, perinaphthenes, benzthiophenes
5-7	99.5	0.5	...	160	18.8	← Like Fraction 4 →												
8-9	99	1	...	740	10.3	xxx[d]		xxx		(s)		(s)		(s)	x	x	x	Diaromatics like fractions 4-7 phenanthrenes, benzthiophenes
10-11	30	70	...	1000	14.6		xx	xx		(s)		(s)		(s)	x	x	0	Phenanthrenes
12	....	99.5	0.5	65	21.8	xxx	← Broad nondescript →								xx	x	x	....
Total					96.0													

[a]From Ref. 69.
[b,c,d](As in Table 11.)
[f]n-Hexane.

TABLE 13

Composition of the Hydrocarbon Fractions Obtained by the Application of the API-60 Separation Method to a 370° C–535° C Distillate[a]

Saturates		Monoaromatic		Diaromatics		Polyaromatics	
Rings[b]	Wt-%	Series[d]	Wt-%	Series[d]	Wt-%	Series[d]	Wt-%
0	27.7	-12	17.8	-12, -25, -165	11.8	-26, -28, -165, -305, -65_2, -205_2, -245_3, ?	14.3
1	22.3	-14	15.0	-14, -45, -185	17.6	-28, -45, -185, -325, -85_2, -225_2, -265_3	12.0
2	15.8	-16	10.3	-16	19.4	-16, -30, -65, -205, -105_2, -245_2, ?	11.0
3	12.2	-18	6.7	-18	16.1	-18, -32, -85, -225, -125_2, -265_2	17.3
4	12.0	-6, -20, -105	14.4	-20, -105	16.2	-20, -34, -105, -245, -285_2	16.3
5	5.7	-8, -22, -125	17.5	-22, -125	11.3	-22, -125, -265, -305_2	15.7
6	3.6	-10, -145	18.3	-24, -05, -145	7.6	-24, -05, -145, -285	13.4
MA[c]	0.7						

[a]From Ref. 44.
[b]Naphthenic rings.
[c]Monoaromatics.
[d]Mass spectral series.

Minishull [31], Haley [40], Bynum et al. [41], and Dougan [42], investigations of the effect of air blowing on asphalts [31, 116], and the change of asphaltene molecular weights after steam stimulation of oil wells [118].

Most GPC runs of residua or asphalts are made on a preparative scale. Gram quantities are needed because often the fractions are used for further analysis by chemical and physical methods or they are further separated by other methods. In many cases, MW distributions of heavy petroleum fractions are obtained by direct MW measurements on all or some of the GPC fractions. Preparative GPC does not have to be inferior to analytical GPC in resolution. It will generally take more time, but well-packed large columns can have the same or better efficiency than analytical ones with small diameters.

General GPC technique and apparatus have been adequately described in the literature [2, 119-123]. Here we want to point out a few factors in methodology that have special application in the GPC of heavy petroleum samples. First of all, care must be taken to avoid oxidation. In our laboratory, we purge the solvent reservoirs with nitrogen. Before the solvent enters the pump, it passes through a 4-in. long heated section of 1/4-in. stainless steel tubing attached to the bottom of the reservoir in such a way that vapor and air bubbles rise through the bulk of the solvent into the nitrogen atmosphere. If the temperature of the tubing is adjusted to give very light boiling, e.g., one vapor bubble every 2 to 3 sec, no problems with air or nitrogen in the solvent are encountered. Furthermore, if the solvent is used up in one or two days, no significant change in solvent composition (in case of mixed solvents) is observed. The solvent composition is easily checked by determining its refractive index and can be restored to the original composition if necessary. A reflux condenser attached to the top of the reservoir may be required for very low-boiling solvents.

A second factor is the gel. We prefer the heavily cross-linked polystyrenes such as the Styragels and Poragels that Waters Associates or the Biobeads from Bio-Rad. Compared with the softer lightly cross-linked gels, these materials give more stable packings and allow considerably faster separations, though at the expense of resolution. Porous glass and silica, on the other hand, strongly adsorb asphaltenes and other heavy polar compounds. Silylation of these packings does not seem to alleviate or even lessen this problem. Some adsorption occurs also with the organic gels but much less than with the inorganic ones.

Very important for good results with petroleum components is the proper solvent, especially in the context of adsorption. In our experience, chloroform + 5% methanol or benzene + 5% methanol give the best results in solvency, suppression of adsorption, and ease of evaporation. (see also Ref. 124). Even better solvents are trichlorobenzene and α-methylnaphthalene + 5% methanol; but they have the disadvantage of a high boiling point, and recovery of the fractions is cumbersome. Since these solvents cannot be evaporated, they have to be separated from the solute on a short, small-pore GPC column. Methylene chloride containing 5% methanol would also be a good choice except for its very low boiling point, which would make it necessary to refrigerate the solvent reservoir or at least use a cold finger to avoid excessive losses.

Glass columns are very helpful because they allow us to see what is going on. Heavy petroleum compounds are usually black or brown and can therefore be observed in their progression along the column. Uneven flow or the presence of air, in case of leakage, is easily recognized; and immediate corrective action can be taken. Also, we can see early in the separation if a sample is badly adsorbed. In such a case, we may decide to backflush the sample out to save the column; or we may want to clean up the column right after the run by rinsing it with a more powerful solvent, e.g., a mixture of α-methylnaphthalene and pyridine. Whenever strong adsorption is suspected, a precolumn should be used, preferably one also made of glass. It is easier and cheaper to clean or discard a small precolumn than the larger main column.

In work with glass columns, pump rates are restricted in that only limited pressure can be applied, depending on column diameter and wall thickness. W. Lame's formula

$$P = 2S \frac{(OD - ID)^2}{(OD + ID)^2} \approx 2S \frac{W}{ID}$$

with OD = outer diameter, ID = inner diameter, and S = 68 atm (1000 psi) has been suggested for estimating a safe upper pressure limit for Pyrex columns (safety factor of 10). Under the assumption of medium wall thickness, this means that the pressure should not exceed 150 psi for 1/2-in. (12.7-mm) column diameter and 100 psi for 1-in. (25.4-mm) diameter. Avoiding exposure to bright sun or other light containing a substantial amount of UV frequencies helps to prevent photoreactions, which might change the sample and aggravate adsorption on the gel.

Even with the best precautions, some adsorption of asphaltenes cannot be avoided; and after a number of runs, the columns take on brown or grey shades. In our experience, this gradual discoloration usually does not impair performance, especially if the columns are regularly and generously rinsed between runs. Eventually, however, heavily stained column packings will increase the adsorption tendency of a sample and must be discarded. Again, a glass column makes it much easier to recognize when this point has been reached, compared with an opaque metal column.

Glass columns cannot be packed by the usual high-pressure method. However, very good results are obtained with most polystyrene gels in the following way. An auxiliary column fitted with a large funnel or an open flask on top is attached to the main column. The latter is filled to about one-half of its length with acetone. A slurry of the gel in enough acetone to have about one-half as much supernatant liquid as swollen gel is poured all at once into the two columns. The valve at the bottom is opened to allow the solvent to drain. While the gel is settling, the columns are turned back and forth around their longitudinal axis about once every 2 to 3 sec. The turning action agitates the gel when it settles on the packed surface and thus helps to pack it more tightly. Furthermore, it serves to even out any irregularities arising from a slight deviation from the vertical position of the columns, or more important, from the eddies caused by the gel falling through the solvent. When the gel has settled, the level of the solvent is lowered to the top of the gel. The solvent flow is then stopped, excessive gel and the auxiliary column are removed, and the main column is closed with its upper end fitting, which is connected to the solvent pump. About one column volume of acetone is then pumped rapidly downwards. If some settling should occur, either the fitting is readjusted or more gel is added. When no more settling takes place, the solvent is changed stepwise from acetone to the intended solvent, usually chloroform + 5% methanol. The polystyrene gel swells slightly more in the good final solvent than in the poor initial solvent acetone. Consequently, the packing becomes tight and quite regular, to judge from the high efficiency usually achieved, viz., 800-3000 theoretical plates per foot (2600-10,000 TP/meter) as measured with benzene. The back pressure remains reasonably small with 30-40 psi at a flow rate of 600 ml/hr in a column of 1-in. diameter and 36-in. length. With this packing technique, it is impossible for the bed to shrink when a change is made from chloroform to a poorer solvent. No solvent suitable for petroleum components will shrink the polystryene gel more than the initially used acetone.

2. Polynuclear Ring Systems

We have stated that GPC separates by molecular size and not by chemical structure. The discriminating parameter is, in good approximation, the hydrodynamic volume [39, 126], or for small molecules, the molar volume [127, 128]. The hydrodynamic volume of polymers and other large molecules can be expressed as the product of intrinsic viscosity $[\eta]$ and weight-average molecular weight M_w. The plot of molar volume or of hydrodynamic volume against elution volume is a single curve for a given column irrespective of solvent and sample type [39, 121]. This is very important is general and for petroleum compounds in particular because of their great variety of chemical structure. However, in the case of polynuclear ring systems, this principle is seemingly violated by an anomaly, the cause of which has not been clearly identified yet but is probably related to partitioning or adsorption of these compounds on the gel [97, 129-133]. According to Oelert et al. [97, 129-132], the plot of molar volume vs elution volume yields three different relations for the three groups: paraffinic and monoring hydrocarbons, catacondensed aromatic hydrocarbons, and pericondensed aromatic hydrocarbons. This is shown in Fig. 23. The catacondensed compounds follow a curve with decreasing MW for increasing V_e as is expected in GPC, but with a considerably steeper slope than the curve for the paraffins. The pericondensed hydrocarbons, however, increase in MW with increasing V_e, indicating a predominant sorption or partitioning rather than an exclusion effect. Oelert et al. [132] established similar relations for N- and O-containing compounds. Note that the samples used in these studies were model compounds, presumably synthetic ones, and generally without alkyl substituents. This means that they constituted extreme cases that are not necessarily representative of the majority of polynuclear aromatic compounds present in petroleum. Whatever the

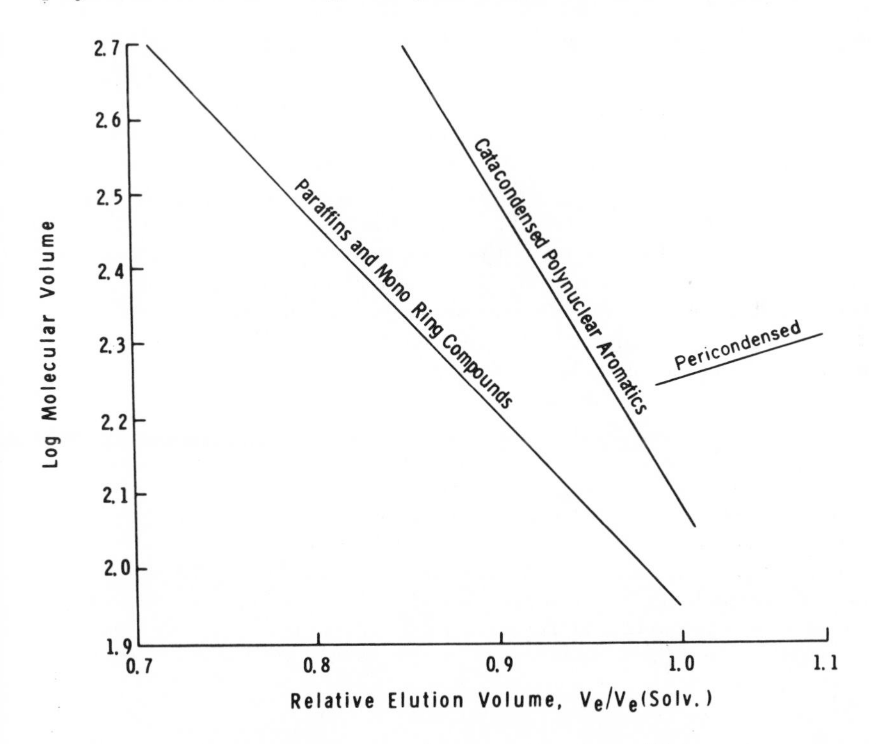

FIG. 23. Calibration curves, log M vs V_e of paraffinic, catacondensed, and pericondensed hydrocarbons. (From Ref. 130. Courtesy Separation Sci.)

alkyl-substituted samples included in the comparison, they eluted at lower volumes than their parent compounds. This is to be expected not only because of their greater molar volume but also because of their higher solubility and because of a shielding effect, both of which minimize adsorption.

Bergmann et al. [133] observed the same elution anomalies with numerous model compounds, including n-alkanes, cata- and pericondensed aromatics, hydropyrenes, and heterocyclics. However, they could practically eliminate the differences in elution behavior by carrying out the GPC with trichlorobenzene (TCB) as solvent. They performed most runs at 140° C, but they found that TCB overcame the anomalies even at room temperature. This was the case even though their model compounds did not contain any alkyl substitution and included molecules as large as coronene (300 MW) and ovalene (396 MW). These authors observed an increasing elution anomaly with decreasing sample solubility in different solvents. Such a relation points to partitioning as the underlying mechanism; although it does not preclude adsorption.

It is probably fair to state that adsorption and partitioning <u>can</u> be important in the GPC of polynuclear ring systems. Since these principles counteract the exclusion effect, we must be aware of their possible existence in GPC and try to avoid them. Their effect will be smaller with increasing alkyl substitution of the molecules-which is known to increase their solubility and would also tend to shield their aromatic rings- and with increasing solvent power and temperature. Oelert et al. [44] suggested taking advantage of the sorption effect for separations based on chemical structure by using carefully selected solvents and gels. While this may lead to simple and effective fractionations in specific cases [45, 97], we believe that a more generally valid method would combine GPC under conditions of optimum exclusion with adsorption chromatography on deactivated alumina. With this latter combination, the separations by molecular size and by chemical structure can be more easily kept from interfering with each other and should therefore give better results. Gel permeation chromatography of asphalts and crude oil residua has generally yielded fractions of decreasing molecular weight and color intensity [25-36, 40-44]. Yet, investigation of high- and low-MW asphaltenes by NMR, elemental analysis, and density measurements indicated higher degrees of aromaticity and ring fusion in the low-MW fraction [29], quite possibly as a consequence of some adsorption on the gel columns. In plots of these properties and of S, N, and O contents versus MW or elution volume, V-shaped curves are obtained with distinct minima in the medium ranges [29, 124, 134].

3. Calibrating and Determining MW Distributions

Most GPC runs of nonpetroleum samples are performed in analytical columns with milligram amounts. Calibration curves are used to convert, often by computer, elution volumes into molecular weights. This technique depends on the knowledge of the relation between hydrodynamic volume, in case of polymers, or molar volume, in case of small molecules, and molecular weight. The relation of hydrodynamic volume $[\eta]M$ with M, or of intrinsic viscosity $[\eta]$ with M, depends on the configuration of the molecules [121]. Once this relation has been determined for a polymer across the whole MW range, the MW distribution of any sample of this polymer type can be determined by GPC without any further measurement by using a calibration curve.

However, the situation is different with heavy petroleum samples. These are not homologous mixtures differing only in MW. Even within a given crude oil or asphalt, we have a tremendous variety of different configurations, ranging from largely paraffinic wormlike molecules to compact, nearly spherical or disk-shaped polynuclear ring systems. We should not expect these constituents to follow the same $[\eta]$-M relation. Likewise, petroleum samples of different origin should have different $[\eta]$-M relations. This has indeed been found [26, 29]. In Fig. 24, $[\eta]$ is plotted versus M for various asphaltene fractions derived from four crude oils with different chemical composition. Three curves result reflecting the different compactness of the molecules though not necessarily their paraffinic, naphthenic, and aromatic character. Differences between the asphaltene and the resin fractions of one asphalt are seen in Fig. 25. These differences are paralleled in the MW-elution volume calibration curves.

Another problem encountered with petroleum samples is the variation of the detector response with the type of material monitored. Refractive index, UV, and visible light absorption all depend greatly on the aromaticity, the degree of ring fusion, and the size and nature of the polynuclear aromatic ring systems. Variations by factors of 100 and more have been observed in our laboratory [116] between the light absorption of different GPC fractions of asphaltenes as well as of whole asphalts. Snyder [124] reported similar results, although only in qualitative terms. Even IR absorption in the C-H stretch vibration range at 2900 cm^{-1} becomes unreliable for highly aromatic samples. The only true automatic mass detection is by flame ionization. Flame ionization detectors with chain or wire mechanisms for sample transportation have been very problematical in the past. More recently they seem to have been refined to the point that now they can be used for routine measurements [44]. For certain applications, RI detectors also may give satisfactory results (see Sect. IIA2).

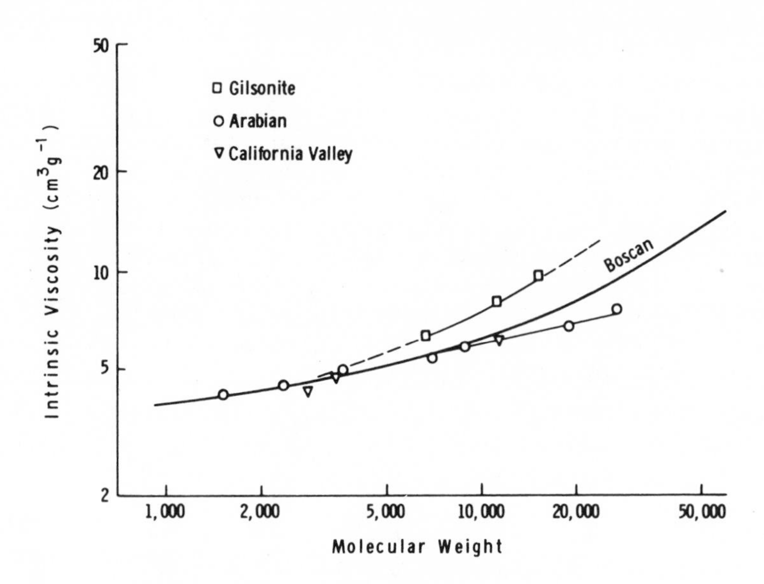

FIG. 24. [η]-Molecular-weight relation of various asphaltenes. (From Ref. 29. Courtesy Bitumen, Teere, Asphalte, Peche.)

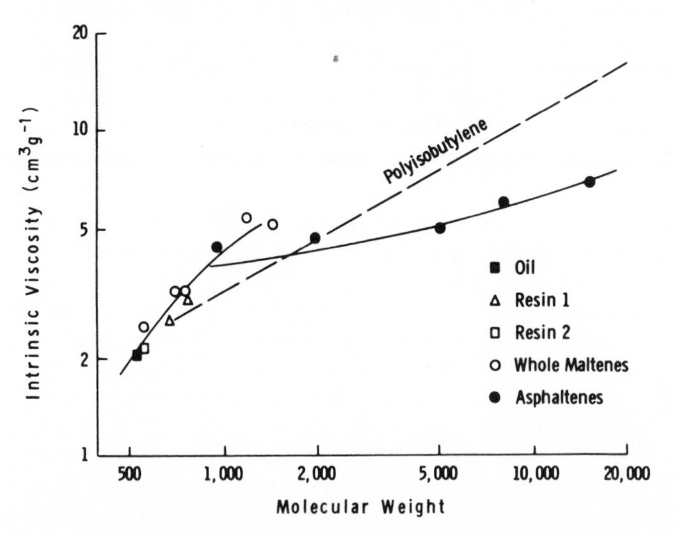

FIG. 25. [η]-Molecular-weight relation of oil, resin, and asphaltene fractions of a steam-refined Boscan asphalt. (From Ref. 29. Courtesy Bitumen, Teere, Asphalte, Peche.)

In general, however, we feel that, except for routine measurements on well-characterized or fairly uniform samples, the MW distribution of heavy petroleum fractions as determined by GPC should not be based on analytical runs and on the use of a universal calibration curve without additional measurements. The only safe method is to fractionate enough material (about 0.5 g or more) on a preparative GPC column and to measure the MWs of several fractions. These MWs plotted against the elution volume will result in a specific calibration curve that can be used for the inter- and extrapolation to all remaining fractions of that particular sample, and if necessary, to elution volumes in between fractions. Corrections for axial dispersion (peak broadening) are usually unnecessary for heavy petroleum samples because of their broad MW distributions.

For quick and easy assessment of the quantitative MW composition of a sample, the results are displayed by an integral MW distribution curve such as that in Fig. 26. Here the cumulative weight percentages of the fractions are plotted versus MW or log MW. Often the differential MW distribution is preferred because it gives a more graphic and easily comprehensible illustration of the MW composition. An example is Fig. 27, which shows the differential distribution curve (DDC) of the same sample as that depicted in Figure 26. The ordinate in Fig. 27 is the slope of the curve in Fig. 26. The ordinate in Fig. 27 is the slope of It can be calculated from the equation [29, 137].

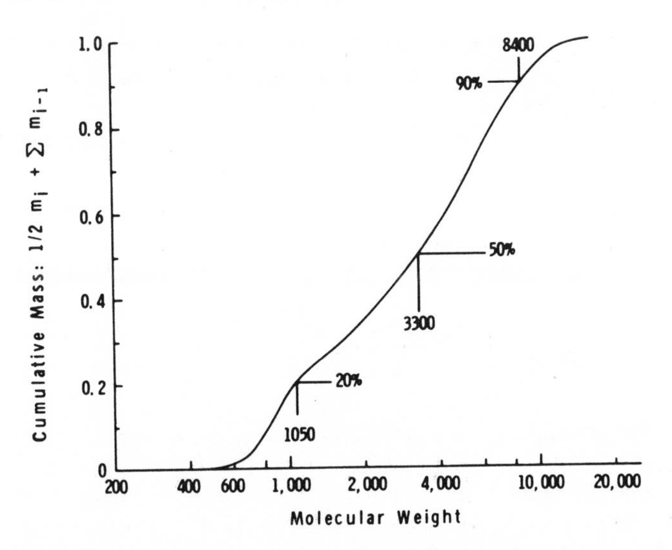

FIG. 26. Integral molecular-weight-distribution curve of a model mixture of three components. (From Ref. 29. Courtesy Bitumen, Teere, Asphalte, Peche.)

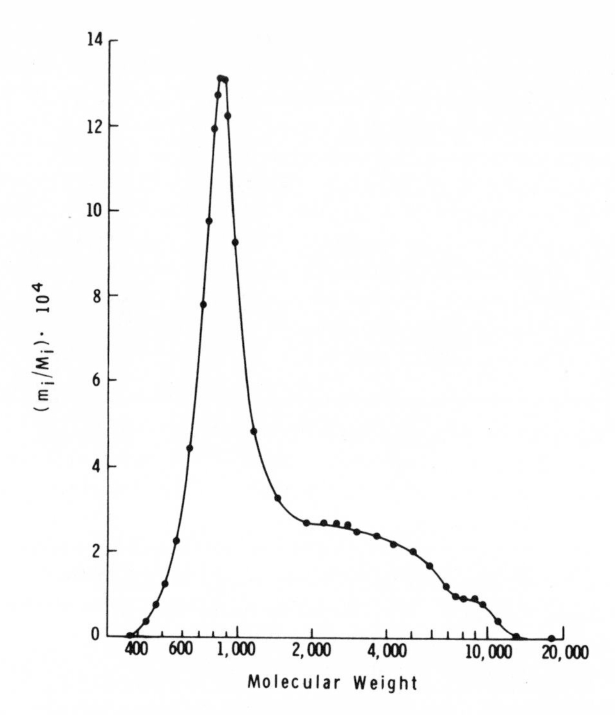

FIG. 27. Differential molecular-weight distribution of a model mixture of three components (same as in Fig. 26). (From Ref. 29. Courtesy Bitumen, Teere, Asphalte, Peche.)

$$\frac{d\sum_{i=1}^{j} m}{dM_i} = \lim_{Mj-1 \to M_j} \left[\frac{m_j}{M_j - M_{j-1}}\right] = \lim_{V_{e(j-1)} \to V_{e(j)}} \left[\frac{m_j}{V_{ej} - V_{e(j-1)}}\right] x \lim_{M_{j-1} \to M_j} \left[\frac{V_{ej} - V_{e(j-1)}}{M_j - M_{j-1}}\right] \tag{8}$$

by using a computer or even a desk calculator. Here M_j is the mass percentage of fraction j; M_j, its MW; and V_{ej} its elution volume. The first factor on the right-hand side of Eq. (8) defines the concentration of fraction j, the second one the reciprocal slope of the calibration curve at point j. Thus, the DDC is a plot of the product of (normalized) concentration and reciprocal slope of the calibration curve of fraction j against its MW.

It is interesting to compare the DDC of a sample with its elution curve. In Fig. 28, we see the elution curve of the same sample as that used for Figs. 26 and 27. The peaks and valleys observed in the elution curve (Fig. 28) are reflected in the DDC although in a grossly distorted fashion. This comparison demonstrates how difficult it is to estimate the MW distribution from the appearance of an elution curve without the proper evaluation. The primary reason for the distortion is the logarithmic relation between MW and V_e. An additional contribution to the distortion is the nonlinearity of the calibration curves. The greatest uncertainty in the shape of the DDC is near the peaks and the valleys. These regions require the most data points, therefore. Here it is often good practice to interpolate between fractions, whereas in other regions one may even leave one or several fractions out of the calculation.

4. GPC and Structural Analysis

Done and Reid [43] first used GPC as a rapid tool for classifying or recognizing petroleum samples. Albaugh and Talarico [44] expanded this method by monitoring the eluate with three detectors in series. A flame ionization detector registers the total mass in good approximation. Ultraviolet absorption and refractive index measurements allow some deductions about the chemical composition. The elution volume itself is a function of MW. With this combination, a fair amount of qualitative and semiquantitative information can be obtained rapidly and automatically. The amount of useful information is multiplied by applying this technique to fractions obtained by alumina chromatography or by one of the more elaborate separation schemes. (See fig. 4.)

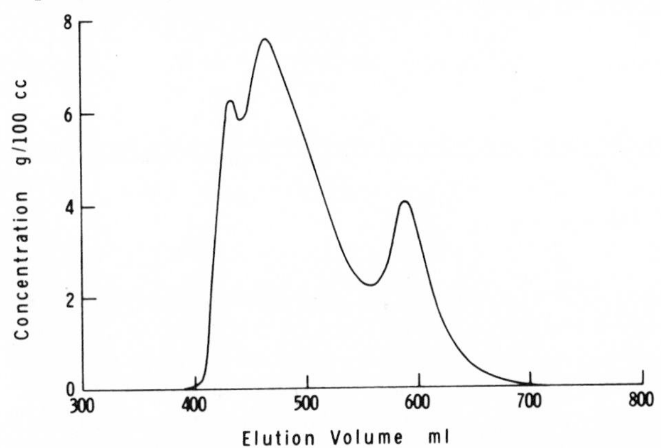

FIG. 28. Elution curve of a model mixture of three components (same as in Figs. 26 and 27). (Courtesy Bitumen, Teere, Asphalte, Peche.)

More quantitative results are derived when fractions obtained by GPC and possibly other methods are subjected to structural analyses. Heavy distillates and the lighter portions of residua can be analyzed by mass spectrometry; the nonvolatile compounds require less straightforward measurements. Examples for both types of approaches are given in a paper by Coleman, Hirsch, and Dooley [37]. Nuclear magnetic resonance data provide structural information on GPC fractions of the nonvolatiles, mass spectrometry on GPC fractions of the distillate cuts. Molecular-weight determinations round out the picture.

Oelert [138-143] and Haley [40] applied group-type analysis to GPC fractions of petroleum and coal and studied their chemical structure as a function of molecular weight. They based their analyses on elemental composition, IR, and NMR data. NMR and elemental analysis of C, H, and S was used by Dickson, Wirkkala, and Davis [134] for the characterization of GPC fractions from a crude oil residuum. Molecular weights were determined by vapor-phase osmometry and compared with "unit weights" obtained from NMR. Altgelt and Hirsch [136] fractionated a whole asphalt by GPC. After further separation by n-pentane extraction and alumina chromatography [28, 29], the fractions were characterized by "integrated structural analysis." By this term the authors describe the combination of NMR, elemental analysis, MW, and density determinations with a computer program that reconciles all experimental data and evaluates them in terms of structural parameters [137].

These three approaches yield only statistical averages as results. Only average values of aromaticity, ring size, degree of ring fusion, etc., are obtained and obtainable. But because of this very fact, these methods give the most information if applied to the narrowest fractions. In practice, one has to weigh the effort spent on the separation and the effort spent on the characterization against the needed information. Often one will find sufficient data in the fairly quick results accessible from a preparative GPC run combined with a group analysis of some of the fractions. Jewell et al. [55] discuss the combination of their SARA separation with GPC, GLC, and spectroscopic techniques for the characterization of residual oils.

Coleman, Dooley, Hirsch, and Thompson [51] employed GPC in a different way. They used it for the interpretation of mass spectral data in terms of molecular structures--specifically, for the distinction between the larger straight-chain compounds and the more compact ring compounds. For example, naphthenologs, i.e., saturated fused six-membered ring compounds, elute at the same colume as alkyl-substituted molecules, which are lower in MW by 1.0 to 1.5 carbons per ring number. The combination of GPC elution volumes with mass spectral MWs allowed this team to make structural assignments more easily and with greater certainty and ease.

C. Ion-Exchange Chromatography

1. General Considerations

Ion-exchange chromatography is widely used in sophisticated analyses of petroleum fractions for the isolation and preliminary separation of acidic and basic components in crude oil. It has the advantage of greatly improving the quality of a complex operation; a disadvantage is the very time-consuming nature of the separation. Although the first synthetic ion-exchange resin was reported as early as 1935 by Adams and Holmes [144], its first application to petroleum chemistry appears to have been the work by Munday and Eaves [145], which was published 24 years later. Considerable progress has been noted in the past two decades on the physical forms of the resin, and with it we have also observed an increase in the applicability of the above technique to many different problems.

Cation-exchange chromatography is now used primarily to isolate the basic constituents in a petroleum fraction. Most of these are nitrogen compounds, mainly pyrroles and indoles. Snyder and Buell [146, 147] have studied these compounds extensively. Other weak bases found in petroleum are the pyrindines, pyridines, carbazoles, and amides, including the quinolines [148, 149]. Much work has been carried out on these compounds since they have a deleterious effect in many petroleum refining processes. They reduce the activity of cracking and hydrocracking catalysts and contribute to gum formation, color, odor, and poor storage properties of the fuel [148]. Not all basic compounds isolated by cation exchange chromatography are nitrogen compounds. A study by Okuno, Latham and Haines [150] showed that sulfoxides are also removed in this process. These compounds do not occur naturally in crude oil, but they are readily formed in some oils by mild air oxidation.

Anion-exchange chromatography is used to isolate the acidic components from petroleum hydrocarbon fractions. Most of these compounds are either carboxylic acids or phenolic. Naturally occurring carboxylic acids in petroleum have been the subject of many studies about the origin of petroleum [62, 65] and life on earth. Work prior to 1959 has been covered in excellent detail by Lochte and Littmann [151]. More recent work on these compounds has been reported by Seifert et al. [65, 69], Whitehead [62, 152], Burlingame [153], and others [62, 150, 154, 155].

2. Nature of the Resins

Ion-exchange resins are prepared from aluminum silicates, synthetic resins, and polysaccharides. The most widely used resins have a skeletal structure of polystyrene cross-linked with varying amounts of divinylbenzene. The first products of this kind were gel-type resins. These polymers, which date back to 1945, are still used on a wide scale. They do not have true pores but a loose gel structure of cross-linked polymer chains through which the sample ions must diffuse to reach most of the exchange sites. Since ion-exchange resins are usually prepared as beads that are several hundred micrometers in diameter, most of the exchange sites are located at points quite distant from the surface. Because of the polyelectrolytic nature of these organic resins, they can absorb large amounts of water and swell to volumes considerably larger than the dried gel. Ion exchange is usually associated with water and aqueous solutions of polyelectrolytes, but this does not always have to be the case. Other solvents besides water can be used to induce swelling. The liquid present in the pore structure enhances the diffusion of the ionic species throughout the whole ion-exchange resin particle. The size of the species that can diffuse through the particle is determined by the intermolecular spacing between the polymeric chains of the three-dimensional polyelectrolyte resin. This spacing is generally referred to as the apparent porosity of the gel. Even with lightly cross-linked resins, the maximum apparent porosity is only about 40 Å.

In ion-exchange chromatography of heavy petroleum fractions, we have to consider two additional aspects. Chromatography has to be carried out in a nonaqueous medium, and the average molecular dimensions of many of the compounds to be separated are considerably larger than 40 Å.

The swelling of the cross-linked polyelectrolyte and the rate of exchange depend on the dielectric properties of the solvent; the rate is larger for solvents with a high dielectric constant. Since diffusion depends on the viscosity of the solvent η, Kunin suggested using the ratio of the dielectric constant ϵ to the viscosity for evaluating the effect of solvent parameters on relative ion-exchange performance [156]. Some values of $\epsilon:\eta$ are shown in Table 14.

These data are useful in the choice of a solvent system, especially for chromatography with gel-type resins. For kinetic reasons, solvents with a large value of $\epsilon:\eta$ are preferred. Table 14 shows that rates of exchange in nonaqueous media are generally lower than in an aqueous environment. Lower flow rates are therefore necessary in nonaqueous ion-exchange operations. The data in Table 14 can also be used to transpose data from one solvent system to another. There is a rough linear relation between the capacity of a resin and the value of $\epsilon:\eta$.

An important development of more recent vintage, especially for chromatography in nonaqueous media, involves the macroporous, or macroreticular, resins. These polymers can be described as small, spherical, tough or rigid sponges that have discrete pores ranging from 100 Å to 1200 Å. This structure allows large molecules to diffuse freely through the system and results in a more complete interchange than is

TABLE 14

Solvent Parameters in Ion-Exchange Chromatography at 25° C[a]

Solvent	ϵ	η(poise)	$\epsilon : \eta$
Water	78.5	0.00893	8790
CH_3OH	32.6	0.00545	5980
C_2H_5OH	24.2	0.0109	2220
Benzene	2.28	0.006	380
Methylene chloride	8.80	0.00421	2090
Acetone	20.5	0.00304	6740
Pyridine	12.0	0.00882	1360
Hexane	1.89	0.0033	580
Dodecane	2.01	0.013	155
Lubricating oil	2.5	0.3	8.3
Acetic acid	7.14	0.0116	616
Acetic anhydride	20.5	0.008	2560

[a]From Ref. 21.

possible with the gel-type resins. Because of the greater pore diameter and the more readily accessible exchange sites, solvent selection is not restricted to materials that swell the resin. Macroreticular resins are more resistant to organic fouling than the gel-type resins. Organic fouling can be viewed as adsorption of high-molecular-weight-organic compounds that are not readily desorbed during regeneration [157]. The exchange capacity of these resins is about 20-30% lower than the exchange capacity of the conventional microreticular gel, but this minor disadvantage is greatly outweighed by the advantages. Macroporous ion-exchange beads for specifically nonaqueous applications are commercially available. For the fractionation of petroleum components, they are the only resins of practical use.

Ion-exchange resins can be bought in varying mesh sizes and meshranges. Smaller particles or narrow-mesh fractions or both distinctly improve chromatographic separations as, e.g., reported by Mondino [158] for ion-exchange chromatography, and also in many treatises on general modern gas and liquid chromatography [2, 84, 86, 159]. Using these small-diameter particles may necessitate special packing techniques and special modifications in the procedures described in this chapter.

Narrow-particle-size distribution is not so important if only a gross separation of the acidic or basic fraction from the petroleum hydrocarbon matrix is intended. Although most commercially available resins are used directly as obtained, particle sizing can further improve chromatographic separations. Several sizing techniques have been described in the literature [160-162].

The spherical shape and rigidity of macroreticular ion-exchange resins improves packing and flow characteristics of the columns. For the same pressure drop at a given flow velocity, spherical resins of much smaller size than irregular particles can be used, generally with a concomitant increase in efficiency.

3. Conditioning the Resins

Acidic and basic compounds in crude oil are present only in trace amounts. Artifacts can be easily introduced into the system unless extreme cleanliness is observed for reagents and materials. Therefore, the ion-exchange resins are carefully cleaned and conditioned before use. The conditioning procedure depends on the ionic nature of the resin. Cation-exchange resins are first washed several times with 10 wt-% methanolic KOH. Often the washing is done right in the loosely packed column by percolating it with about 10 vol of the KOH solution. The material is then thoroughly rinsed with distilled water until the effluents are neutral to litmus. This step can be very time-consuming. In some cases the KOH treatment can be omitted, and the resin is simply washed with water followed by acetone and dried under vacuum. The resin is subsequently activated by carefully contacting it with a solution of 10 vol-% HCl in methanol. Heat is released during this reaction, and caution is prescribed, especially if the material is contained in a column. In this case, it is best to begin with a 1% HCl solution in methanol and to follow that with the 10% HCl solution. The activated resin is then rinsed with distilled water until the washings are neutral to litmus. The final purification consists of 24-hr Soxhlet extractions with methanol, benzene, and finally pentane. After the resin is dried in a vacuum oven at slightly above-ambient temperatures, it is finally ready for use.

Most anion-exchange resins are furnished in the water-moist chloride form for maximum stability in storage and shipping. Before use, they are washed several times with a 10 vol-% aqueous HCl solution in methanol. This removes any extractable amines that might be present. The material is then rinsed thoroughly with distilled water until the effluent has a pH of 6 or higher. Activation is carried out by treating the resin with a 10 wt-% KOH solution in methanol. Again, caution is necessary in this step because of the exothermic nature of the reaction. With some resins, higher hydroxide regeneration levels can be obtained by inserting an intermediate bicarbonate conversion step prior to the treatment with dilute alkali. In this case, the resin is first flushed with about 20 bed volumes of 10% $NaHCO_3$ solution, followed by 5 bed volumes of deionized water, and finally 20 bed volumes of 10 wt-% NaOH solution.

After activation, the resin is washed with deionized water until the effluents are essentially neutral to phenolphthalein. Flow rates during these operations should be maintained at about 4 bed volumes per hour. The resin is then extracted in a Soxhlet for 24 hours to remove adsorbed organic material using first methanol, then benzene, and finally pentane. The resin can be dried in a vacuum oven at slightly above ambient temperatures before use.

4. Experimental Aspects

Ion-exchange chromatography is carried out in two or more steps. In the first one, the "exhaustion" cycle, the reactive components are retained on the column and separated from the petroleum hydrocarbon matrix. In the subsequent steps, the polar compounds are displaced from the ion-exchange resin with successively stronger solvents.

Columns 24 or more inches long are usually needed for nonaqueous processes since often the selectivity of the sample material is low. For packing, an empty auxiliary column should be attached to the top of the main column to allow for a 50% bed expansion. The resin is first swollen with the solvent to be used. After packing, the bed is back-flushed with the same solvent at a high enough flow rate to expand it by about 50%. This condition is maintained for 10-15 minutes, after which the flow is stopped and the resin beads are allowed to settle by gravity. This process eliminates all the air pockets. It also tends to classify the resin as a function of height, leading to a lower pressure drop along the column. After classification, the bed is compacted by allowing 2-3 bed volumes of solvent to flow down through the column at a flow rate of approximately 2 bed volumes per hour. Now the auxiliary column is removed and the main column is connected to the inlet system.

In preparative ion-exchange chromatography, it is often advisable to use a water-jacketed column that is connected to a thermostat. This reduces detrimental temperature gradients resulting from the chemical exchanges occurrring in the column, and it allows heating, which accelerates the exhaustion cycle. Most cation-exchange resins are stable to about 120° C. Hydroxide-form quaternary anion-exchange resins, however, should not be exposed to temperatures above 60°C. Even prolonged exposure to temperature above 45° C will lead to loss of some capacity. Other types of anion-exchange resins may be heated to about 75°C. For one-time use, much higher temperatures are permissible.

Flow rates in ion-exchange chromatography are in the order of 2 bed volumes per hour, i.e., much lower than in most other techniques. They are determined by the affinity of the resin for the material to be retained. For cases with slow equilibrium, e.g., when the solvent has a low value of $\epsilon:\eta$, even lower flow rates are necessary to use bed capacities more effectively.

As is general with petroleum fractions, the separations should be carried out under nitrogen.

The amount of sample charged should, as a rule, correspond to about 1 meq of acid per gram of resin. The sample is dissolved in approximately 10 times its volume of pentane, or in the case of wax-containing stocks, of cyclohexane. Insoluble components such as asphaltenes or wax are ordinarily separated by an appropriate method before the ion-exchange step. After the solution is charged to the column by running it through several times, the unreacted hydrocarbons are washed out with the pure solvent by eluting with 5-10 times the volume of the sample solution. In this exhaustion cycle, flow rates should be around 2 bed volumes per hour.

Removal of the acids from the anion-exchange column is carried out by eluting in succession with benzene, methanol, and finally CO_2-saturated methanol. Since the acidity of these eluents increases in the order shown, compounds of increasing acid strength are removed from the column in this process. These three fractions can then be combined or analyzed separately for further characterization.

The basic constituents are displaced from the cation-exchange column by elution with, e.g., benzene, methanol, and finally an 8 vol-% isopropylamine solution in methanol. Prior to the methanol wash, the resin bed is first loosened by forcing nitrogen gas up the column. This precautionary measure prevents column breakage due to the swelling of the resin in the methanol and to the exothermic reaction of the resin with the amine. The three cuts obtained in this process can be combined or analyzed separately.

A sharper separation of the acidic fractions into different classes was achieved in a multistep operation described by Seifert and Howells [65]. These authors combined a preliminary extraction step with three ion-exchange chromatographic separations, as described in Sect. IIB1. With this approach, there is the possibility of the formation by hydrolysis of more interfacially active acid material than originally present in the crude oil.

The first chromatographic separation in Seifert and Howell's scheme was carried out on the weak basic ion-exchange resin, Duolite A-6 (Diamond Alkali Company), by elution with a 2:1 mixture of benzene:ethanol until the eluate became colorless. After solvent removal, the eluted part was rechromatographed through a column of Amberlite IRA 401 with a 2:1 mixture of benzene:ethanol. The eluate yielded a fraction rich in phenols. The retained acids in the Amberlite column were recovered with a 2:2:1 mixture of ethanol-benzene-acetic acid and identified as a mixture of phenols and carboxylic acids.

The strong acids retained on the Duolite column were displaced with ethanol-benzene-acetic acid. After removal of the solvent with n-heptane to azeotrope out the acetic acid, the recovered material was chromatographed again on a Duolite A-6 column. Elution with a 2:1 mixture of benzene-ethanol resulted in a fraction consisting mainly of carboxylic acids and its derivatives; elution with ethanol-benzene-acetic acid mixture finally rendered a fraction that consisted essentially only of carboxylic acids.

A combination of sophisticated instrumental techniques, including mass spectroscopy, showed that the last fraction contained thousands of different petroleum carboxylic acids. A highly fluorescent fraction isolated by preparative thin-layer chromatography from this cut yielded a number of acids with molecular weights between 200 and 700 and a maximum in the 300-400 range [72] and with structures comprising terpenoids, polynuclear saturates, mono-, and polynuclear aromatics, heterocyclics, and naphthenoaromatics. The most abundant species contained two, three, and five saturated rings and fused polynuclear structures. Many of these compounds were hitherto not known to be present in petroleum.

The use of anion-exchange resins to separate petroleum acids into subfractions of different acidity has also been reported by Jewell, Latham, McKay, and others [48, 63]. Their separation scheme is different from that of Seifert et al. Jewell et al. make extensive use of Amberlyst A-29 resin in the hydroxide and in the carbonate forms. The acid concentrates are first passed through the resin in the carbonate form where the stronger acids are retained. The eluate is introduced into a second column packed with the resin in the hydroxide form where weaker acids are retained. The effluent from this column contains the weakest acids.

This separation scheme is especially useful for obtaining concentrates of nitrogen compounds and weakly acidic phenols. Its disadvantages are the lengthy operations and a 15-20% weight loss by irreversible adsorption on a kaolin column.

5. High-Resolution Ion-Exchange Chromatography

With the advent of high-resolution liquid chromatography and the development of specialty column packings, ion-exchange chromatography can also be used as an analytical technique. In classical ion-exchange chromatography, a big contribution to the time required for analyses and to the band broadening is due to the resistance to mass transfer in the stationary phase. Equilibration times can be greatly reduced by using small resin particles. Porous ion-exchange resins in very narrow mesh size ranges around 10 μ are now available commercially. Although the resultant columns are more efficient per unit length, they also show sharply increased pressure drops for the same flow velocity. This may be of minor concern because modern liquid chromatographs have provisions to generate high inlet pressures.

Another elegant solution to the conflicting requirements of fast equilibration and low-pressure drops has been attained in the modern pellicular packings, where the ion-exchange resin is deposited as a thin coat on the surface of a spherical fluid-impermeable core [2, 163]. These beads range about 30-50 μm in diameter with a resin coating of only a few micrometers thickness. The shallow depth of the active layer permits rapid equilibration, and the relatively large impermeable core contributes to packing stability and a low resistance to flow. Although the new column packings are capable of carrying out rapid separations, their capacity is necessarily much lower than the capacity of conventional resins. The greater rapidity and much higher resolution these resins afford usually far outweigh the disadvantage of decreased capacity. These column packings are now available commercially from several sources.

D. Gas Chromatography

Gas chromatography has proved to be an exceptional and versatile instrumental tool for analyzing compounds that are of low molecular weight and that can be volatilized without decomposition. However, because of these constraints, the principal applicability in petroleum chemistry has been to hydrocarbons of low-to-medium boiling range. The use of this technique for direct component analysis in the heavy ends of petroleum is fraught with many problems. The number of possible components of a certain molecular weight range increases geometrically with increasing molecular weight; at the same time, a corresponding sharp decrease is observed in the differences in physical properties between isometric structures. This makes it very difficult for gas chromatography to separate and identify single components in heavy petroleum streams unless they are present in rather large quantities. Complete component analysis is generally out of the question, unless the complexity of the sample has been reduced by other fractionation techniques.

Because of the substantial molecular weights involved, long residence times are inevitable. This is compounded by the necessity of using long columns to obtain adequate resolution. Increasing the column temperature would alleviate the situation to a certain degree, but the gain is offset by the increased danger of thermal degradation.

Despite these problems, gas chromatography is still widely used in special applications in analyzing higher-molecular-weight petroleum hydrocarbons. Three primary areas of application are the conventional analytical procedures, pyrolysis gas chromatography, and inverse gas liquid chromatography.

1. Conventional Analytical Procedures

Conventional gas chromatography has a place in the analysis of heavy petroleum fractions, viz., for the qualitative and quantitative determination of simple constituents, mostly hydrocarbons, that are present in larger amounts than most of the other components.

Because of the high molecular weights involved, resolution as ordinarily expected from GC has to be sacrificed if meaningful chromatograms are to be obtained with reasonable column temperatures and in a practical length of time. Short columns are almost mandatory. Although in many cases packed columns can be used, the most versatile column for these higher-molecular-weight hydrocarbon mixtures appears to be a short capillary column [164]. Besides exhibiting an exceptional degree of resolution, it gives practically quantitative elution with many heavy materials that are retained in ordinary packed columns, even when very light loading levels are employed. Gouw, Jentoft, and Whittemore [164] demonstrated the advantage of the capillary column over a packed one by analyzing a heavy lubricating oil (260 SUS at 210° F) to which n-eicosane had been added as internal standard. Only 40% of this oil was eluted from a regular 3-m by 1/8-in. OD column, which had been packed with 0.5% SE-30 silicone on 80-100 mesh Chromosorb G-HP. The operating conditions included temperature programming from 150°C to 300°C at 6° C/min and a helium flow rate of 25 ml/min. The same material, chromatographed under similar conditions on a short OV-101 capillary column, eluted with 90% yield. Indeed, using a somewhat higher-than-usual gas velocity and temperature programming from 280°C to 350° C at 10°C/min, Gouw et al. successfully chromatographed a microcrystalline wax with a 190-195°F melting range on this column [164]. The heaviest species, nC_{58} (boiling point 1125° F), eluted in 11 minutes; and a complete baseline resolution was still observed for nC_{28} from nC_{29}, the lowest n-hydrocarbon homologs present in the sample.

Thermally more stable stationary liquid phases make it possible to use longer columns and obtain somewhat better resolutions. Poly-m-carborane siloxane polymers, which have been observed to possess exceptional thermal stability, can be deposited in thin homogeneous films on etched and silylated glass capillaries or on glass beads. These stationary phases allow temperatures up to 350° C. Bleeding is observed to exist above 250° C but can apparently be suppressed by special treatment. In a series of communications, Novotny et al. [165] showed the applicability of these columns to the chromatography of complex high-boiling mixtures and high-molecular-weight waxes up to nC_{60}. A disadvantage of these polymers is that they have high melting points, which necessitate high initial column temperatures. This makes the columns unsuitable for resolving lower-molecular-weight components.

The separations described can be used for analyzing single constituents in a heavy petroleum cut, provided that these components are present in fairly large quantities. In heavy petroleum hydrocarbon stocks, it is not uncommon to encounter waxy streams with considerable amounts of n-hydrocarbons. The chromatogram of such a sample shows a poorly defined background pattern over the whole boiling range, with sharp n-hydrocarbon peaks superimposed at regular intervals. See, for example, Fig. 30. A fair estimate of the amount of these compounds can be obtained by using the background pattern of the chromatogram as the baseline of these peaks. Sojak and Bucinska, e.g., determined in this way residual n-alkanes in the desorbates from the dewaxing of gas oils [166].

Another class of compounds that yield conspicuous peaks in gas chromatograms of crude oil samples are the isoprenoids. The most pronounced of these compounds have 13, 14, 15, 16, 18, 19, and 20 carbon atoms. On a nonpolar column, the C_{19} isoprenoid, pristane, elutes at approximately the same place as n-hexadecane; the C_{20} isoprenoid, phytane, has a retention time close to that of n-octadecane. The lower-molecular-weight isoprenoids form distinct peaks separate from those of the n-hydrocarbons. The higher-molecular-weight compounds of this class, together with the C_{17} isoprenoid, occur in much lower quantities, if at all, in nature and are generally not distinguishable from the background pattern of the chromatogram. Because of their lower stability these compounds are generally not observed in manufactured petroleum products.

Conspicuous peaks in a chromatogram do not necessarily have to be identified if the object of the chromatographic analysis is to obtain a "fingerprint"-type analysis of the sample. Marschner et al. [167]

chromatographed a number of bitumens on a 2-m or 6-m-long column packed with OV-107 on Chromosorb, temperature-programmed from 20°C to 400°C, and observed that entire groups of biological marker hydrocarbons, such as the diterpanes, n-alkanes, steranes, triterpanes, and tetraterpanes, tend to be either present or absent.

A more recent, very important development in gas chromatography is its combination with a mass spectrometer as the detector. Gas chromatography/mass spectrometry (GC/MS) has proved to be a powerful tool for identifying many compounds at very low levels in a wide-range-boiling matrix.

By combination of the two techniques in one instrument, the onerous trapping of fractions from the GC column is avoided, and higher sensitivities can be attained. In passing through the GC column, the sample is separated more or less according to its boiling point. Although insufficient component resolution is observed in most cases, the eluting compounds at any time are usually closely related to each other in boiling point and molecular weight or both and are free from interfering lower- and higher-molecular-weight species. Because of the much reduced complexity of the GC fractions, mass spectrometer scans carried out at regular intervals yield simpler spectra, from which compound classes can now more easily be determined. Modern interfacing techniques between the gas chromatograph and the mass spectrometer inlet system result in essentially no degradation of the resolution attained in the column. In some cases, it is possible to get complete identification of compounds that without MS are partly or completely submerged in the background pattern of the chromatograph. With this technique, Gallegos identified 16 steranes, terpanes, and branched paraffins in the branched cyclic hydrocarbon fraction of Green River shales [168]. These compounds range from the C_{16} isoprenoid of molecular weight 226 to the $C_{40}H_{76}$ perhydro-β-carotenes. The latter have estimated boiling points well above 950° F. These compounds are especially important in geochemical studies because they are considered to be biological marker hydrocarbons indicating the origin and maturity of the shale. Similar studies have been reported by Burlingame [153], Eglinton [155], and others [62].

Another area in which GC/MS has been found to be especially useful is in research on petroleum waxes [169]. Mass spectrometric methods for analyzing higher-molecular-weight hydrocarbons have been reviewed by Kajdas and Tuemmler [170].

A very useful recent variation in GC/MS techniques is the so-called mass chromatography. In this approach, the output of the mass spectrometer is continuously scanned for relatively few selected m/e fragments. This can be done mathematically [154] or by using separate detectors

[171]. If a monitor is adjusted to react only to a typical m/e fragment, a signal will be recorded only if a compound belonging to a certain homologous series is eluted off the chromatographic column. For instance, fragments with m/e 71 are monitored for the emergence of paraffins; m/e 77 is distinctive of alkylbenzenes; naphthenes and monoolefins can be detected by scanning m/e 99 fragments; cyclic diolefins yield m/e 65 fragments; and m/e 67 fragments are associated with dinaphthenes, diolefins, or naphthenic olefins. The technique can also be used to monitor the emergence of specific compounds with known or predictable mass spectra. By displaying the output of several monitors parallel or superimposed to the regular flame ionization detector output from the gas chromatograph or from the total ion monitor of the mass spectrometer, one can establish the compound classes of many known peaks in the chromatogram. Hites and Biemann [154] described application of this technique to analyzing the methyl esters of a non-steam-distillable acidic extract of Green River oil shale.

In another application, gas chromatography is used to determine the boiling range of hydrocarbon fractions by the technique called simulated distillation [172-175]. This approach requires a good chromatographic system with a highly reproducible injector temperature control and column programming option. A calibration curve is established relating the retention times of a synthetic mixture of n-hydrocarbons boiling in the range of interest to their atmospheric boiling points. Figure 29 shows the chromatogram of a C_4-C_{42} n-hydrocarbon mixture; except for the first members of the series, the observed boiling points all lie on an essentially straight line.

The sample is chromatographed under identical conditions as the standard hydrocarbons and cumulative peak areas are determined at regular intervals during elution of the sample. Each cumulative peak area is divided by the total area of the chromatogram to obtain the fraction of material eluted or "distilled" at each point. The simulated distillation curve is now obtained by plotting these values against the temperature at the corresponding points as derived from the calibration curve. Taking points at regular intervals from initial elution point to total elution allows a complete simulated distillation plot to be constructed. With the fraction distilled on the horizontal and the "boiling points" on the vertical axis, quantitative information on the amount of material boiling in a certain temperature range can be derived by this approach. If a portion of the sample is retained on the column, an internal standard has to be used to determine the percentage of sample eluted. Other approaches are using a combination of back-flushing and combustion or using the area of a portion of the crude as the internal standard [176]. Figure 30 shows a simulated distillation curve of a 370-535°C distillate of a South Swan Hills, Alberta, Canada, crude oil with the calculated amounts of material in different temperature ranges [177].

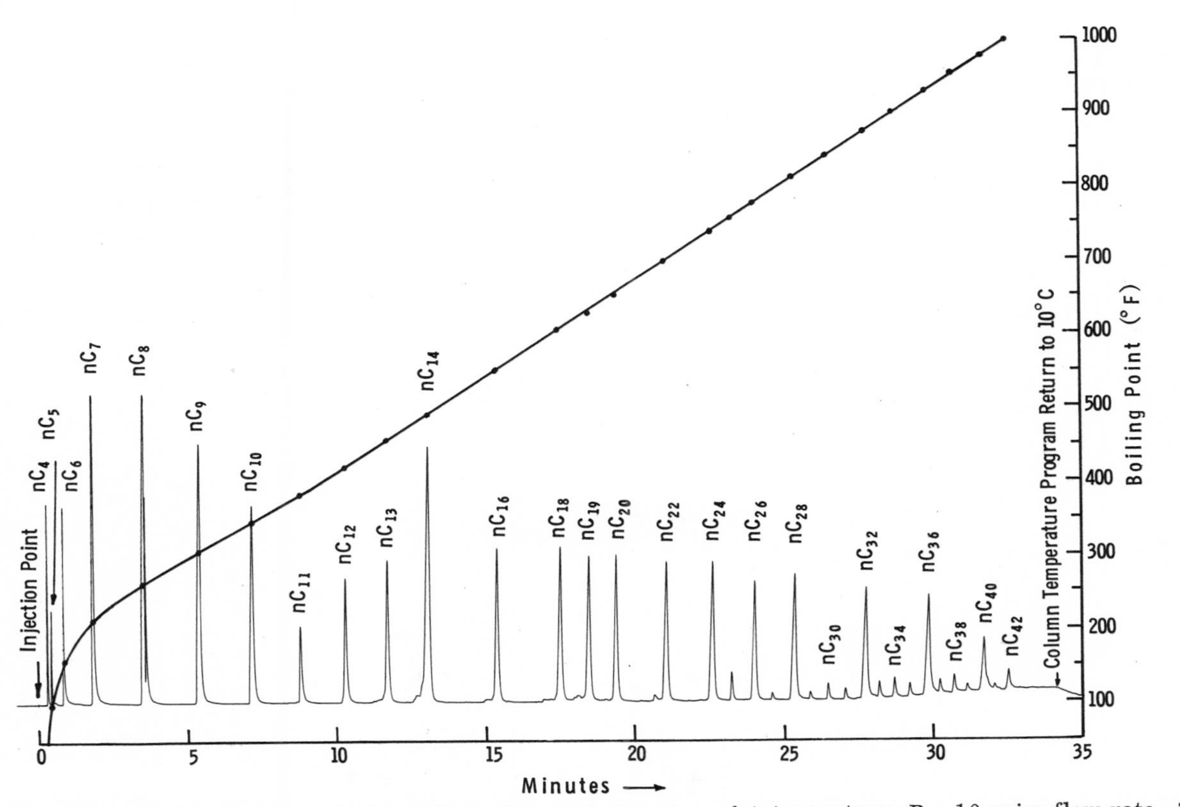

FIG. 29. Synthetic mixture of n-hydrocarbons from n-butane to n-dotetracontane; P_i, 10 psig; flow rate, 3.5 ml He/min.; T, 10 - 350°C at 10°C/min. Note the linear relation between boiling point and retention time for all compounds eluting after n-decane. (From Ref. 164. Courtesy of American Chemical Society.)

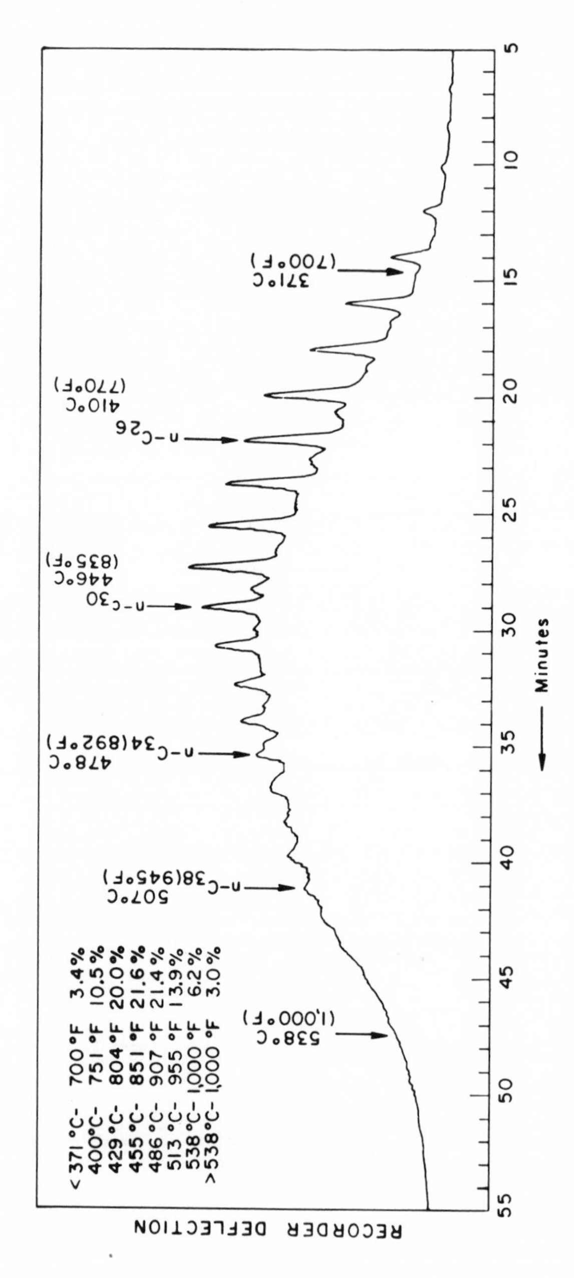

FIG. 30. Simulated distillation by gas-liquid chromatography of a 370–535°C distillate of South Swan Hills, Alberta, Canada, crude oil. (From Ref. 177. Courtesy of API Research Project 60.)

Computer programs are available to carry out the calculations and to plot out the curve [174]. Simulated distillation is now widely used to establish the boiling curve of high-molecular-weight distillation fractions. For example, this approach is routine for most high-molecular-weight distillation cuts obtained in the API 60 research project. Some of the problems encountered with narrow, high-boiling hydrocarbon fractions have been described recently [175]. In this case, a graphical approach is recommended over the usual procedure.

2. Pyrolysis Gas Chromatography (PGC)

Pyrolysis gas chromatography can be used for rapid though undetailed information on the gross composition of heavy petroleum fractions. Most of the published information has little comparative value because fragmentation patterns depend not only on the sample itself but also on sample size, pyrolysis chamber geometry, pyrolysis temperature, heating rate, and pyrolysis time. In PGC, the sample investigated is pyrolyzed in a separate chamber; the pyrolysis fragments are then introduced into a gas chromatography system for analysis. Work before 1966 has been covered in excellent detail in a critical review by Levy [178].

Two main types of pyrolysis units are in use: pyrolyzers, where heat is applied as a pulse of a relative short duration, and units where a continuous source of energy is present. In the pulse mode approach, microgram quantities of sample are deposited on a filament that is heated in a very short time to the pyrolysis temperature. In the continuous mode the sample, usually in milligram amounts, is introduced into a hot pyrolysis chamber. Secondary reactions are more apt to occur in the latter approach, and chromatograms obtained by these two techniques differ quite distinctly from each other.

The rate of heating is actually far more important to the nature of the pyrolysis products than the more commonly reported maximum temperatures of the pyrolyzing element [179]. An elegant approach of the pulse mode class involves the curie point pyrolyzers, which use a ferromagnetic filament heated in a radio-frequency field. The temperature of this filament increases rapidly and stabilizes at the curie point of the material. This temperature is a function of the composition of the filament.

While in these techniques the pyrolysis temperature is maintained at 300-700°C for seconds or even minutes, the advent of the laser has led to pyrolysis techniques in which the sample is irradiated by a narrow, high-intensity laser pulse. Surface temperatures to 8000°K and higher can be generated in a period ranging from 10^{-9} to 10^{-1}sec. The irradiated region is very small permitting, e.g., localized pyrolysis patterns of materials from different areas of a geological specimen. Because of the extremely rapid heating and cooling of the sample, fewer secondary reactions occur, resulting in more characteristic and reproducible chromatograms; greater distinctions can therefore be made between similar substances than with the previously indicated techniques. Application of laser pyrolysis chromatography to petroleum products has been reported by Folmer and Azaragge [180], Ristau and Vanderborgh [181], and others [182].

In another variation that claims somewhat more reproducible pyrograms, the sample is vaporized prior to introduction into the pyrolysis unit. A noteworthy paper of vapor-phase pyrolysis is that by Cramers and Keulemans [183]; they describe a carefully designed microflow reactor that gives uniform distribution of the temperature along a gold-plated capillary of 1000-mm length. Another interesting paper is that of Wolf, Walker, and Fanter, who studied the pyrolysis patterns of 83 C_6-C_{10} compounds [184, 185]. Higher precision in first-order pyrolysis reactions than is generally observed in gas-phase kinetics was claimed for the samples pyrolyzed in a wall-less, gas-phase reactor [186]. Walker [187] has reviewed design problems and operational parameters such as sample size, sample preparation, pyrolysis temperature, temperature rise time, and repeatability.

Knotnerus [188] studied pyrolysis patterns of single compounds and applied this knowledge to the analysis of asphaltic bitumens. His data indicate that PGC can be used to differentiate between strongly paraffinic and cyclic materials and even identify predominant structures. Thus, he found in his asphalts an average of three isoparaffin side chains and one isolated ring per asphalt molecule.

Perry [189] has also carried out asphalt analysis by PGC. He related differences in three asphalts to the polynuclear structures in these heavy materials. In an analogous study, Karr et al. [190] used PGC to obtain information on the chemical structure of coal tar pitches. Asphaltic materials were also pyrolyzed by Ariet and Schweyer, who observed certain differences in the pyrograms from different asphalts [191]. However, these authors questioned the applicability of this method as a practical classification tool to bitumen from different sources. Poxon and Wright [192] pyrolyzed several asphalts and gilsonites by Curie point pyrolysis gas chromatography at different temperatures. The observed differences in the chromatograms were very small, but the method appeared to be applicable to differentiating samples from various sources if a computer was available to carry out the statistical analyses of the results.

Not wholly unexpected is the extensive use of PGC by geochemists to correlate crude oil with source rock and to derive geochemical characterization parameters from oil-bearing strata. LePlat [193] carried out an extensive study on hundreds of rock and earth samples. He observed that it was possible to differentiate the types of samples from the obtained chromatograms. The area of the chromatograms turned out to be proportional to the amount of material that can be extracted by $CHCl_3$.

3. Inverse Gas-Liquid Chromatography

In the technique of inverse gas-liquid chromatography, which Petersen et al. [194-196] developed, the asphalt under study is used as the stationary phase; and a number of volatile test compounds are chromatographed on this column. The interaction coefficient I_p determined for these compounds is a measure of certain qualities of the liquid phase. The coefficient I_p is defined as the difference in the Kovats Retention Index between the test compound and the hypothetical n-paraffin of the same molecular weight. I_p is therefore indicative of the chemical interaction of the solute with the stationary phase. In later papers, a more fundamental operator I_g, the specific interaction coefficient, was used to correct for the void volume of the system.

Robertson and Moore [197] used inverse gas-liquid chromatography to study variations in the performance of asphalts from different crude sources. Inverse gas-liquid chromatography has also been applied to asphalt oxidation studies [198, 199]. The asphalt under investigation can be oxidized prior to packing on the chromatograph column; it can also be oxidized in situ in the GLC column. Results on coating-grade asphalts show that after a 24-hr oxidation period, the I_ps for phenol and for propionic acid show a good correlation with the durability of the asphalts. In a subsequent study, the results from inverse gas chromatography have been correlated with data obtained by the Kleinschmidt chromatographic fractionation [22], which yields five fractions, i.e., asphaltenes, white oils, dark oils, asphaltic resins, and a small fifth one. In the 25 samples studied, the I_p for formamide, corrected for nitrogen and acid content of the asphalt, shows a very good correlation with the sum of the asphaltenes and asphaltic resins. The phenol I_p on oxidized road asphalts was found to correlate well with the changes in viscosity as determined by a microfilm durability test [197].

Inverse gas chromatography seems to have merits as a screening tool but not so much as a research method, since most of the results can be obtained by other more direct physical test methods.

E. Miscellaneous

1. Coordination Chromatography

With certain compound groups, advantage can be taken of their reactions with specific chemicals to separate them from a mixture. Examples are the retention and separation of polynuclear aromatics by alumina coated with 2,4,7-trinitrofluorenone (TNF)[200] or by silica treated with caffeine [200], picric acid [201], TNF [202], trinitrobenzene [202], and other compounds [201, 202], and the removal of olefins by silver nitrate on silica [203], and of nonbasic nitrogen compounds by ferric chloride supported on kaolin [4, 46, 53].

The only one of these methods that has been applied to heavy petroleum fractions is the complexing of nitrogen compounds with ferric chloride on kaolin. It was introduced by Jewell and Snyder [50] and has become an integral part of the separation scheme that the API-Research Project 60 group [4, 46, 53] developed.

The ferric chloride is deposited on the clay mineral from a saturated methanolic solution. After being washed with methanol and benzene, the adsorbent is ready for packing. The petroleum sample is introduced into the column as a solution in hexane or pentane. A rapid color change to green or blue is observed as the reaction takes place. The unreacted hydrocarbons are exhaustively eluted with hexane or pentane. This may be quite difficult and time-consuming. Traces of the colored complex in the eluate can be removed by contacting with a strong anion-exchange resin. The complexed N compounds are desorbed from the $FeCl_3$-kaolin column by 1,2-dichloroethane and subsequently dissociated on a column of a strong anion-exchange resin (e.g., Amberlyst A-29) in the hydroxide form. The metal salt remains on the resin while the free nitrogen compounds are recovered from the eluate. For quantitative removal of the N compounds, the sample may have to be complexed several times with fresh ferric chloride adsorbent. Generally, the sample must be free of basic and acidic compounds.

2. Supercritical Fluid Chromatography

Classically, a substance is considered to be a gas when it is heated beyond its critical temperature. In supercritical fluid chromatography, the mobile phase is a substance maintained at a temperature a few degrees above its critical point. The physical properties of this substance are intermediate to those of a liquid and of a gas at ambient conditions; hence, it is preferable to designate this condition as the supercritical phase. In a chromatographic column, the supercritical fluid usually has a density about one-third to one-fourth of that the the corresponding liquid when used as the mobile phase; the diffusivity is about 1/100 that of a gas and about 200 times that of the liquid. The viscosity is of the same order of magnitude as that of the gas. For chromatographic purposes, such a fluid has more desirable transport properties than a liquid. In addition, the high density of the fluid results in a 1000-fold better solvency than a gas. This is especially valuable for analyzing high-molecular-weight compounds. Rijnders observed, e.g., that dimyricyl phthalate, a compound with the formula $C_{68}H_{126}O_4$ and a molecular weight of 1006, is eluted in a few minutes from a column 2 m long [204]. Jentoft and Gouw chromatographed a mixture of polystyrene oligomers and eluted a 3386-molecular-weight oligomer in 70 minutes through a 4-ft by 1/4-in. column [205].

Although supercritical fluid chromatography appears to be an excellent technique for analyzing heavy ends of petroleum hydrocarbons, very little has been published about it. This is, in part, due to some of the experimental problems with the technique and the reluctance of many investigators to try this approach.

Figure 31 shows a chromatogram Sie and Rijnders obtained in 1967 by supercritical fluid chromatography of a coal tar used in bitumen blending [206]. In this chromatogram, pronounced peaks are observed, which are attributed to phenanthrene, anthracene, fluoranthene, pyrene, and other polynuclear aromatic hydrocarbons with up to 7 and 8 rings. The authors indicate also that work is being carried out on other high-boiling hydrocarbon mixtures [204].

In a later paper, Rijnders [207] described the separation of a 345-350° C petroleum hydrocarbon distillate on a 3.5-m-long column packed with alumina at 203° C and 50 atm, with n-pentane as the mobile phase. The eluate was separated into 10 consecutive fractions that were further characterized by spectrometric techniques. Distinct differences were observed among the obtained cuts. Fraction 1 contained substituted monoaromatics; Fraction 2 contained the same class of compounds plus benzthiophene derivatives; Fraction 3 consisted of benzthiophene derivatives plus substituted naphthalenes; Fractions 4 and 5 were essentially only substituted naphthalenes; Fraction 6 contained predominantly dibenzthiophenes; Fractions 7 and 8 were mixtures of dibenzthiophene and phenanthrene derivatives; and Fraction 9 consisted of mostly substituted phenanthrenes.

Supercritical fluid chromatography at present has not attained the level of development that allows an easy combination with other techniques, especially mass spectrometry. If the direct combination of these two techniques becomes a practical reality, we expect a surge of applications in the higher-molecular-weight field. This area is not covered by present-day gas chromatography/mass spectrometry techniques because of the temperature limitations in the gas chromatography column. In the meantime, however, the discontinuous approach should pose no problems because fraction collection from supercritical fluid chromatography is quite straightforward [208].

A primary advantage of this technique is related to the extreme selectivity that can be achieved by adjusting the operating parameters. Sie and Rijnders showed, e.g., that with n-pentane as the mobile phase and 23 wt% PEG 6000 on 120/140 mesh Sil-O-Cel as the stationary

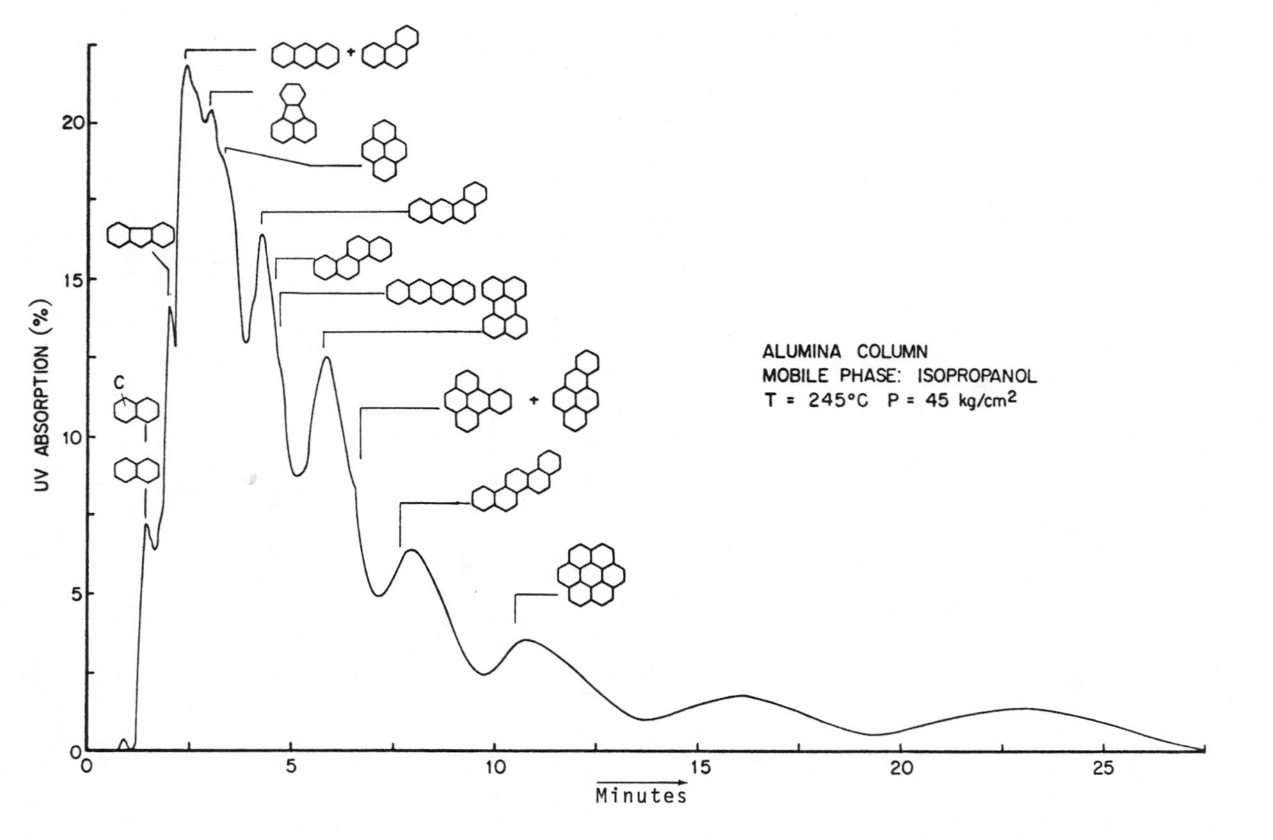

FIG. 31. Supercritical fluid chromatogram of coal tar. (From Ref. 206. Courtesy of Separation Science.)

substrate, dimyricyl phthalate is eluted well before diphenyl phthalate, notwithstanding that the more paraffinic ester has 48 more atoms [209]. This separation was carried out at 213°C and a pressure of 44.9 kg/cm^2. This example indicates that supercritical fluid chromatography may be an important technique in the area of high-molecular-weight group-type separations.

3. Reversed-Phase Partition Chromatography

To our knowledge, Helm [210] is the only one who published on the application of reversed-phase partition chromatography (RPPC) to heavy petroleum samples. He separated a molecular distillation cut of an asphalt by this technique, using nitromethane saturated with cyclohexane as the mobile phase and cyclohexane on Fluoropak 80 as the stationary phase. After the column was conditioned with the mobile phase, the sample, dissolved in an equal amount of cyclohexane, was placed on the column, and elution was initiated. When the concentration in the effluent became constant at less than 1% sample per fraction, the RPPC residue remaining in the column was washed out with cyclohexane and finally with benzene. In this way, the sample was recovered quantitatively.

Infrared analysis of the fractions indicated early elution of the more polar and aromatic fractions, followed by others of lower polarity and aromaticity. Subsequent chromatography of some of the RPPC fractions on silica gave better resolution than silica chromatography without prior RPPC.

Experiments with this technique in our laboratory indicated that quantitative recovery of the sample is not achieved with heavier material i.e., with asphaltenes and even resins of molecular weights higher than about 1000. These heavy samples are not sufficiently soluble in the nitromethane. This was demonstrated in a nonchromatographic solubility study [116]. In summary, it may be stated that RPPC has its merits. However, with petroleum samples, it is limited to relatively low molecular weights (up to 600); and even in this range, it does not seem to give great resolution.

4. Paper and Thin-Layer Chromatography

Because of the limited resolution of paper and thin-layer chromatography, they have been used in petroleum applications mainly as a gross separating tool for quick routine analyses or as a fingerprint technique [211] applied to prefractionated samples. One example for the latter is Seifert and Teeter's [72] fractionation by preparative TLC of a carboxylic acid fraction of high interfacial activity isolated from a California crude oil. They prepared multiple TLC plates from silica gel, 5% $CaSO_4$, and an inorganic fluorescent indicator. As developer solvent, they used a mixture of acetic acid, methanol, and benzene (0.5:1:98.5). Analysis of a strongly fluorescent fraction from this separation by mass spectrometry revealed about 1500 compounds, many of which belonged to homologous series.

There are many papers on the analysis by PL and TLC of polynuclear aromatic hydrocarbons in petroleum fractions [212]. The advent of modern, high-resolution liquid chromatography techniques has, however, supplanted many of the applications that were previously carried out by these two-dimensional techniques. This is confirmed by the observation that the number of published applications of TLC for the analysis of hydrocarbons has decreased considerably in the last few years [213].

Most of the papers regarding applications of PC and TLC to heavy petroleum fractions are seemingly unrelated to each other. Peurifoy et al. [214] used TLC to separate a middle distillate oil with an average molecular weight of 373 into fractions rich in monoaromatics, polyaromatics, and resins. A three-stage discontinuous layer gradient plate, consisting of silica gel G, alumina G, and Florisil G, was developed for this purpose. With a mixture of cyclohexane, benzene, and ethyl acetate (105:1:1) as the solvent, improved separations over those with single adsorbent plates were obtained. This paper also reported on the fractionation of resins from a molecular distillate of straight-run residues with average molecular weights to 959. Thin-layer chromatography, LC, and GC with subsequent characterization of the fractions by infrared spectroscopy were used for operation control in manufacturing lubricants [215].

Thin-layer chromatography of commercial waxes has been reported by a number of investigators. Reutner [216] used silica gel G and a variety of developing solvents and observed that the technique does not yield sufficient resolution to distinguish paraffinic hydrocarbon waxes from different sources. TLC is, however, quite useful for discriminating between petroleum waxes and waxes from natural sources such as carnauba wax, candellila wax, and Brazilian beeswax.

To distinguish between n-paraffins and branched chain hydrocarbons in waxes, Sucker [217] impregnated his TLC plates with urea. Urea is a "molecular sieve" in that it can form cagelike inclusion compounds with straight-chain hydrocarbons but not with branched ones. Dietsche [218] subsequently studied the applicability of this technique to analyzing petroleum-derived waxes. TLC layers were prepared from a mixture of 15 g urea, 30 g silica gel G, 5 g calcined $CaSO_4$, and 3 g of a 75% aqueous solution of sorbitol, the latter to prevent crystallization of the urea. The separation between n-paraffins from isoparaffins becomes less sharp with increasing molecular weight. Higher-molecular-weight isoparaffins with a low degree of branching or with a straight-chain moiety of at least 20 carbon atoms are also retained by the stationary substrate. The mobility of the compounds is influenced by the temperature, the polarity of the developing solvent, and the molecular weight of the components in the sample.

In another application, paper chromatography has been used for quantitative determination of asphaltenes in crude oils or in residua. Bzdega et al. [219] measured the circular zone produced on Whatman No. 1 paper by a sample of the oil dissolved in heptane and ethanol and developed with a mixture of benzene-ethanol.

A comparable application was described by Lantos and Lantos [220], who developed a simple paper chromatography technique to determine the amount of free carbon and oxidized material in used lubricating oils.

Petroleum and coal tar oils and residues can be identified in sewer effluents by TLC, as Mathews [221] has described. He established several systems based on TLC plates with Kieselguhr G, alumina T, or silica gel T, and developing solvents such as petroleum ether 40-60° C, acetone, ethanol, toluene, and chloroform. By choosing from several alternative procedures, one can differentiate among several grades of heavy lubricating oils, greases, fuels, coal tar and coal tar products, residual and crude petroleum products, and other nonpetroleum hydrocarbon oils and waxes that may be present.

ACKNOWLEDGMENTS

The authors are grateful to Drs. R. E. Jentoft, H. H. Oelert, W. K. Seifert, and L. R. Snyder, and W. K. Siefert for reading the manuscript and for helpful comments. They also thank Drs. G. L. Cook, W. E. Haines, and C. G. Thompson for advising them on the status of their work at the American Petroleum Institute.

REFERENCES

1. L. R. Snyder, Acc. Chem. Res. 3, 290 (1970).

2. "Modern Practice of Liquid Chromatography," ed. J. J. Kirkland, Wiley-Interscience, New York, 1971.

3. "Handbook of Chromatography," ed. G. Zweig and J. Shenna, vol 2, CRC Press, Cleveland, Ohio, 1972.

4. W. E. Haines and L. R. Snyder, Proc. 8th World Petrol. Congr. 6, 223 (1971).

5. L. R. Snyder and B. E. Buell, Anal. Chem. 40, 1295 (1968).

6. L. R. Snyder, Anal. Chem. 38, 1319 (1966).

7. J. L. Oudin, Inst. Francais Pétrole Rev. 25, 470 (1970).

8. B. Tissot, Y. Califet-Debyser, D. Geroo, and J. L. Oudin, Amer. Assoc. Petrol. Geol. Bull. 55, 2177 (1971).

9. D. E. Anders and W. E. Robinson, Geochim. Cosmochim. Acta 35, 661 (1971).

10. D. E. Anders, F. G. Doolittle, and W. E. Robinson, Geochim. Cosmochim. Acta 37, 1213 (1973).

11. M. Poulet and J. Roucaché, Inst. Francais Pétrole Rev. 25, 127 (1970).

12. W. R. Middleton, Am. Chem. Soc. Div. Petrol. Chem. Prepr., A-45, April 1958.

13. W. R. Middleton, Anal. Chem. 39, 1839 (1967).

14. L. W. Corbett, Anal. Chem. 41, 576 (1969).

15. L. W. Corbett and R. E. Swarbrick, Am. Chem. Soc. Div. Petrol. Chem. Prepr., B-161, (1966).

16. G. O'Donnell, Anal. Chem. 23, 894 (1951).

17. R. L. Griffin, W. C. Simpson, and T. K. Miles, Am. Chem. Soc. Div. Petrol. Chem. Prepr., A-13, April 1958.

18. G. U. Dineen, J. R. Smith, R. A. Van Meter, C. S. Albright, and W. R. Anthoney, Anal. Chem. 27, 185 (1955).

19. W. E. Robinson and J. J. Cummins, Chem. and Eng. Data, 5, 74 (1960).

20. W. E. Robinson and D. L. Lawlor, Fuel 40, 375 (1961).

21. M. Bestougeff, Compt. Rend. Acad. Sci. 577 (1966).

22. L. R. Kleinschmidt, J. Res. Nat. Bur. Standards 54, 163 (1955).

23. R. F. Marshner, L. J. Duffy, and J. C. Winters, Am. Chem. Soc. Div. Petrol. Chem. Prepr. 18 (3), 572 (1973).

24. L. J. Duffy, reported at the summer meeting of the API Research Project 60 Advisory Group at Laramie, Wyoming, 1971.

25. K. H. Altgelt, Makromol. Chem. 88, 75 (1965).

26. K. H. Altgelt, Am. Chem. Soc. Div. Petrol. Chem. Prepr. 13 (3), 37 (August 1968).

27. K. H. Altgelt, J. Appl. Polymer Sci. 9, 3389 (1965).

28. K. H. Altgelt, Separ. Sci. 5, 855 (1970).

29. K. H. Altgelt, Bitumen, Teere, Asphalte, Peche, 21, 475 (1970).

30. W. B. Richman, Assoc. Asphalt Technol. Proc., 106 (1968).

31. J. A. Minshull, Aust. Road Res. Board 1968, Pap. 428.

32. J. F. Branthaver and J. M. Sugihara, Am. Chem. Soc. Div. Petrol. Chem. Prepr., B-105 (1966).

33. R. J. Rosscup and H. P. Pohlmann, Am. Chem. Soc. Div. Petrol. Chem. Prepr., 12, A-103 (1967).

34. T. Edstrom and B. A. Petro, J. Polym. Sci., C, 21, 171 (1968).

35. J. P. Dickie and T. F. Yen, Anal. Chem. 39, 1847 (1967).

36. M. H. B. Hayes, M. Stacey, and J. Standley, Fuel 51, 27 (1972).

37. H. J. Coleman, D. E. Hirsch, and J. E. Dooley, Anal. Chem. 41, 800 (1969).

38. J. C. Moore, J. Polym. Sci., Part A, 2, 835 (1964).

39. Z. Grubisic, P. Rempp, and H. Benoit, J. Polym. Sci., Part B, 5, 753 (1967).

40. G. A. Haley, Anal. Chem. 43, 371 (1971).

40a. G. A. Haley, Anal. Chem. 44, 580 (1972).

41. D. Bynum, Jr., R. N. Traxler, H. L. Parker, and J. S. Ham, J. Inst. Petr. 56, 147 (1970).

42. C. E. Dougan, Bureau of Highways, Res. Proj. 175-210, 1970.

43. J. N. Done and W. K. Reid, Separ Sci. 5, 825 (1970).

44. E. W. Albaugh and R. C. Talarico, J. Chromatogr. 74, 233 (1972).

45. H. H. Oelert, J. Chromatogr. 53, 241 (1970).

46. C. J. Thompson, H. J. Coleman, J. E. Dooley, and D. E. Hirsch, Oil and Gas J., 1971.

47. D. E. Hirsch, R. I. Hopkins, H. J. Coleman, F. O. Cotton, and C. J. Thompson, Anal. Chem. 44, 915 (1972).

48. D. M. Jewell, J. H. Weber, J. W. Bunger, H. Plancher, and D. R. Latham, Anal. Chem. 44, 1391 (1972).

49. W. E. Haines, C. C. Ward, and J. M. Sugihara, Preprint API, Div. Refining, May, 1971.

50. D. M. Jewell and R. E. Snyder, J. Chromatog. 38, 351 (1968).

51. H. J. Coleman, J. E. Dooley, D. E. Hirsch, and C. J. Thompson, Anal. Chem., 45, 1724 (1973).

51a. D. E. Hirsch, J. E. Dooley, and H. J. Coleman, API RP 60, Published Report No. 27. Burea of Mines Report on Investigations, 1974.

52. J. W. Bunger, Personal Communication (1974).

53. D. M. Jewell, E. W. Albaugh, B. E. Davis, and R. G. Ruberto, Am. Chem. Soc. Div. Petrol. Chem. Prepr., 17 (4), F 81 (1972).

54. H. H. Oelert, Personal Communication, 1974.

55. D. M. Jewell, E. W. Albaugh, B. E. Davis, and R. G. Ruberto, Ind. Eng. Chem. Fundam. 13, 279 (1974).

56. D. M. Jewell, R. G. Ruberto, and B. E. Davis, Anal. Chem. 44, 2318 (1972).

57. L. R. Snyder, Anal. Chem. 41, 1084 (1969).

58. L. R. Snyder, Anal. Chem. 37, 713 (1965).

59. J. E. McKay and D. R. Latham, Anal. Chem. 45, 1050 (1973).

60. H. H. Oelert, D. Severin, and H. J. Windhager, Erdol und Kohle, 26, 397 (1973).

61. H. N. M. Stewart, R. Amos, and S. G. Perry, J. Chromatogr. 38, 209 (1968).

62. See; for instance: G. L. Speers and E. V. Whitehead: Crude Petroleum in Organic Geochemistry, eds. G. Eglington and M. T. J. Murphy, Springer Verlag, New York, 1969; P. M. Gardner and E. V. Whitehead, Geochim. et Cosmochim. Acta 36, 259 (1972); P. C. Anderson, P. M. Gardner, and E. V. Whitehead, D. E. Anders and W. E. Robinson, Geochim. et Cosmochim. Acta 33, 1304 (1969); I. R. Hills and E. V. Whitehead, Nature 209, 977 (1966).

63. J. F. McKay, D. M. Jewell, and D. R. Latham, Separ. Sci. 7, 361 (1972).

64. T. E. Cogswell, J. F. McKay, and D. R. Latham, Anal. Chem. 43, 645 (1971).

65. W. K. Seifert and W. G. Howells, Anal. Chem. 41, 554 (1969).

66. W. K. Seifert, Anal. Chem. 41, 562 (1969).

67. W. K. Seifert, R. M. Teeter, W. G. Howells, M. J. R. Cantow, Anal. Chem. 41, 1638 (1969).

68. W. K. Seifert and R. M. Teeter, Anal. Chem. 42, 180 (1970).

69. W. K. Seifert and R. M. Teeter, Anal. Chem. 42, 750 (1970).

70. W. K. Seifert, E. J. Gallegos, and R. M. Teeter, J. Am. Chem. Soc. 94, 5880 (1972).

71. W. K. Seifert, Pure and Appl. Chem. 34, 633 (1973).

72. W. K. Seifert and R. M. Teeter, Anal. Chem. 41, 786 (1969).

73. L. R. Snyder, B. E. Buell, and H. E. Howard, Anal. Chem. 40, 1303 (1968).

74. L. R. Snyder, Anal. Chem. 41, 314 (1969).

75. H. V. Drushel and A. L. Sommers, Anal. Chem. 37, 1319 (1965).

76. H. V. Drushel, Am. Chem. Soc. Div. Petrol. Chem. Prepr. 17 (4) F92 (1972).

77. A. Giraud and M. A. Bestougeff, J. Gas Chromat. 5, 464, (1967).

78. L. R. Snyder, Principles of Adsorption Chromatography, Marcel Dekker, New York, 1968.

79. L. R. Snyder, J. Chromatogr. 5, 430 (1960).

80. L. R. Snyder, Anal. Chem. 33, 1527 (1961).

81. L. R. Snyder, Anal. Chem. 33, 1535 (1961).

82. L. R. Snyder, Anal. Chem. 33, 1538 (1961).

83. L. R. Snyder, Anal. Chem. 34, 771 (1962).

84. J. C. Giddings, Dynamics of Chromatography, Part 1, Principles and Theory, Dekker, New York, 1965.

85. B. L. Karger in "Modern Practice of Liquid Chromatography," ed. J. J. Kirkland, Wiley Interscience, 1971.

86. L. R. Snyder and J. J. Kirkland, Introduction to Modern Liquid Chromatography, Wiley, 1974.

87. L. R. Snyder, ASTM, STP, 389 (1965).

88. L. Hagdahl and R. T. Holman, J. Am. Chem. Soc. 72, 701 (1950).

89. R. J. P. Williams, L. Hagdahl, and A. Tiselius, Arkiv. Kemi, 7, 1 (1954).

90. H. Engelhardt and N. Weigand, Anal. Chem. 45, 1149 (1973).

91. Du Pont Sales Brochure for "Sorbax" SIL.

92. L. R. Snyder and B. E. Buell, J. Chem. Eng. Data 11, 545 (1966).

93. L. R. Snyder, J. Chromatogr. 28, 300 (1967).

94. B. J. Mair, P. T. R. Hwang, and R. G. Ruberto, Anal. Chem. 39, 839 (1967).

95. M. Wilk, J. Rochlitz, and H. Bende, J. Chromatog. 24, 414 (1966).

96. H. H. Oelert, Erdöl und Kohle, 22, 536 (1969).

97. H. H. Oelert, Z. Anal. Chem. 244, 91 (1969).

98. L. R. Snyder and D. L. Saunders, J. Chromatogr. Sci. 7, 195 (1969).

99. R. P. W. Scott and P. Kucera, Anal. Chem. 45, 749 (1973).

100. J. Hildereand and R. Scott, "Solubility of Non-Electrolytes," Reinhold, New York, 1949.

101. L. M. Hansen, I & EC, Prod. Res. Devel. 5, 2 (1969).

102. L. R. Snyder in J. J. Kirkland (Editor), Modern Practice of Liquid Chromatography, Wiley-Interscience, New York, 1971, Ch. 4.

103. C. M. Hansen, J. Paint Technol. 39, 104 (1967).

104. C. M. Hansen, J. Paint Technol. 39, 511 (1967).

105. J. D. Crowley, G. S. Teague, Jr., and J. W. Lowe, Jr., J. Paint. Technol. 38, 269 (1966).

106. J. D. Crowley, G. S. Teague, Jr., and J. W. Lowe, Jr., J. Paint. Technol. 39, 504 (1967).

107. E. B. Bagley, T. P. Nelson, and J. M. Scigliano J. Paint. Technol. 43, 35 (1971).

108. L. R. Snyder, J. Chromatog. 92, 223 (1974).

109. L. R. Snyder and J. J. Kirkland, "Introduction to Modern Liquid Chromatography," Wiley-Interscience, New York, 1974

110. L. Rohrschneider, J. Chromatogr. 22, 6 (1966).

111. L. Rohrschneider, J. Chromatogr. 39, 383 (1969).

112. L. R. Snyder, J. Chromatogr. 11, 195 (1963).

113. L. R. Snyder, J. Chromatogr. 63, 15 (1971).

114. L. R. Snyder and H. D. Warren, J. Chromatogr. 15, 344 (1964).

115. See also L. R. Snyder's Comments in Anal. Chem. 46, 1384 (1974).

116. K. H. Altgelt, unpublished work.

117. L. R. Snyder and W. F. Roth, Anal. Chem. 36, 128 (1964).

118. C. A. Stout and S. W. Nicksic, Separ. Sci. 5, 843 (1970).

119. A. Lambert, Brit. Polym. J., 3, 13 (1971).

120. K. H. Altgelt and L. Segal, eds. Gel Permeation Chromatography, Dekker, New York, 1971.

121. K. H. Altgelt: Theory and Mechanics of GPC, Adv. in Chromatog. 7, 3 (1970).

122. J. Cazes, J. Chem. Educ. 43, 567 (1966).

123. M. J. R. Cantow, R. S. Porter, and J. F. Johnson, Rev. Macromolec. Chem. 1, 393 (1966).

124. L. R. Snyder, Anal. Chem. 41, 1223 (1969).

125. H. M. Nutt, Fusion 5, 9 (1958). See Also K. E. Bett and D. M. Newitt in "Chemical Engineering Practice," edts. H. W. Cremer and T. Davis, Vol. 5, pp 205-206, Academic Press, Inc., New York, 1958.

126. See Also W. W. Yau, J. Polymer Sci. A-2, 7, 483 (1969); H. Coll, Separ. Sci. 5, 273 (1970); J. V. Dawkins, J. W. Maddock, and D. Coupe, J. Polymer Sci., A-1, 5, 1391 (1967); A. C. Oanu, A. Broido, A. Barrall, II, and A. C. Javier-Son Am. Chem. Soc. Div. Polymer Chem. Prepr. 12, (September 1971).

127. J. G. Hendrickson and J. C. Moore, J. Polymer Sci. A-1, 4, 167 (1966).

128. A. Lambert, Analytica Chem. Acta, 53, 63 (1971). Many other references given there.

129. H. H. Oelert, Erdöl und Kohle 22, 19 (1969).

130. H. H. Oelert, D. R. Latham, W. E. Haines, Separ. Sci., 5, 657 (1970).

131. J. H. Weber and H. H. Oelert, Separ. Sci. 5, 669 (1970).

132. H. H. Oelert and J. H. Weber, Erdöl und Kohle 23, 484 (1970).

133. J. G. Bergmann, L. J. Duffy, R. B. Stevenson, Anal. Chem. 43, 131 (1971).

134. F. E. Dickson, R. A. Wirkhala, B. E. Davis, Separ. Sci. 5, 811 (1970).

135. W. W. Yau and S. W. Fleming, J. Appl. Polymer Sci. 12, 2111 (1968).

136. K. H. Altgelt and E. Hirsch, Separ. Sci. 5, 855 (1970).

137. E. Hirsch and K. H. Altgelt, Anal. Chem. 42, 1330 (1970).

138. H. H. Oelert, Z. Analyt. Chem. 231, 81 (1967).

139. H. H. Oelert, Z. Analyt. Chem. 231, 105 (1967).

140. H. H. Oelert, Brennstoff-Chem. 11, 3 (1967).

141. H. H. Oelert, Brennstoff-Chem. 12, 1 (1967).

142. H. H. Oelert, Adv. in Org. Geochem. 1971, 629

143. G. Albers, L. Lenard, and H. H. Oelert, Fuel 53, 47 (1974).

144. B. A. Adams and E. L. Holmes, J. Soc. Chem. Ind. 54, 1 (1935).

145. W. A. Munday and A. Eaves, Proc. 5th World Petrol. Congr. New York, Sect. V, Paper 9 (1959).

146. L. R. Snyder and B. E. Buell, Anal. Chem. 36, 767 (1964).

147. L. R. Snyder and B. E. Buell, Anal. Chim. Acta, 33, 285 (1965).

148. H. V. Drushel and A. L. Sommers, Anal. Chem. 38, 19 (1966).

149. K. A. Kvenvolden, J. Am. Oil Chem. Soc. 44, 628 (1967).

150. I. Okuno, D. R. Latham, and W. E. Haines, Anal. Chem. 39, 1830 (1967).

151. H. L. Lochte and E. R. Littmann, "Petroleum Acids and Bases," Chemical Publishing Co., New York (1955).

152. I. R. Hills, E. V. Whitehead, D. E. Anders, J. J. Cummins, and W. E. Robinson, Chem. Commun. 752 (1966).

153. A. L. Burlingame, P. Haug, T. Belsky, and M. Calvin, Proc. Nat. Acad. Sci. 54, 1406 (1965).

154. R. A. Hites and K. Biemann, Anal. Chem. 42, 855 (1970).

155. Sister Mary Murphy, T. J. A. McCormick, and G. Eglinton, Science 157, 1040 (1967).

156. R. Kunin, Amber-Hilites (Rohm and Haas Co.) 126, 1 (1972).

157. D. Dowling and R. Hetherington, Rohm and Haas Co., Phila., "Macroreticular Anion-Eschange Resins," presented at the International Water Conference, Pittsburgh, Pa. October 1963.

158. A. Mondino, J. Chromatogr. 50, 260 (1970).

159. T. H. Gouw and R. E. Jentoft, in "Guide to Modern Methods of Instrumental Analysis," ed. T. H. Gouw, Wiley-Interscience 1972.

160. P. B. Hamilton, Anal. Chem. 30, 914 (1958).

161. C. D. Scott, Anal. Biochem. 24, 292 (1968).

162. K. Tesařík and M. Nesčasová, J. Chromatog. 75, 1 (1973).

163. J. J. Kirkland, Anal. Chem. 43, (12), 36A (1972).

164. T. H. Gouw, I. M. Whittemore, and R. E. Jentoft, Anal. Chem. 42, 1394 (1970).

165. M. Novotny, R. Segura, and A. Zlatkis, Anal. Chem. 44, 9 (1972).

166. L. Sojak and A. Bucinska, Ropa Uhlie, 10, 572 (1968); Anal. Abstracts (London), 18, 1759 (1970).

167. R. F. Marschner, L. J. Duffy, and J. C. Winters, Am. Chem. Soc. Div. Petrol. Chem. 18, (3) 57 (1973).

168. E. J. Gallegos, Anal. Chem. 43, 1151 (1971).

169. C. Kajdas, Chem.-Techn. Ind. 65, 777 (1969).

170. C. Kajdas and R. Tuemmler, Chem.-Techn. Ind. 65, 259 (1969).

171. E. J. Gallegos, Private Communication.

172. L. S. Green, L. J. Schmauch, and J. C. Worman, Anal. Chem. 36, 1512 (1964).

173. ASTM D 2887-70T, "1971 Annual Book of ASTM Standards," Part 17, p 1072, American Society for Testing and Materials, Phila., Pa. 19103.

174. T. H. Gouw, Ruth L. Hinkins, and R. E. Jentoft, J. Chromatogr, 28, 219 (1967).

175. T. H. Gouw, Anal. Chem. 45, 987 (1973).

176. F. Trussell and L. J. Miculas, Am. Chem. Soc. Div. Petr. Chem. Prepr. 12 (3), 517 (1973).

177. American Petroleum Institute Research Project No. 60, Report No. 12, "Characterization of the Heavy Ends of Petroleum," 12, 15 (1972).

178. R. L. Levy, Chromatogr. Rev. 8, 49 (1966).

179. F. Farre-Rius and G. Guiochon, Anal. Chem. 40, 998 (1968).

180. O. F. Folmer, Jr., and L. V. Azaragga, J. Chromatogr. Sci. 7, 665 (1969).

181. W. T. Ristau and N. E. Vanderborgh, Anal. Chem. 43, 702 (1971).

182. N. W. Vanderborgh and W. T. Ristau, American Lab. 5 (5), 41 (1973).

183. C. A. M. G. Cramers and A. I. M. Keulemans, J. Gas Chromatogr. 5, 58 (1967).

184. Clarence J. Wolf and J. Q. Walker, Oil and Gas Journal, March 1969.

185. D. L. Fanter, J. Q. Walker, and Clarence J. Wolf, Anal. Chem. 40, 2174 (1968).

186. J. E. Taylor, D. A. Hutchings, and K. J. French, J. Am. Chem. Soc. 91, 2215 (1969).

187. J. Q. Walker, Chromatographia 5, 547 (1972).

188. J. Knotnerus, Ind. Eng. Chem. Prod. Res. and Development, 6, 43 (1967).

189. S. G. Perry, J. Gas Chromatogr., 2, 54 (1964).

190. C. Karr, J. R. Comberiati, and W. C. Warner, Anal. Chem. 35, 1441 (1963).

191. M. Ariet and H. E. Schweyer, Ind. Eng. Chem., Prod. Res. Development 4, 215 (1965).

192. D. W. Poxon and R. G. Wright, J. Chromatogr. Sci. 6, 142 (1971).

193. P. LePlat, J. Gas Chromatogr. 5, 128 (1967).

194. T. C. Davis and J. C. Petersen, Proc. Ass. Asphalt Paving Technologists, 36, 1 (1967).

195. T. C. Davis and J. C. Petersen, Anal. Chem. 39, 1852 (1967).

196. F. A. Barbour, S. M. Dorrence, and J. C. Petersen, Anal. Chem. 42, 668 (1970).

197. J. C. Robertson and J. R. Moore, Proc. Ass. of Asphalt Paving Technol. 40, 438 (1971).

198. T. C. Davis and J. C. Petersen, Anal. Chem. 38, 1938 (1966).

199. W. E. Haines, Riv. Combust. 25, 311 (1971).

200. A. Berg and J. Lam, J. Chromatogr. 16, 157 (1964).

201. H. Kessler and H. Muller, J. Chromatogr. 24, 469 (1966).

202. R. G. Harvey and M. Halonen, J. Chromatogr. 25, 294 (1966).

203. See the review by G. Jurriens, Riv. Ital. Sostanze Grasse 42, 116 (1965).

204. G. W. A. Rijnders, 5th International Symposium on Separation Methods in Column Chromatography, (1969). Chimia, Supplement.

205. R. E. Jentoft and T. H. Gouw, J. Chromatogr Sci. 8, 138 (1970).

206. S. T. Sie and G. W. A. Rijnders, Separ. Sci. 2, 755 (1967).

207. G. W. A. Rijnders, Chem. Ing. Techn. 42, 290 (1970).

208. R. E. Jentoft and T. H. Gouw, Anal. Chem. 44, 681 (1972).

209. S. T. Sie and G. W. A. Rijnders, Separ. Sci. 2, 729 (1967).

210. R. V. Helm, Am. Chem. Soc. Div. Petr. Chem. Prepr. 14 (1) 13 (1969).

211. T. T. Martin, Am. Chem. Soc. Div. Petr. Chem. Prepr. 18 (3) 562 (1973).

212. T. J. Mayer, Anal. Chem. 43, 176R (1971).

213. T. J. Mayer, Anal. Chem. 45 (5), 183R (1973).

214. P. V. Peurifoy, M. J. O'Neal, and L. A. Woods, J. Chromatogr. 51, 227 (1970).

215. B. Mostert and H. Bohnes, Schmiertechnik 17, 64 (1970).

216. F. Reutner, Fette. Seifen. Anstrichm. 70, 162 (1968).

217. H. Sucker, Fette, Seifen. Anstrichm. 70, 849 (1968).

218. W. Dietsche, Fette. Seifen. Anstrichm. 72, 778 (1970).

219. J. Bzdega, T. Rambuszek, and A. Lewandowski, Chem. Anal. (Warsav 13, 767 (1968).

220. F. S. Lantos and J. Lantos, Lubric. Eng. 27, 184 (1971).

221. P. J. Matthews, J. Appl. Chem. 20, 87 (1970).

Chapter 4

DETERMINATION OF THE ADSORPTION ENERGY, ENTROPY, AND FREE ENERGY OF VAPORS ON HOMOGENEOUS SURFACES BY STATISTICAL THERMODYNAMICS

Claire Vidal-Madjar, Marie-France Gonnord, and Georges Guiochon

Laboratoire de Chimie Analytique Physique
Ecole Polytechnique
Paris, France

I. INTRODUCTION

In this article a review of the statistical theory of adsorption at zero surface coverage is presented and an attempt is made to generalize the method for hydrocarbon molecules of any shape and size. A semiempirical interaction potential is used to calculate the partition function of a molecule in an adsorbed state and this, in conjunction with fundamental physical properties of the adsorbate and solid adsorbent (molecular polarizability, magnetic susceptibility, atomic radius), can be used to calculate the several thermodynamic functions. These complex calculations are greatly facilitated if a computer is used.

A series of models are discussed; these models describe the behavior of an adsorbate on an adsorbent and the corresponding expansion of the interaction potential into terms of the adsorption partition function. They are fashioned to represent the movements of the molecule in its adsorbed state, with cognizance taken of molecular dimensions and shape.

Also discussed are the methods of measuring adsorption thermodynamic functions at zero surface coverage and hence the determination of adsorption energy, entropy, and free energy by gas-solid chromatographic methods.

There cannot be complete understanding of physical adsorption until we are able to calculate the equilibrium constant and the thermodynamic functions of adsorption for any molecule. As is so often the case, it has been possible to obtain a rigorous solution for the adsorption of the hydrogen molecule only on a "homogeneous" surface. However, it is now possible to overcome many of the recognized difficulties by adopting a semiempirical potential similar to that which describes molecular interactions. Moreover, the attendant problems of processing and calculating can be safely entrusted to a computer.

The principles of calculating thermodynamic functions of adsorption were enunciated for diatomic molecules (and benzene) by Hill [1, 2], but to date they have not been used much [3]. By means of quantum mechanics, different mathematical methods, and different models for the adsorption potential, White and Lassettre [4] and Evett [5] have made the calculations for ortho- and para-hydrogen and have shown how these results account for the separation of these compounds. Katorski and White [6] developed a complete theory of adsorption of homo- and heteronuclear isotopes of hydrogen that gave results in excellent agreement with experimental data derived from gas-solid chromatography [7].

Freeman and Hagyard [8] derived a hybrid partition function that allowed calculation of the adsorption coefficients of hydrogen isotopes on homogeneous surfaces. Their model gave results that were in good agreement with the relative retentions of the symmetrical isomers measured by gas-solid chromatography on synthetic zeolites [9].

Kiselev and Poshkus have carried out a semiempirical calculation of the adsorption coefficient of inert gases [10-12], benzene [11, 13, 14], nitrogen and ethane [15], and hydrogen and deuterium [16] on graphitized carbon black. Poshkus, Kiselev and Afreimovich [17-20] made similar calculations for light alkanes (CH_4, C_2H_6, C_3H_8, C_4H_{10}, C_5H_{12}, C_6H_{14}), taking into account the internal rotations of the molecules by considering the vapors as mixtures of the stable conformers in thermodynamic equilibrium.

A simplifying assumption is introduced in this work [10-20] that allows considering molecules much more complex than hydrogen. It is that the adsorption potential can be derived by describing the interaction between the molecule and the adsorbent as the sum of the individual interactions that arise between each atom of the adsorbed molecule and each carbon atom of graphite, with all these binary interactions calculated by a semi-empirical potential and summing for all the atoms involved. Because of the complexity of calculation, several effects had to be neglected and certain simplifications had to be made [11, 12]. Furthermore, it was necessary to introduce limiting constants in order to demonstrate the correlations between theoretical calculations and experimental data [17, 21].

More recently, Poshkus and Kiselev have calculated the ratio of the adsorption coefficients of isotopic pairs of molecules--CH_4, CD_4; C_2H_6, C_2D_6; C_2H_4, C_2D_4; C_6H_6, C_6D_6; C_6H_{12}, C_6D_{12}-- on graphitized carbon black, together with the differences between the differential enthalpy and entropy of adsorption of these molecules [22, 23]. They have shown that the isotopic effect is mostly accounted for by the difference in the adsorption potential of the isotopic molecules, while the contribution of the quantum effects of vibration is noticeable only at low temperatures and only for the lighter molecules. These conclusions concur with those of Di Corcia and Liberti, who have studied the adsorption of similar pairs on various adsorbents [24].

The results are interesting since they illustrate the possibility of relating the thermodynamic functions of adsorption to such general properties as diamagnetic susceptibility, polarizability, atomic weight, and the Van der Waals radii of the atoms of the adsorbent and the adsorbate.

The aim of this work is to show how some refinements of the relevant theory together with some very simple assumptions regarding the movements of a molecule in an adsorbed state allow one to derive easily the thermodynamic functions of adsorption. Comparing the calculated values with experimental results will then facilitate the refinement of the model of adsorption and provide a direct insight to the behavior of the adsorbed molecules.

The methods of calculation and the various models used to account for the behavior of spherical, planar or quasiplanar, and linear or quasilinear molecules are discussed here and the reasons for selecting these different types of models are examined. This involves the calculations of the equilibrium constant of adsorption, the free energy, the entropy and the enthalpy of adsorption, the heat capacity of the adsorbed molecules, and the gas chromatographic volumes. All these calculations are carried out for adsorption at zero surface coverage. As gas chromatography is the best method now available for determining adsorption data at very low surface coverage (down to 10^{-4} of a monolayer or lower), the derivation of thermodynamic functions from retention data is also described. In several other papers, we have presented these calculations for various aromatic hydrocarbons and compared the results with experimental data. A study on the adsorption of alkanes and alkenes incorporating the treatment of the internal rotation will be published later.

II. STATISTICAL THERMODYNAMICS OF ADSORPTION

The theory will be derived for a quasirigid molecule and it will be assumed that physical adsorption does not affect the modes of internal vibration of the molecule nor their frequency. Later, we shall have to study molecules exhibiting rotational isomerism or semifree internal rotation. The problem will always be solved by using the quasirigid molecular model, with such molecules considered as mixtures of quasirigid rotamers in equilibrium.

The only way to solve the problem of the interaction of a molecule with a homogeneous surface would be to calculate the different energy levels of the system by solving the Schrödinger equation, considering each electron-electron interaction separately, and then applying the conventional results of statistical thermodynamics. The complexity of the calculations grows exponentially with the size of the atoms or molecules involved because of the greater numbers of electrons that must be considered. Up to now, such a study has been made for ortho- and para-hydrogen [4-6, 8, 9] and for some isolated atoms adsorbed on a graphite surface [25]. Even for such simple adsorbates, the molecular orbital theory, which is a tempting means of predicting adsorption behavior, has had to be simplified by the extended Hückel theory or CNDO approximation [25]. It just cannot be used for studying the adsorption of molecules more complex than H_2.

A. The Potential Energy of Adsorption

As the present state of the theory does not permit a quantum ab initio calculation, Kiselev and his coworkers [26, 27] simplified the potential equation by introducing a semiempirical expression like the Lennard-Jones or the Buckingham potential (Eqs. (1) and (2)):

$$\phi_i^{LJ} = -\sum_j C_{ij1} r_{ij}^{-6} - \sum_j C_{ij2} r_{ij}^{-8} + B_i^{LJ} \sum_j r_{ij}^{-12} \tag{1}$$

$$\phi_i^{B} = -\sum_j C_{ij1} r_{ij}^{-6} - \sum_j C_{ij2} r_{ij}^{-8} + B_i^{B} \sum_j \exp \frac{-r_{ij}}{\rho} \tag{2}$$

In these equations, r_{ij} is the distance between the ith atom center of the adsorbate molecule and the jth atom center of the adsorbent, and ρ is the repulsion constant taken as equal to 0.28 Å for all graphite-hydrocarbon systems and for neon, argon, and krypton adsorption on graphite [27]. The potential energy of the molecule is taken as the sum of the potential energies of its separate atoms ϕ_i. The potential energy of individual atoms is thus calculated taking into account a dipole-dipole term and a dipole-quadrupole term in the potential of the dispersion-attraction forced inclusive of repulsion forces such as r_{ij}^{-12} or exp $-r_{ij}/\rho$. The quadrupole-quadrupole term is not taken into consideration for it is less than 1% of the dipole-dipole term. The constants of attraction were calculated by the Kirkwood-Muller equation [28]:

$$C_{ij1} = -6\, mc^2 \frac{\alpha_i \alpha_j}{\dfrac{\alpha_i}{\chi_i} + \dfrac{\alpha_j}{\chi_j}} \tag{3}$$

and an analogous expression for C_{ij2} [29]:

$$C_{ij2} = \frac{45\, h^2}{35\, \pi^2\, m} \alpha_i \alpha_j \left[\frac{1}{(2\alpha_j \chi_i)/(\alpha_i \chi_j) + 1} + \frac{1}{(2\alpha_i \chi_j)/(\alpha_j \chi_i) + 1} \right] \tag{4}$$

where h is the Planck constant, m the electron mass, and C the light velocity; α_i and χ_i are the polarizability and diamagnetic susceptibility of the ith atom, and α_j and χ_j those of the adsorbent atom.

The same adsorption potential expressions have been used for graphitized carbon black [26, 27, 30] and for other adsorbents such as boron nitride [31], the phthalocyanines [32], zeolites [33, 34, 35] and potassium chloride [36]. Calculation is easy when the crystalline structure of the adsorbent can be strictly defined.

Table 1 gives the constants used for all our calculations. The constants for the repulsion forces were calculated for each type of atom at the equilibrium distance between the adsorbate atom and the graphite lattice,

TABLE 1

Parameters Used for Calculating Force Constants for Adsorption Potential

	C graphite	C	H
Polarizability ($\mathring{A}^3$)	0.937 [37]	0.96 [18]	0.43 [18]
Diamagnetic susceptibility (10^{-5} $\mathring{A}^3$)	1.05 [37]	1.23 [18]	0.37 [18]
Atomic radius ($\mathring{A}$)	1.70 [37]	1.70 [18]	1.30 [18]
C_{i1} ($\mathring{A}^6$ x kcal/mole)	....	381.33	139.02
C_{i2} ($\mathring{A}^8$ x kcal/mole)	....	595.01	268.12
B^{LJ} x 10^{-3} ($\mathring{A}^{12}$ x kcal/mole)	....	658.29	117.12
B^{B} x 10^{-3} ($\mathring{A}^{12}$ x kcal/mole)	....	52.09	10.19

at which the repulsion and attraction forces cancel; this equilibrium distance z_0 is given by the sum of the van der Waals radii. The potential energies are calculated for the infinite surface of graphite by summation over the nearest 10,000 carbon atoms in the five first planes of the graphite network (cf. Fig. 1), following the principles described in a previous paper [38].

Figure 2 shows the adsorption potential curves for hypothetical C and H atoms adsorbed on a graphite surface, obtained using either the Lennard-Jones or the Buckingham potential equations. The adsorption potential of a molecule can then be calculated as a function of the distance of its mass-center to the adsorbent surface atoms and of the molecule orientation, by summation of all atomic contributions. The equilibrium position for the whole molecule is derived as the position corresponding to the minimum of this potential.

B. The Classical Partition Function in the Adsorbed State

The classical Hamiltonian of an adsorbed molecule can be written

$$H = \frac{p_x^2}{2M} + \frac{p_y^2}{2M} + \frac{p_z^2}{2M} + \frac{u^2}{2I_X} + \frac{v^2}{2I_Y} + \frac{w^2}{2I_Z} + \phi \tag{5}$$

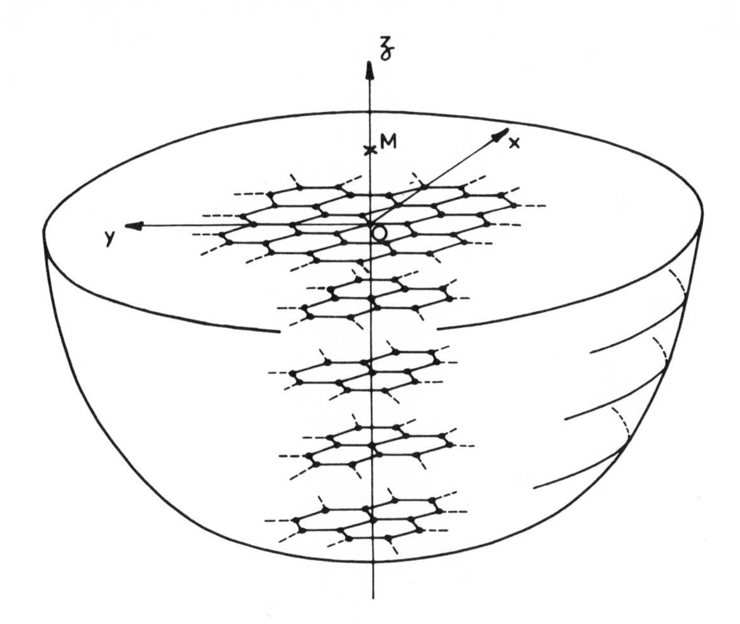

FIG. 1. This figure shows disposition of the carbon atoms in the crystal of graphite, which is used to calculate the adsorption potential. The potential in M is calculated after the sum of the potentials created there by all the atoms inside the half-sphere shown.

ϕ is the interaction potential between the adsorbent and the adsorbed molecule; I_X, I_Y, I_Z are the main moments of inertia of the molecule, and M its mass; p_x, p_y, p_z are the components of the linear momentum; and p_θ, p_ϕ, p_ψ, the components of the angular momentum, with [39]

$$u = p_\theta \cos\psi + \frac{\sin\psi}{\sin\theta}(p_\phi - p_\psi \cos\theta)$$

$$v = p_\theta \sin\psi + \frac{\cos\psi}{\sin\theta}(p_\phi - p_\psi \cos\theta) \tag{6}$$

$$w = p_\psi$$

In our calculations, the position of the mass center O of the molecule is defined in a three-dimensional plot xyz; the plane x, y (z = 0) is the plane of the adjacent carbon atom centers in the outer basal plane of the graphite. The main axes of inertia of the molecule are denoted OX, OY, and OZ, and the Eulerian angles of the molecule in the xyz reference system are θ, ψ, ϕ (Fig. 3).

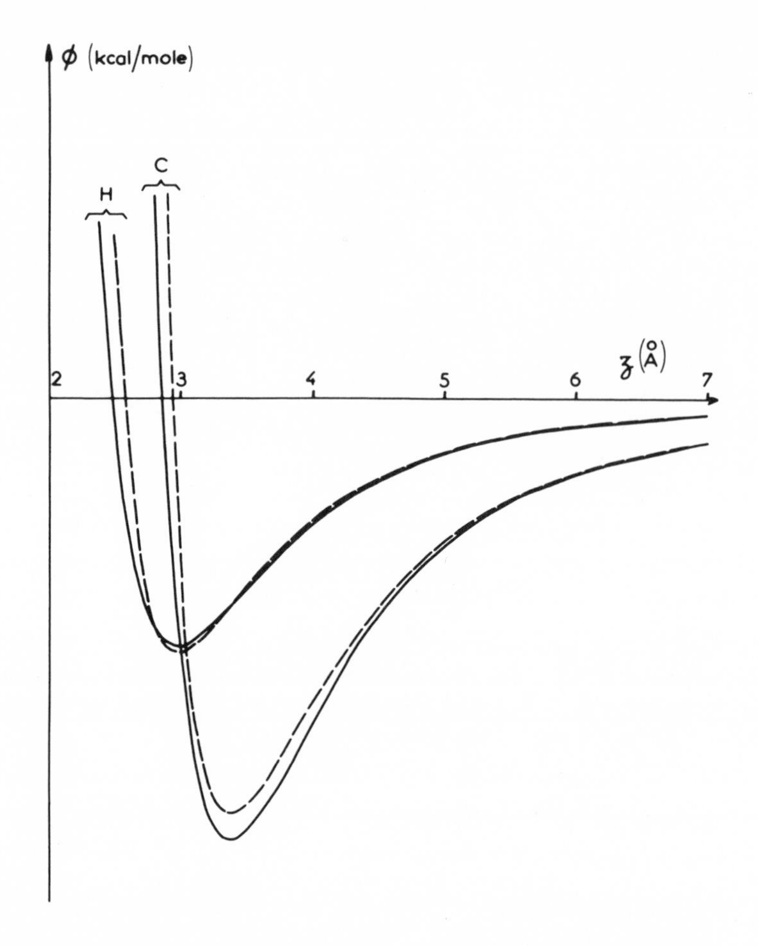

FIG. 2. Variation of the adsorption potential of the carbon and hydrogen atoms with their distance to the graphite surface. Solid lines show Buckingham adsorption potential. Dashed lines show Lennard-Jones adsorption potential.

Hill [1], then Kiselev and Poshkus [10-14], have given an approximate solution to this problem, adopting a method similar to the one Pitzer and Gwinn [40] used to solve the problem of the steric hindrance of internal rotation. They have assumed that the quantum partition function can be inferred from the classical partition function by multiplying it by a quantum factor that is a function of the frequencies of the harmonic vibrations of the adsorbed molecule.

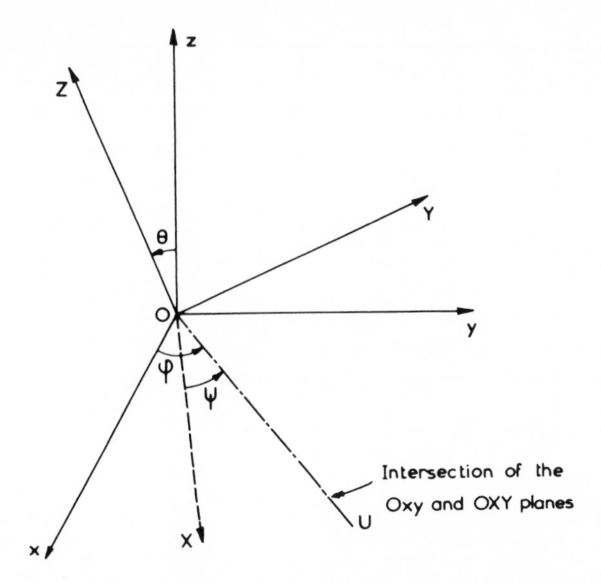

FIG. 3. Schematic representation of the rotation movements of a rigid molecule OXYZ in a fixed reference system Oxyz. OU intersection of the Oxy and OXY planes. For the calculations to be more clearly and easily understandable, sometimes OZ is considered as nearly perpendicular to the adsorbent surface and sometimes as nearly parallel to that surface.

Unfortunately, it is not possible to solve the Schrödinger equation directly for the classical Hamiltonian (Eq. (5)) of the adsorbed molecule, in order to calculate the energy levels and to treat them statistically.

If no account is taken of any intraatomic or intramolecular factors, i.e., if we assume that the electronic and the vibration partition functions are unchanged by adsorption, we can calculate the partition function in classical mechanics for some simple models that represent the behavior of the adsorbed molecule. Calculations can be made as precise as necessary for each model, and comparison with the experimental results will help in selecting the model that describes the best results, and thus eventually in its refinement to better describe the movements of the molecule in the adsorbed state [41].

C. Calculating the Classical Partition Function in the Adsorbed State: Resolution Models

Each of the models discussed below corresponds to a definitive, possible behavior of an adsorbed molecule but all are only approximations. The only rigorous way to carry out calculations would be to calculate energy levels by quantum mechanics. In all these models the molecule is assumed to move freely in any direction parallel to the homogeneous graphite surface: for molecules larger than methane, the potential is not a function of x and y [38].

1. Spherical and Quasispherical Molecules

Model A. For such molecules the rotation around the three axes of inertia remains unchanged. If we assume that the movements of rotation of the molecule around its three main axes of inertia are not hindered when the molecule is sorbed in the space above the crystal surface [15, 16], the partition function in classical mechanics is given by

$$Z_{a\,\mathrm{class}} = \frac{1}{h^{\sigma}} \int \cdots$$

$$\int \exp \frac{-H}{kT}\, dx\, dy\, dz\, \sin\theta\, d\theta\, d\varphi\, d\psi\, dp_x\, dp_y\, dp_z\, du\, dv\, dw \tag{7}$$

where H is given by Eq. (5) and σ is the symmetry number of the molecule. Equation (7) can be easily reduced to

$$Z_{a\,\mathrm{class}} = \frac{A}{\sigma} \left[\frac{2\pi M kT}{h^2}\right]^{3/2} \left[\frac{2\pi I_X kT}{h^2}\right]^{1/2} \left[\frac{2\pi I_Y kT}{h^2}\right]^{1/2} \left[\frac{2\pi I_Z kT}{h^2}\right]^{1/2} 2\pi J_1 \tag{8}$$

where

$$J_1 = \int_{\theta} \int_{\psi} \int_{z} \exp \frac{-\phi}{kT}\, dz\, \sin\theta\, d\theta\, d\psi \tag{9}$$

with $\phi \leq 0$, $0 \leq \theta \leq \pi$ $0 \leq \psi$ 2π, the integration made for all the space available to the molecule and for all orientations of the molecule within that space.

The integration of Eq. (9) can be carried out directly by numerically calculating the adsorption potential for all the orientations of the molecule on the graphite surface and by separately integrating every variable z, θ, and ψ. This model introduces no further approximations and it is the usual mode of operation in classical mechanics. However, it cannot account for the behavior of flat or quasilinear molecules whose rotation is hindered by the surface.

Model B. In Model B, we assume harmonic oscillations in a perpendicular sense to the carbon surface. So ϕ is approximated by a harmonic expansion of z near z_1, the equilibrium distance of the mass center for given values of θ and ψ (z_1 is a function of θ and ψ):

$$\phi(z, \theta, \psi) = \phi_0 + \frac{1}{2}(z - z_1)^2 \ \phi''_{z_1^2}(\theta, \psi) \tag{10}$$

where $\phi''_{z_1^2}$ stands for the value of the second derivative $\partial^2\phi/\partial z^2$ for $z = z_1$.

The frequency of the assumed vibration is

$$\nu_z(\theta, \psi) = \frac{1}{2\pi}\sqrt{\frac{\phi''_{z_1^2}(\theta, \psi)}{M}} \tag{11}$$

The integration of J_1 must be carried out as for Model A. The physical significances of both models A and B are very similar since the rotation of the molecule around its axis remains free, thus the results are numerically very close.

2. Planar or Quasiplanar Molecules

The adsorption potential varies inversely with the distance from the adsorbent surface, so these molecules normally have a preferential orientation. When they are near the surface, their plane is roughly parallel to the adsorbent surface and they oscillate around this equilibrium position. Two models have been derived to account for the behavior of planar or quasiplanar molecules. In both models, the mass-center of the molecule is moving with a harmonic vibration perpendicular to the adsorbent surface. The molecule oscillates, with harmonic movement, around the two main axes of inertia, which are in the adsorption plane instead of rotating freely. The average position of the molecular plane is parallel to the adsorbent surface. The two models differ only by the limits of the freedom of rotation around the third axis.

Model C. In Model C, rotation around the third axis, which is perpendicular to the molecular plane and corresponds to the largest moment of inertia, is free. The reference system of the molecule (cf. Fig. 3), OXYZ, is chosen here so that OX and OZ are the two main axes of inertia in the molecular plane. The frequency of the oscillation around OX is ν_θ and around OZ is ν_ψ. The interaction potential can be expanded in powers of z, θ, and ψ near the equilibrium position ($z = z_0$, $\theta = \pi/2$, $\psi = 0$):

$$\phi(z, \theta, \psi) = \phi_0 + \frac{1}{2}\phi''_{z^2}(z - z_0)^2 + \frac{1}{2}\phi''^{\pi/2}_{\theta^2}\cos^2\theta + \frac{1}{2}\phi''^{0}_{\psi^2}\sin^2\psi \tag{12}$$

where $\phi''^{0}_{z^2}$ is the value of ϕ''_{z^2} for $z = z_0$ and the derivatives

$$\phi''^{\pi/2}_{\theta^2} = \sum_{i=1}^{n} (Z_i^2 + Y_i^2)\left(\frac{\partial^2 \phi_i}{\partial z^2}\right)_{z = z_0}$$

$$\phi''^{0}_{\psi^2} = \sum_{i=1}^{n} (X_i^2 + Y_i^2)\left(\frac{\partial^2 \phi_i}{\partial z^2}\right)_{z = z_0} \tag{13}$$

are calculated for the whole molecule, n being the number of atoms X_i and Z_i the coordinates of these atoms in the molecule plane. The calculation is detailed in Appendixes I and II.

The integral J_1 (cf. Eq. (9)) is given by

$$J_1 = 2\left(\frac{2\pi kT}{\phi''^{0}_{z^2}}\right)^{1/2} \left(\frac{2\pi kT}{F_1}\right) \exp\frac{-\phi_0}{kT} \tag{14}$$

where

$$F_1 = \left(\phi''^{\pi/2}_{\theta^2}\right)^{1/2} \left(\phi''^{0}_{\psi^2}\right)^{1/2} \tag{15}$$

The frequencies of the three movements of oscillation of the molecule are

$$\nu_z = \frac{1}{2\pi}\sqrt{\frac{\phi''^{0}_{z^2}}{M}}$$

$$\nu_1 = \frac{1}{2\pi}\sqrt{\frac{\phi''^{\pi/2}_{\theta^2}}{I_X}} \tag{16}$$

$$\nu_2 = \frac{1}{2\pi}\sqrt{\frac{\phi''^{0}_{\psi^2}}{I_Z}}$$

A similar result can be obtained by choosing a different coordinate system with OX, OY in the molecular plane and OZ perpendicular to the plane. The formulas derived are different, as the oscillation in θ can be made around either OX or OY, but the numerical results are the same. The system used has been chosen because it gives more general formulas, although it might seem more complicated.

Model D. In Model D, the rotation around the axis perpendicular to the surface is hindered. The value of the potential barrier V_0 is taken as the difference between the potential energies of the molecule at equilibrium distance when either main axis of inertia located in the molecular plane is perpendicular to the adsorbent surface. In the following calculations, θ will be defined as the angle of the molecular plane with the adsorbent surface and OZ is the main axis perpendicular to the molecular plane (cf. Fig. 3). This model has been introduced in Ref. 41 to explain the large ratio of the adsorption coefficients of phenanthrene and anthracene on graphitized carbon black.

The adsorption potential is now given by the following expansion:

$$\phi(z, \theta, \psi) = \phi_0 + \frac{1}{2} \phi''^{0}_{z^2} (z - z_0)^2 + \frac{1}{2} \phi''^{0}_{\theta^2} \sin^2 \theta + V_0 \sin^2 \psi \tag{17}$$

J_1 is then written:

$$J_1 = 4 \left(\frac{2\pi kT}{\phi''^{0}_{z^2}} \right)^{1/2} \left(\frac{kT}{\phi''^{0}_{\theta^2}} \right) \left(\frac{\pi kT}{V_0} \right)^{1/2} \exp \frac{-\phi_0}{kT} \tag{18}$$

There are now four modes of vibration. The frequencies of the one vibration in z, perpendicular to the lattice surface, and of the two oscillations of the plane of the molecule around the main axis of inertia in that plane are still given by equations similar to Eq. (16). The frequency associated with the hindered rotation of the plane of the molecule around the axis of inertia perpendicular to this plane is ν_ψ:

$$\nu_\psi = \frac{1}{2\pi} \sqrt{\frac{2 V_0}{I_Z}} \tag{19}$$

3. Linear and Quasilinear Molecules

In the strong anisotropic field of adsorption, linear and quasilinear molecules have a preferential orientation with their axis of lowest moment of inertia roughly parallel to the surface. Let θ be the angle between OZ and the perpendicular to the adsorbent surface Oz. As in the B, C, and D models, we shall assume a harmonic oscillation of the mass-center of the molecule perpendicular to the adsorbent surface.

The movements of the whole molecule in the mass-center system can be described by the sum of the independent movements of rotation around each of the three main axes of inertia of the molecule. These movements can remain as free rotation in the gas phase, or can become either hindered rotation or harmonic oscillation. Many combinations can be imagined but a large number are impossible or improbable: If OZ is the axis of inertia of lowest moment and OY the one of largest moment (cf. Fig. 3), the equilibrium position of the molecule will be with the axes OX and OZ approximately parallel to the adsorbent surface; the rotation around OX will be more hindered than the rotation around OZ, which itself will be more hindered than the rotation around OY. Table 2 shows the possible combinations.

If we consider free rotations only, no extra terms are introduced in the potential expansion of Model B (Eq. (10)). The force constants corresponding to the harmonic oscillations are $(\phi''_{\theta^2})_{eq}$ and $(\phi''_{\psi^2})_{eq}$. The potential barriers of hindered rotations around OX and OZ, V_θ and V_ψ, are calculated as the difference between the adsorption potentials of the molecule when it is placed either with the OXZ plane parallel to the adsorbent surface or with the corresponding axis perpendicular to the graphite lattice. The third potential barrier V_ϕ has been calculated as for Model D: This is the difference between the extreme values of the potential energy of the molecule during its rotation around OY when this axis is parallel to the adsorbent surface (θ or ψ = 0° and 90°).

The models corresponding to the acceptable combinations of these different movements are listed in Table 2. Since the potential barrier V_ψ is usually very low for a quasilinear molecule, it is practical to consider either a free rotation in ψ or a hindered one as equivalent.

Five models remain for consideration.

Models E: Free rotation around OZ and OY. There are two E models, depending whether the movement around OX is a hindered rotation or an oscillation. The corresponding adsorption potentials of the molecule are the following near the equilibrium position ($z = z_0$, $\theta = \pi/2$):

$$\phi_{E_a} = \phi_0^0 + \frac{1}{2}\phi''^{0}_{z^2}(z - z_0)^2 + V_\theta \cos^2\theta \tag{20}$$

$$\phi_{E_b} = \phi_0^0 + \frac{1}{2}\phi''^{0}_{z^2}(z - z_0)^2 + \frac{1}{2}\phi''^{\pi/2}_{\theta^2}\cos^2\theta \tag{21}$$

Kiselev [15, 16] used these two models; he noted a large difference between the sets of results given by the two potential expressions and he later considered only the first expression for linear molecules such as hydrogen, deuterium, and nitrogen.

For model E_a, the integral J_1 (cf. Eq. (9)) is given by

TABLE 2

Movements of Adsorbed Molecules around Their Main Axis of Inertia[a]

	Free rotation around OY			Hindered rotation around OY		
Movement around OX (θ) / Movement around OZ (ψ)	Free rotation	Hindered rotation	Oscillation	Free rotation	Hindered rotation	Oscillation
Free rotation	A[b] B[b]	E_a	E_b	No[d]	No	No
Hindered rotation	No			No		
Oscillation	No	F_a	C(or F_b)[c]	No		D[c]

[a]Oscillation around OY leads to only one possible combination: oscillation around the three axis. This combination is quite improbable for linear molecules.

[b]These movements of rotation are more or less perturbed by presence of adsorption fields.

[c]Models C and D are improbable for quasilinear molecules since they involve an oscillation around OZ for which the potential barrier is small.

[d]"NO" indicates combinations that are not acceptable. They cannot be, since rotation around axis OX is more hindered than around axis OZ, which itself is more hindered than around axis OY.

$$J_1 = 2\pi \left(\frac{2\pi kT}{\phi''^{o}_{z}}\right)^{1/2} \left(\frac{2\pi kT}{2V_\theta}\right)^{1/2} \exp\frac{-\phi_o}{kT} \tag{22}$$

Vibration frequencies are easily deduced from (22). An equivalent relation exists for Model E_b,

Models F: Free rotation around OY; oscillation around OZ. As noted above, due to the small potential barrier for the rotation around OZ, the numerical results obtained are the same whether free or hindered rotation is assumed around OZ. If an oscillation is assumed, the potential is given by

$$\phi_{F_a} = \phi_o^{\pi/2} + \frac{1}{2}\phi''^{o}_{z^2}(z - z_o)^2 + V_\theta \cos^2\theta + \frac{1}{2}\phi''^{o}_{\psi^2}\sin^2\psi \tag{23}$$

or

$$\phi_{F_b} = \phi_o^{\pi/2} + \frac{1}{2}\phi''^{o}_{z^2}(z - z_o)^2 + \frac{1}{2}\phi''^{\pi/2}_{\theta^2}\cos^2\theta + \frac{1}{2}\phi''^{o}_{\psi^2}\sin^2\psi \tag{24}$$

depending on whether the movement around OX is an oscillation or a hindered rotation. The integral J_1 for both models F is equivalent to Eq. (14). The frequencies associated with Model F_a for instance are then

$$\nu_z = \frac{1}{2\pi}\sqrt{\frac{\phi''^{o}_{z}}{M}} \qquad \nu_\theta = \frac{1}{2\pi}\sqrt{\frac{2V_\theta}{I_X}} \qquad \nu_\psi = \frac{1}{2\pi}\sqrt{\frac{\phi''^{o}_{\psi^2}}{I_Z}} \tag{25}$$

Model F_b is equivalent to Model C considered above for quasiplanar molecules. Equations (12) and (24) are identical.

Models D: Hindered rotation around OY. This last model, D, which assumes in addition a hindered rotation around the third axis of inertia, is identical to Model D described above, for which we have assumed, for the sake of simplicity, that OZ is the axis of largest moment of inertia, OY being now the axis of smallest moment or axis of quasilinearity of the molecule.

The difference between the E_a and F_a models lies in the term $\phi''^{o}_{\psi^2}$, which indicates that the molecule is not rigorously linear but has become quasiplanar, like a very narrow rectangle, and that the assumption of a free rotation around the vertical axis is not exactly convenient. The results obtained with these five models will be discussed in a subsequent paper.

Variation of the adsorption potential with θ. It should be easy to determine whether a model should assume an oscillation or a hindered rotation for a given molecule but often it is not. Because of the strong anharmonicity of the curve $\phi = f(\theta)$, as exemplified in Figure 4, the force constant for the oscillation can be extremely large whereas the potential barrier is relatively low: the potential decreases sharply at first when the atoms that are near the surface recede from it; when they are more than 1 Å away, the potential does not vary much with the distance. This can explain the results Kiselev et al. obtained for linear molecules using the E_a and E_b models [15, 16].

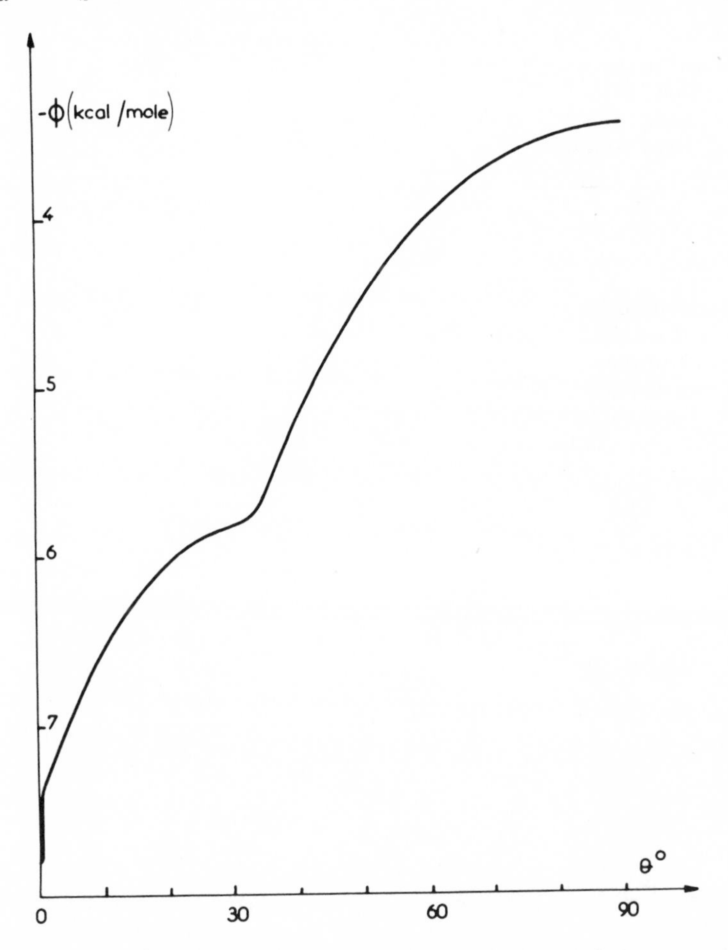

FIG. 4. Variation of the adsorption potential of cis-2-butene versus the angle θ of the axis of smallest moment of inertia of the molecule with the plane of the adsorbent lattice.

Choice of the best model is thus a specific matter and for a given molecule may also depend on temperature; under certain conditions the progressive transformation of an oscillation into a hindered rotation could result in an abnormal variation of the thermodynamic functions of adsorption.

4. The thermodynamic functions of adsorption

Assuming that the gas phase is ideal and that the surface coverage is so small that there is no adsorbate-adsorbate interaction, we can write the adsorption isotherm equation in a virial form, relating the partition functions of a molecule in the adsorbed phase Z_a and in the gas phase Z_g [3, 12]:

$$\Gamma = \frac{1}{A} \frac{Z_a - Z_g}{Z_g/V} \frac{p}{kT} \tag{26}$$

where Γ is the number of adsorbed molecules per unit surface area of adsorbent, A the surface area of the adsorbent, and V the volume of the gas phase. At zero surface coverage, the adsorption coefficient is equal to the first virial coefficient:

$$U_A = \frac{V}{A} \frac{Z_a - Z_g}{Z_g} \tag{27}$$

The partition function of a molecule in the gas phase can be written very easily if we assume the molecule to have only translational and rotational degrees of freedom. As only the ratio $Z_a:Z_g$ is involved in Eq. (27) and as we have assumed the internal vibrations to be unaffected by adsorption, their contributions to U_A cancel out.

$$Z_g = \frac{8\pi^2 V}{\sigma} \left(\frac{2\pi M kT}{h^2}\right)^{3/2} \left(\frac{2\pi I_X kT}{h^2}\right)^{1/2} \left(\frac{2\pi I_Y kT}{h^2}\right)^{1/2} \left(\frac{2\pi I_Z kT}{h^2}\right)^{1/2} \tag{28}$$

A combination of the Eqs. (8), (27), and (28) gives the retention volume per unit surface area in the case of Model A, if the symmetry number and the intramolecular properties of the molecule are not modified by the adsorption phenomenon:

$$U_A = \frac{1}{4\pi} \iiint \left[\exp \frac{-\phi}{kT} - 1\right] \quad dz \ \sin\theta \ d\theta \, d\psi \tag{29}$$

Generally ϕ/kT is large enough to permit us to neglect unity in Eq. 29.

Similar equations are obtained for each model considered. These equations are given in Table 3.

TABLE 3

Relations Giving the Adsorption Coefficient Corresponding to the Different Models of the Adsorption Potential

Model	Adsorption coefficient
A	$U_A = \frac{1}{4\pi} \iiint \exp \frac{-\phi}{kT} \; dz \sin\theta \, d\theta \, d\phi$
B	$U_A = \frac{1}{4\pi} \iint \left(\frac{2\pi kT}{\phi''_{z^2}(\theta,\psi)}\right)^{1/2} \exp\left(-\frac{\phi_0(\theta,\psi)}{kT}\right) \sin\theta \, d\theta \, d\psi$
C	$U_A = \frac{1}{2\pi} \left(\frac{2\pi kT}{\phi''^{0}_{z^2}}\right)^{1/2} \left(\frac{2\pi kT}{\phi''^{0}_{\psi^2}}\right)^{1/2} \left(\frac{2\pi kT}{\phi''^{\pi/2}_{\theta^2}}\right)^{1/2} \exp \frac{-\phi_0}{kT}$
D	$U_A = \frac{1}{\pi} \left(\frac{2\pi kT}{\phi''^{0}_{z^2}}\right)^{1/2} \left(\frac{kT}{\phi''^{0}_{\theta^2}}\right) \left(\frac{2\pi kT}{2V_0}\right)^{1/2} \exp \frac{-\phi_0}{kT}$
E_a	$U_A = \frac{1}{2} \left(\frac{2\pi kT}{\phi''^{0}_{z^2}}\right)^{1/2} \left(\frac{2\pi kT}{2V_\theta}\right)^{1/2} \exp \frac{-\phi_0}{kT}$
E_b	$U_A = \frac{1}{2} \left(\frac{2\pi kT}{\phi''^{0}_{z^2}}\right)^{1/2} \left(\frac{2\pi kT}{\phi''^{\pi/2}_{\theta^2}}\right)^{1/2} \exp \frac{-\phi_0}{kT}$
F_a	$U_A = \frac{1}{2\pi} \left(\frac{2\pi kT}{\phi''^{0}_{z^2}}\right)^{1/2} \left(\frac{2\pi kT}{\phi''^{0}_{\psi^2}}\right)^{1/2} \left(\frac{2\pi kT}{2V_\theta}\right)^{1/2} \exp \frac{-\phi_0}{kT}$
F_b	$U_A = \frac{1}{2\pi} \left(\frac{2\pi kT}{\phi''^{0}_{z^2}}\right)^{1/2} \left(\frac{2\pi kT}{\phi''^{0}_{\psi^2}}\right)^{1/2} \left(\frac{2\pi kT}{\phi''^{\pi/2}_{\theta^2}}\right)^{1/2} \exp \frac{-\phi_0}{kT}$

The thermodynamic functions of adsorption and the adsorption coefficient at zero surface coverage can be derived by the conventional equations of statistical thermodynamics [15] once a reference state has been selected. The reference state chosen assumes that the number of molecules in the adsorbed state per unit surface area of the adsorbent equals the number of molecules in the gaseous phase per unit volume. The thermodynamics quantities are as follows.

1. Retention volume and adsorption coefficient:

$$U_A = \frac{1}{4\pi} J_1 \tag{30}$$

2. Differential free energy of adsorption:

$$\overline{\Delta A} = -RT \ln \frac{J_1}{4\pi} \tag{31}$$

3. Differential energy of adsorption:

$$\overline{\Delta U} = RT \frac{J_2}{J_1} \tag{32}$$

4. Differential entropy of adsorption:

$$\overline{\Delta S} = R \ln \frac{J_1}{4\pi} + R \frac{J_2}{J_1} \tag{33}$$

5. The specific heat capacity of the adsorbed molecule can be derived from the following equation, which gives the difference between the heat capacity of the adsorbed molecule and of the molecule in the gas phase:

$$\Delta C_V = R \left[\frac{J_3}{J_1} - \left(\frac{J_2}{J_1} \right)^2 \right] \tag{34}$$

In these equations [15],

$$J_n = \iiint \left(\frac{\phi}{kT} \right)^{n-1} \exp \frac{-\phi}{kT} \, dz \, \sin\theta \, d\theta \, d\psi \tag{35}$$

The quantum correction factor that Pitzer and Gwinn [40] introduced and Kiselev and Poshkus [16, 17] calculated has not been taken into account, for its influence is very small: in all the cases we have studied, this factor is between 0.999 and 0.99 [41].

III. EXPERIMENTAL MEASUREMENT OF THERMODYNAMIC FUNCTIONS

Gas chromatography offers an accurate and fast method for determining the thermodynamic functions of adsorption. The net retention volume of a compound is the volume of carrier gas flowing through the column during the time between injection of a pulse of that compound and elution of the mass center of the corresponding band, which is corrected for decompression and for gas holdup of the column. The adsorption isotherm is obtained by gas solid chromatography from the change of the net retention volume V_R^o (cm³) with the amount of solute in the gas phase, n_g(g/cm³), according to the relation [42-44]:

$$n_s = \frac{1}{A} \int_o^{a_g} V^o{}_R \, dn_g \tag{36}$$

A is the surface area of the adsorbent and n_s is the amount adsorbed per unit surface area (g/m²), in equilibrium with a_g. The net retention volume is obtained in practice from the gaseous volume of the column V_g, and the column capacity factor ($k' = (t_R - t_m)/t_m$), through the relation

$$V_R^o = k' \, V_g \tag{37}$$

t_R is the retention time of the solute and t_m is the retention time of an unretained compound (i.e. methane). Methane is not completely inert over graphitized thermal carbon black, even at a temperature well above ambient. Around 100°C, the column capacity factor k', with conventional packed column, is about 0.005. This effect is important only when very accurate measurements are carried out. Theoretical and experimental results on the adsorption of methane on graphitized carbon black will be published soon.

At zero surface coverage, the retention volume per unit surface area, $U_A = V_R^o/A$, is equal to the slope of the isotherm at the origin [44]; for practical purposes the isotherm is nearly linear in this region. When the isotherm is plotted as n_s vs p (partial pressure of the adsorbate), the slope of the linear portion at the origin, K (Henry coefficient), is simply related to U_A:

$$K = \frac{U_A}{RT} \tag{38}$$

A change in retention volume with the amount of sample injected is observed when the isotherm is not linear in the sample size range studied. If the amount of solute injected into the column is small enough however, the adsorption isotherm can be considered as linear.

The maximum concentration in the gas phase at the column outlet is given by the following equation,

$$n_g^M = \frac{m\sqrt{N}}{V_R\sqrt{2\pi}} \tag{39}$$

where m is the amount of solute injected; N is the number of theoretical plates; and V_R is the retention volume (i.e., product of outlet flow-rate by retention time). The retention volume V_R is not corrected for the gas holdup of the column so that

$$\frac{V_R^\circ}{V_R} = \frac{k'}{1 + k'} \tag{40}$$

The maximum concentration n_s^M in the adsorbed phase at the column outlet is accordingly

$$n_s^M = n_g^M \frac{V_R^\circ}{A} \tag{41}$$

Then, combining Eqs. (39) to (41) gives

$$n_s^M = \frac{m\sqrt{N}}{A\sqrt{2\pi}} \frac{k'}{1 + k'} \tag{42}$$

For a 2-m-long column filled with graphitized carbon black (A = 28.4 m^2), k' = 7.65 for benzene at 373°K. With a 0.1-μg sample, the maximum density of benzene molecules adsorbed at the outlet end of the column is

$$n_s^M = 0.39 \cdot 10^{-7}\ g/m^2$$

Near the injection point of the column (column length = 2 cm),

$$n_s^{M^2} = 0.39 \cdot 10^{-6}\ g/m^2$$

This value is obtained from Eq. (42), which shows that n_s^M is inversely proportional to the square root of the column length [45]. The surface occupied by a molecule of benzene can be estimated to be 47.2 Å^2 [46]:

the amount needed to form a complete monolayer is 2.73 10^{-4} g/m^2 . The maximum surface coverage at the column outlet is thus 1.4 10^{-4}, and at 2 cm from the injection point, 1.4 10^{-3}. This example shows that the surface coverage at which gas chromatographic measurements can be made with precision is much lower than the surface coverage that can be studied by static experiment, and that the former technique is meaningfully operative within the initial linear part of the adsorption isotherm. Gas-solid chromatography is then an important method of studying adsorption thermodynamics without adsorbate-adsorbate interactions.

The change in the differential free energy of the adsorption associated with the transfer of a molecule from the gaseous state to the adsorbed state, at constant temperature is [47]

$$\overline{\Delta A} = RT \log \frac{a_g}{a_s} - RT \log \frac{a_g^o}{a_s^o} \tag{43}$$

where a_g and a_s are the activities in the gasous and adsorbed phases and a_g^o and a_s^o represent the activities in the tridimensional and bidimensional standard states. If the standard state is chosen so that $a_g^o/a_s^o = 1$, then, when both gaseous and adsorbed phases are ideal:

$$\overline{\Delta A} = -RT \log U_A \tag{44}$$

The differential energy of adsorption is obtained from the change of U_A with temperature.

$$\overline{\Delta U} = -R \frac{d \log U_A}{d(1/T)} \tag{45}$$

It can be obtained directly by plotting the column capacity factor k' vs (1/T). The slope of the straight line, proportional to the differential energy of adsorption is determined by a least-square method. In fact, this plot is not exactly linear: the curvature depends on the variation in heat capacity between the adsorbed and the gas molecule, but very accurate measurements of k' make it possible to determine ΔC_v. The differential entropy of adsorption is easily derived from $\overline{\Delta A}$ and $\overline{\Delta U}$.

IV. RESULTS

To illustrate the possibilities of the theory discussed above, we present the results obtained with three molecules: the flat, aromatic benzene and naphthalene and the elongated ethane molecules. More complete and systematic results will be published soon [48]. Table 4 lists the values calculated for benzene after models A, B, and C, for naphthalene after models A, B, C, and D, and for ethane after models A, B, Ea, Eb, Fa, and Fb and compare them with experimental data.

For the molecules of benzene and ethane, the model D cannot be used for there is no potential barrier hindering the rotation of these molecules around their main axis of symmetry, all the positions of the molecules around this axis being equivalent. Then Eq. (17) has no significance.

Figures 5 and 6 show the variations of the retention volumes of benzene, naphthalene, and ethane with the temperature, derived from experimental data and from the different models studied.

The adsorption of the benzene (cf. Fig. 5) molecule is well accounted for by the model C, which takes into account its flat, symmetrical structure and assumes harmonic vibrations of its mass center perpendicular to the graphite surface and harmonic oscillations of the ring around its two main axes of inertia.

For naphthalene, models C and D give similar results, but Model D is slightly better. This model assumes a hindered rotation of the molecule around its main axis of inertia perpendicular to the molecule plane. That the theoretical results derived from models C and D are nearly the same is owing to the size of the molecule, which makes the potential barrier hindering the rotation very low (2 kcal/mole).

For benzene as for naphthalene, it can also be observed that both models A and B, which give the same probability to all positions of the molecule above the adsorbent surface, do not suitably describe the behavior of these planar molecules on the graphite lattice.

Models Eb and Fa seem to be equivalent in describing the behavior of ethane adsorbed on graphite. Physically, the model Eb, which assumes free rotation around the C-C axis of the molecule and around the axis of inertia perpendicular to the C-C bond and the graphite plane, seems to be more natural than the Fa model, which assumes hindered rotation around this last axis and harmonic oscillations around the C-C bond. Such oscillations around a C_3 axis are improbable. For reasons yet unknown however, it seems that the model Fa is about as good as the model Eb in describing the adsorption of quasilinear molecules. This is perhaps because in all the calculations, free rotation of the methyl groups around the C-C bond has been assumed, which is not exact.

These results show that very accurate prediction of the retention volumes can be made for aromatic compounds, the error being between 2% and 4% (cf. Table 4). No adjustment of the constants has been made to obtain these results, which are derived straightforward from the theory

TABLE 4

Comparison between Experimental and Theoretical Retention Volume U_A (cm^3/m^2) Adsorption Energy ($-\overline{\Delta U}$, kcal/mole) and Adsorption Entropy ($-\overline{\Delta S}$, cal/mole °K).

Compound		Experimental	A	B	C	D	E_a	E_b	F_a	F_b
Benzene (100° C)	log U_A	0.304	0.522	0.509	0.301	-	-	-	-	-
	$-\overline{\Delta U}$	9.1	9.73	9.78	9.63	-	-	-	-	-
	$-\overline{\Delta S}$	22.94	23.70	23.90	24.44	-	-	-	-	-
Naphthalene (280° C)	log U_A	0.296	0.539	0.530	0.324	0.254	-	-	-	-
	$-\overline{\Delta U}$	14.8	15.22	15.30	15.03	14.48	-	-	-	-
	$-\overline{\Delta S}$	25.46	25.07	25.24	25.70	25.02	-	-	-	-
Ethane (20° C)	log U_A	- 1.272	- 0.997	- 1.038	-	-	- 0.720	- 1.184	- 1.396	- 1.860
	$-\overline{\Delta U}$	4.3	4.21	4.31	-	-	4.40	4.40	4.11	4.11
	$-\overline{\Delta S}$	21.37	18.92	19.44	-	-	18.31	20.43	20.40	22.53

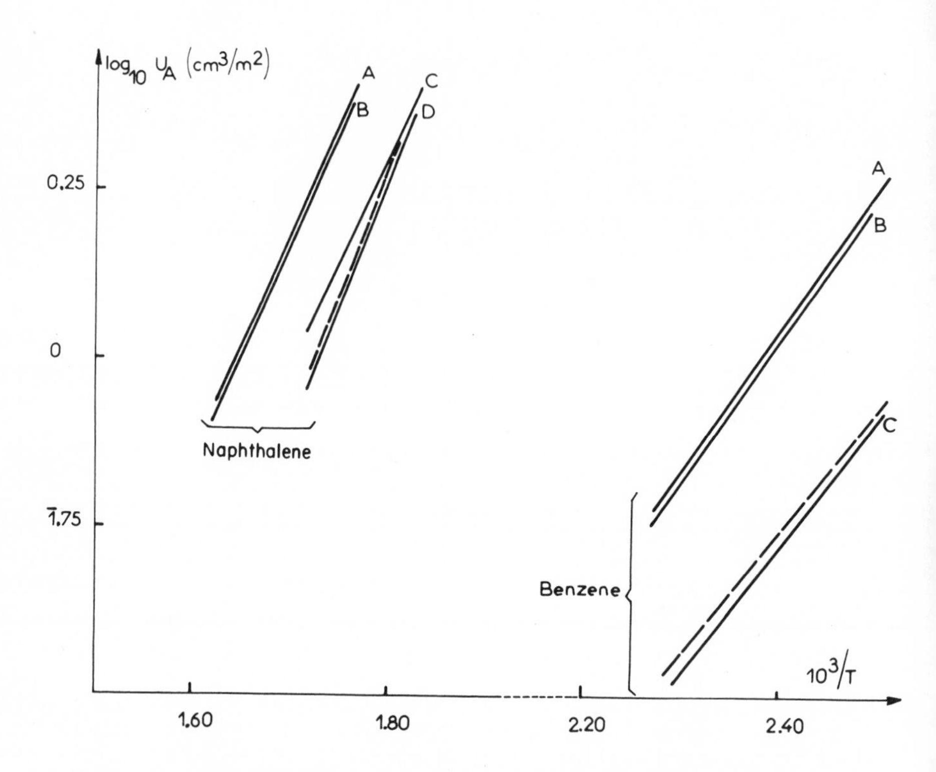

FIG. 5. Variation of the adsorption coefficient as a function of temperature (abscissa: $10^3/T$, T °K). Comparison between calculated values (solid lines) and experimental data (dashed lines) for benzene (models A, B and C) and naphthalene (models A, B, C, and D).

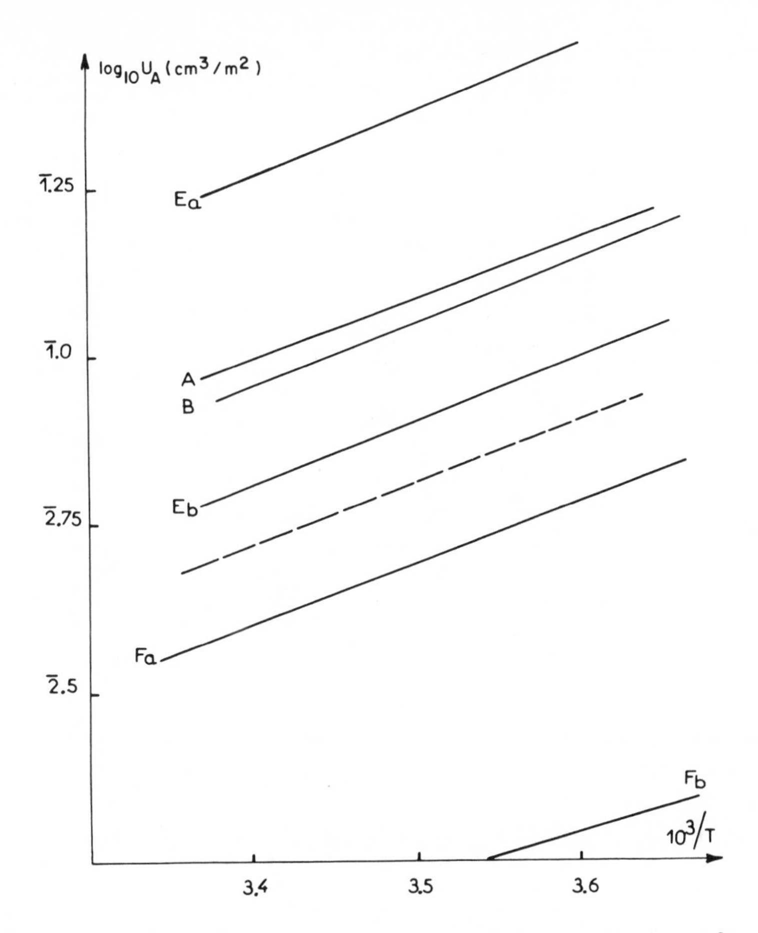

FIG. 6. Variation of the adsorption coefficient as a function of temperature. Comparison between calculated values (solid lines) according to models A, B, Ea, Eb, Fa, and Fb, and to experimental data (dashed lines) for ethane.

and the values of the constants generally accepted (cf. Table 1). We believe that the approximations Kiselev et al. previously found necessary preclude a precise description of the separation of geometrical isomers. With modern computers, these approximations are no longer necessary and as a result much more accurate predictions are possible.

Further results [48] will illustrate the predictive value of molecular statistical thermodynamics applied to gas-solid chromatography, especially for homologous compounds and isomers.

V. CONCLUSION

The principles of the molecular statistical theory of adsorption are very general but their application to a large number of molecules and adsorbents is possible if certain assumptions, which are not too restrictive, are made. These assumptions are listed herewith.

The adsorbent surface is homogeneous.

The adsorption potential is described by the summation of all the binary interactions, between any atom of the adsorbed molecule and any atom of the adsorbent.

Each of these binary interactions is given by the same formula, which applies to the interaction between real gases (Buckingham or Lennard-Jones potentials, Kirkwood-Müller relations, van der Waals radii).

The contribution to the total surface area of the adsorbent of the facets parallel to the various crystallographic planes are known and the surface defects are few.

The quantum partition function can be derived from the classical one by multiplication by a quantic factor that in most cases practically equals unity.

The vibrational state of the molecule is not affected by physical adsorption.

When the molecule has internal rotational degrees of freedom, it can be considered a mixture of the various stable rotamers in equilibrium, which are considered rigid.

Of the three translational degrees of freedom of the molecule in the gas phase, the two that are parallel to the surface remain unchanged, and the one that is perpendicular becomes a harmonic oscillation of the molecule mass-center.

Various models can be derived for the transformation of the rotational degrees of freedom. For spherical or quasispherical molecules they remain unchanged. For other molecules, the strong gradient of the adsorption potential results in a preferential orientation that hinders some rotations that become either oscillations or hindered rotations around the corresponding axis.

The molecule is not polarized by the adsorption. If the molecule has a dipole moment, the contribution due to the polarization of the adsorbent is added to the adsorption potential. This problem however deserves a further study.

Very few adsorbents fulfill these conditions, especially the first, fourth, and last ones.

All these assumptions result in extensive calculations, because for many molecules, several models have to be checked and most of them have different rotational isomers. These calculations can be carried out easily using a modern computer however. As experimental data can be obtained easily and accurately with gas chromatography, a method that allows measurements at very low surface coverages, it is possible to determine the degree of agreement between the theoretical results and the experimental data for molecules of various shapes [41, 48].

APPENDIX I

Calculating the adsorption potential derivatives ϕ''_{z^2} and ϕ''_{θ^2}

The calculations in models C, D, E, and F need the values of the second derivatives of the potential relative to z^2 and θ^2 at the equilibrium position of the molecule on the adsorbent surface. For an atom, the equilibrium distance z_0 is the sum of the van der Waals radii of the graphite carbon atom and of the adsorbate atom. For the molecule, the equilibrium distance z_0 is calculated after the molecular adsorption potential but is practically equal to the sum of the van der Waals radii of the graphite carbon atoms and of the largest groups of the molecule (for example, CH_3 for toluene and CH for benzene). The value of ϕ''_{z^2} at the equilibrium distance z_0 is obtained by differentiation of the potential equation, noting that the first derivative, ϕ'_z, is zero at the equilibrium distance:

$$\phi''^{LJ}_{z^2} = z_0^2 \left[-48 \sum_j C_{ij1}\, r_{ij}^{-10} - 80 \sum_j C_{ij2}\, r_{ij}^{-12} + 168 \sum_j B_i^{LJ}\, r_{ij}^{-16} \right] \tag{I-1}$$

$$\phi''^{B}_{z^2} = z_0^2 \left[-48 \sum_j C_{i1} r_{ij}^{-10} - 80 \sum_j C_{ij2} r_{ij}^{-12} + \sum_j \left(\frac{1}{r_{ij}^2} + \frac{\rho}{r_{ij}^3} \right) \frac{B_i^B}{\rho^2} \exp \frac{-r_{ij}}{\rho} \right] \tag{I-2}$$

At the equilibrium distance z_0, the second derivative ϕ''_{θ^2} equil is related to $\phi''^{o}_{iz^2}$, the second derivative with respect to z^2, for each atom center, at the equilibrium distance z_0, through the relation

$$\phi''_{\theta^2\,\text{equil}} = \sum_i \phi''^{o}_{iz^2} \left(\frac{\partial z}{\partial \theta} \right)^2_{\text{equil}} \tag{I-3}$$

For model D, the equilibrium position is $\theta = 0$, $z = z_0 + D_i \sin\theta$. For Models C, E, and F, the equilibrium position is $\theta = \pi/2$, $z = z_0 + D_i \cos\theta$. D_i is the distance of the atom center to the rotation axis around which the angle of rotation is θ.

One can then write

$$\left(\frac{\partial z}{\partial \theta} \right)_{\text{equil}} = D_i \tag{I-4}$$

In the same way, the second derivative ϕ''_{ψ^2} equil is given by the relation

$$\phi''_{\psi^2\,\text{equil}} = \sum_i \phi''^{o}_{iz^2} \left(\frac{\partial z}{\partial \psi} \right)^2_{\text{equil}} = \sum_i \phi''^{o}_{iz^2} D_i^2 \tag{I-5}$$

if D_i is now the distance of the atom center to the rotation axis around which the angle of rotation is ψ.

It should be realized that for C, E, and F models, in which OX and OZ are the two main axes in the molecular plane, $\phi''^{\pi/2}_{\theta^2}$ at the equilibrium position is given by

$$\phi''^{\pi/2}_{\theta^2\,\text{equil}} = \sum_i (\phi''_{iz^2})_{\text{equil}} \quad (Z_i^2 + Y_i^2) \tag{I-6}$$

$\phi''^{o}_{\psi^2}$ is derived directly from the above calculations by changing θ into ψ and Z_i into X_i:

$$\phi''^{o}_{\psi^2\,\text{equil}} = \sum_i (\phi''_{iz^2})_{\text{equil}} (X_i^2 + Y_i^2) \tag{I-7}$$

Here D_i is the distance to the lowest axis of inertia of the molecule OZ. In the D model, in which OZ is the main axis perpendicular to the molecular plane, the equilibrium position corresponds to $\theta = 0$, where θ is the angle between the plane of the molecule and the plane of graphite. The calculation of $\phi''^{0}_{\theta^2}$ is given by

$$\phi''^{0}_{\theta^2\ \text{equil}} = \sum_i (\phi''_{iz^2})_{\text{equil}} (Y_i^2 + Z_i^2) \tag{I-8}$$

The axis of rotation is the axis of lowest moment of inertia, OX.

A big difficulty arises with the calculation of the contribution to $\phi''^{0}_{\theta^2}$ or ϕ''_{ψ^2} of the hydrogen atoms of the methyl groups that rotate freely around the C-C bond. The calculation is given in Appendix II.

APPENDIX II

Calculating the adsorption potential second derivative ϕ''_{θ^2} for a methyl group

In calculating $\phi''^{0}_{\theta^2}$ for Model D and $\phi''^{\pi/2}_{\theta^2}$ and $\phi''^{0}_{\psi^2}$ for Model C (and for quasilinear molecules), the hydrogen atoms of the CH_3 groups are treated separately. As there is free rotation of the methyl groups around the C-C bond, the integration of Eq. (I-6) for all the positions of the hydrogen atoms of the methyl groups is necessary.

It is sufficient to explain only the calculation of $\phi''^{0}_{\theta^2}$ for Model D; the calculation of ϕ''^{0}_{ψ} (Model C and linear molecules) can thence be derived by calculating the distance D_i to the axis of rotation, which is the axis of lowest moment of inertia (shown previously in Appendix I).

The following calculation is derived for a methyl group bound so that the C-C bond has no special orientation with respect to the main plane of inertia of the molecule, thus it is valid for any molecule.

In Fig. 7, the reference axes system OXYZ (axis parallel or perpendicular to the surface of the adsorbent) is translated into the reference axes system $C\,X_c\,Y_c\,Z_c$, where C is the center of the carbon atom of the methyl group. B is the angle between the axis $\overrightarrow{CZ}$ and the axis $\overrightarrow{Cg}$, passing through the mass-center G of the three hydrogen atoms. A is the angle between the axis $\overrightarrow{CY}$ and the projection $\overrightarrow{Cg_1}$ of $\overrightarrow{Cg}$ on the X_cCY_c plane. The three hydrogen atoms are placed on a circle $\mathcal{C}$ centered in G, with diameter $(2/3)d\sqrt{8}$, where d is the CH distance; obviously CG = d/3.

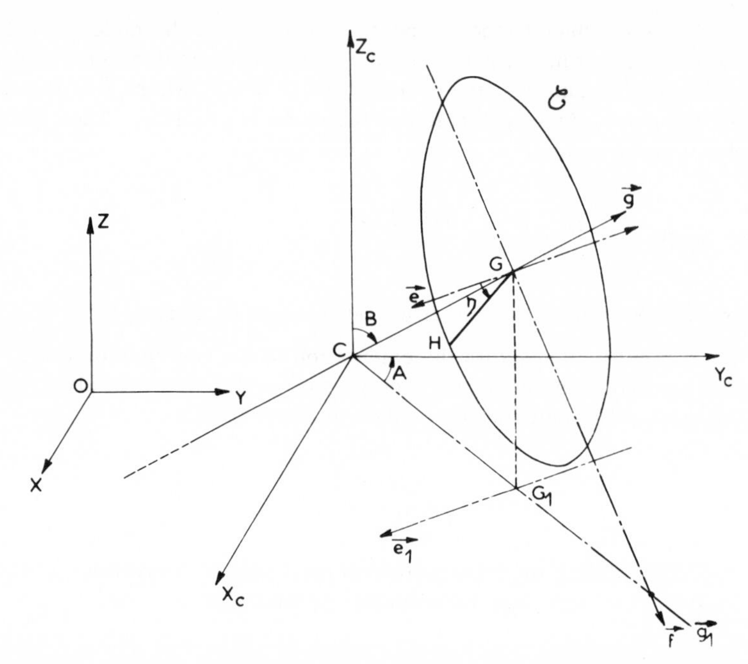

FIG. 7. Schematic representation of a freely rotating hydrogen atom of a CH_3 group. $\overrightarrow{OX}$, $\overrightarrow{OY}$, main axes of the molecule; $\overrightarrow{OX}$, axis of the θ rotation; C, center of the carbon atom of the CH_3 group; G, mass center of the three hydrogen atoms of the CH_3 group; H, current position of the hydrogen atom of the CH_3 group; $\mathcal{C}$, circle described during the free rotation of the three hydrogen atoms around G.

The reference axes system centered in G (Gefg) is defined as follows: $\overrightarrow{Ge}$ is the diameter of the circle $\mathcal{C}$, which is perpendicular to the ZCG plane, and $\overrightarrow{Gf}$ is the diameter of this circle, which is in the ZCG plane. We calculate the position of one of the hydrogen atoms as a function of the angle $\eta(\overrightarrow{Ge}, \overrightarrow{GH})$. Then the other two atoms are on the circle $\mathcal{C}$ at angles $\eta + 2\pi/3$ and $\eta + 4\pi/3$. All positions of the H atoms are catered for by integrating η over 2π.

The molecule rotates by an angle θ around the $\overrightarrow{OX}$ axis, i.e., the axes of lowest moment of inertia, X_c, Y_c, and Z_c, are the coordinates of the carbon atom center in the OXYZ reference system (molecular system) (Figs. 3 and 7).

To calculate the second derivative $\phi''^{\circ}_{\theta^2}$ at the equilibrium position of the molecule, for the hydrogen atoms of the CH_3 group (Model D), it is necessary to know the $(\partial z/\partial \theta)^2$ term in Eq. (I-3). From Eq. (I-4), this derivative must be equal to D_H^2 in the equilibrium position, where D_H is the distance of the hydrogen atom to the axis of rotation, i.e., $\vec{OX}$. This distance is $(Y_H^2 + Z_H^2)^{1/2}$, where X_H, Y_H, Z_H are the coordinates in the OXYZ system (Fig. 7).

In the Gefg system, the coordinates of an hydrogen atom are

$$e = \frac{d\sqrt{8}}{3} \cos \eta$$

$$f = \frac{d\sqrt{8}}{3} \sin \eta \tag{II-1}$$

$$g = 0$$

The matrix of the rotation from the XYZ system to the efg system is the following:

$$\begin{pmatrix} \cos A & -\sin A & 0 \\ \sin A \cos B & \cos A \cos B & -\sin B \\ \sin A \sin B & \cos A \sin B & \cos B \end{pmatrix} \tag{II-2}$$

The transformation is orthogonal and it is only necessary to transpose the matrix to get its inverse.

If we take into account the CG translation, the coordinates of one hydrogen of a methyl group in the C $X_cY_cZ_c$ system can be written

$$X_H^C = \frac{d\sqrt{8}}{3} \cos \eta \cos A + \frac{d\sqrt{8}}{3} \sin \eta \sin A \cos B + \frac{d}{3} \sin A \sin B$$

$$Y_H^C = \frac{-d\sqrt{8}}{3} \cos \eta \sin A + \frac{d\sqrt{8}}{3} \sin \eta \cos A \cos B + \frac{d}{3} \cos A \sin B$$

$$Z_H^C = \frac{-d\sqrt{8}}{3} \sin \eta \sin B + \frac{d}{3} \cos B \tag{II-3}$$

The coordinates Y_H and Z_H are easily derived from Y_H^c and Z_H^c through the OC translation:

$$Y_H = Y_H^c + Y_c$$

$$Z_H = Z_H^c + Z_c \tag{II-4}$$

The mean distance $Y_H^2 + Z_H^2$ for all positions of the hydrogen atom of the CH_3 group is given by

$$D_H^2 = \frac{1}{2\pi} \int_0^{2\pi} (Y_H^2 + Z_H^2)\, d\eta \qquad \text{(II-5)}$$

$$D^2 = \frac{4d^2}{9} (\sin^2 A + \cos^2 A \cdot \cos^2 B + \sin^2 B) + \frac{d^2}{9} (\cos^2 A \sin^2 B + \cos^2 B)$$

$$+ \frac{2d}{3} (Z_c \cos B + Y_c \cos A \cdot \sin B) + Y_c^2 + Z_c^2 \qquad \text{(II-6)}$$

The contribution of one hydrogen atom of a methyl group to the second derivative ϕ''_{θ^2} is derived from Eqs. (I-4), and (II-6). The contribution of the methyl group is three times larger. The calculation has to be carried out for each methyl group.

REFERENCES

1. T. L. Hill, J. Chem. Phys. 16, 181 (1948).
2. J. W. Drenan and T. L. Hill, J. Chem. Phys. 17, 775 (1949).
3. W. A. Steele, "The Gas-Solid Interface," vol. 1, ed. E. A. Flood, Marcel Dekker, New York, 1967. Chap. 10.
4. D. White and E. N. Lassettre, J. Chem. Phys. 32, 72 (1960).
5. A. A. Evett, J. Chem. Phys. 33, 789 (1960).
6. A. Katorski and D. White, J. Chem. Phys. 40, 3183 (1964).
7. M. Mohnke and W. Saffert, "Gas Chromatography 1962," ed. M. Van Swaay, Butterworths, London, 1962, p. 216.
8. M. P. Freeman and M. J. Hagyard, J. Chem. Phys. 49, 4020 (1968).
9. P. L. Gant, K. Yang, M. S. Goldstein, M. P. Freeman, and A. I. Weiss, J. Phys. Chem. 74, 1985 (1970).
10. A. V. Kiselev and D. P. Poshkus, Trans. Faraday Soc. 59, 176 (1963).
11. D. P. Poshkus, Disc. Faraday Soc. 40, 195 (1965).
12. D. P. Poshkus, Russ. J. Phys. Chem. 39, 1582 (1965).
13. A. V. Kiselev and D. P. Poshkus, Dokl. Akad. Nauk SSSR 139, 1145 (1961).
14. A. V. Kiselev and D. P. Poshkus, Trans. Faraday Soc. 59, 428 and 1438 (1963).

15. A. V. Kiselev, D. P. Poshkus, and A. Y. Afreimovich, Russ. J. Phys. Chem. 39, 630 (1965).

16. A. V. Kiselev and D. P. Poshkus, Russ. J. Phys. Chem. 39, 204 (1965).

17. D. P. Poshkus and A. Y. Afreimovich, Russ. J. Phys. Chem. 42, 626 (1968).

18. A. V. Kiselev, D. P. Poshkus, and A. Y. Afreimovich, Russ. J. Phys. Chem. 42, 1345 (1968).

19. A. V. Kiselev, D. P. Poshkus, and A. Y. Afreimovich, Russ. J. Phys. Chem. 42, 1348 (1968).

20. A. V. Kiselev, D. P. Poshkus and A. Y. Afreimovich, Russ. J. Phys. Chem. 44, 545 (1970).

21. D. P. Poshkus and A. Y. Afreimovich, J. Chromatogr. 58, 55 (1971).

22. A. V. Kiselev and D. P. Poshkus, Russ. J. Phys. Chem. 43, 153 (1969).

23. D. P. Poshkus, J. Chromatogr. 49, 146 (1970).

24. A. Di Corcia and A. Liberti, Trans. Faraday Soc. 66, 967 (1970).

25. A. J. Bennett, B. McCarroll, and R. P. Messmer, Phys. Rev. B 3, 1397 (1971).

26. N. N. Avgul, A. V. Kiselev, I. A. Lygina, and D. P. Poshkus, Izv. Akad. Nauk SSSR, Otd. Khim. Nauk, 1196 (1959).

27. N. N. Avgul, A. A. Isirikyan, A. V. Kiselev, I. A. Lygina, and D. P. Poshkus, Izv. Akad. Nauk SSSR, Otd. Khim. Nauk, 1314 (1957).

28. J. G. Kirkwood, Phys. Z. 33, 57 (1932). A. Muller, Proc. Roy. Soc. A 154, 624 (1936).

29. R. Heller, J. Chem. Phys. 9, 154 (1941).

30. A. V. Kouznetsov and G. Guiochon, J. Chim. Phys. 66, 257 (1969).

31. G. Curthoys and P. A. Elkington, J. Phys. Chem. 71, 1477 (1967).

32. A. V. Kouznetsov, C. Vidal-Madjar, and G. Guiochon, Bull. Soc. Chim. France, 1440 (1969).

33. A. V. Kiselev, A. A. Lopatkin, and L. G. Ryaboukhina, Bull. Soc. Chim. France, 1324 (1972).

34. P. Broier, A. V. Kiselev, E. A. Lesnik, and A. A. Lopatkin, Russ. J. Phys. Chem. 42, 1350 (1968) and 43, 844 (1969).

35. G. D. Mayorga and D. L. Peterson, J. Phys. Chem. 76, 1641, 1647 (1972).

36. A. V. Kiselev, A. A. Lopatkin, and E. R. Razumova, Russ. J. Phys. Chem. 44, 82 (1970) and 43, 1004 (1969).

37. R. M. Barrer, Proc. Roy. Soc. A. 161, 476 (1937).

38. C. Vidal-Madjar, L. Jacob, and G. Guiochon, Bull. Soc. Chim. France, 3105 (1971).

39. A. Pacault, "Eléments de thermodynamique statistique," Masson, Paris, 1963.

40. K. S. Pitzer and W. D. Gwinn, J. Chem. Phys. 10, 428 (1942).

41. C. Vidal-Madjar and G. Guiochon, Bull. Soc. Chim. France, 3110 (1971).

42. J. F. K. Huber and A. I. M. Keulemans, "Gas Chromatography 1962," ed. M. Van Swaay, Butterworths, London, 1962, p. 26.

43. A. Saint Yriex, Bull. Soc. Chim. France, 3407 (1965).

44. L. D. Belyakova, A. V. Kiselev, and N. V. Kovaleva, Bull. Soc. Chim. France, 285 (1967).

45. R. L. Gale and R. A. Beebe, J. Phys. Chem. 68, 555 (1964).

46. C. Pierce, J. Phys. Chem. 73, 813 (1969).

47. R. M. Barrer and L. V. C. Rees, Trans. Faraday Soc. 57, 999 (1961).

48. C. Vidal-Madjar, M. F. Gonnord, G. Guiochon, J. Colloid Interface Sci. (1975), to be published.

Chapter 5

TRANSPORT AND KINETIC PARAMETERS BY GAS CHROMATOGRAPHIC TECHNIQUES

Motoyuki Suzuki and J. M. Smith
Institute of Industrial Science
University of Tokyo
Tokyo, Japan
and
Department of Chemical Engineering
University of California at Davis
Davis, California

I. INTRODUCTION

In the last decade, analysis of effluent responses to input disturbances has been widely used to study the behavior of process equipment. There have been two rather different objectives in such experiments: (1) determining the dynamic character of processing units so that control strategies could be devised to optimize performance, and (2) determining rate and equilibrium parameters of the system from the experiments themselves. In the latter case, many of the parameters are independent of the dimensions of the equipment, so that the results from laboratory-scale apparatus are generally applicable. The nature and operation of the laboratory apparatus are the same as for conventional chromatographic experiments. The difference lies in the treatment of the response data. Instead of measuring peak areas, or heights, one uses the retention time and shape of the effluent peak to extract equilibrium and rate parameters.

The purpose of this chapter is to summarize theoretical and experimental studies directed toward this second objective. Two general methods of analyzing data have been used. One is related to chromatography as a separation process and the method is known as the plate, or rate, theory [Giddings, Glueckauf, and Klinkenberg, and their colleagues, 1-4]. In the second, the concentration in the chromatographic column is treated as a continuous function of time and position, described by differential conservation equations and rate expressions. This latter approach has been developed over a long period, beginning with the work of Nusselt [5] and Anzelius [6], and including important contributions by Thomas [7], Rosen [8], Kubin [9, 10], and Kucera [11]. The treatment in this chapter will follow this second approach with tielines at intervals to the plate theory.

The main emphasis is on the usual fixed-bed reactor, or column, which is the normal device for gas chromatography. However, the chromatographic technique can also be applied effectively to other reactor forms, and a summary of developments for slurry and single-pellet devices is given in the last section of the chapter. In a fixed bed of adsorbing particles, the adsorbate-containing input pulse is influenced by several mass transport steps both in the gas and in the porous adsorbent particles. It is difficult to extract reliable values for the rate parameters for all of these steps from the shape of a single effluent pulse. Therefore, pulse-response experiments at different values of such operating conditions as gas velocity, particle size, and temperature are desirable. Particularly, data are helpful at conditions that enlarge the influence of the rate parameter specifically desired. The chapter has been arranged so as first to present the general theory for a monodisperse bed of particles and subsequently to illustrate the use of specific data in evaluating particular rate parameters. The moment method of analysis illustrated here is the one we have used in our work. It should be emphasized that the plate method and modifications of it have been widely employed. References to these procedures are given in the literature review of Section III.

A big limitation to the practical use of the chromatographic technique is that all of the individual steps in the overall process must follow linear rate equations. Section IV describes how this restriction can be circumvented in determinating adsorption rates on solid catalysts. The study of the rates of individual steps on the surface of gas-solid catalytic reactions by chromatographic techniques has been retarded by both theoretical and experimental difficulties. However, the potential benefits are so great--for example, the separation and evaluation of rates of adsorption and surface reaction--that efforts should continue in this area. In Section V, the theoretical problems and progress toward their solution is summarized. The experimental problem is mainly choosing carrier gas-reactant and detector systems such that the effluent peak of reactant can be identified independently of the other components present in the gas.

II. FIXED-BED CHROMATOGRAPHY (MOMENT ANALYSIS)

The equations given in Section A are based on the following model of mass transport: axial dispersion in the interparticle space, mass transfer from gas to particle, intraparticle diffusion, and a first-order, reversible disappearance of tracer component at a localized site in the interior of the particle. The last step could be a reversible, linear adsorption process. The requirement of reversibility means that there is no retention of adsorbable component (tracer) in the bed during the time period of the experiment. Although this restriction is assumed (see Eq. (4)) in this section, it is possible to derive moment equations for irreversible disappearance of the tracer, as discussed in Section VI. It is assumed also that the entire bed is isothermal and that there are no radial gradients.

A. Basic Equations

For the described model and restrictions, conservation equations for the mass of tracer, or adsorbable, component are of the form Masamune and Smith [12] and Kubin [9, 10] and Kucera [11] described. For the gas phase in the bed, where the concentration C of tracer is a function of bed depth z and time t, mass conservation requires

$$E_z \frac{\partial^2 C}{\partial z^2} - u \frac{\partial C}{\partial z} - \frac{3(1-\alpha)}{\alpha R} N_0 = \frac{\partial C}{\partial t} \tag{1}$$

where E_z is the axial dispersion coefficient (cm^2/sec) based on the interstitial cross-sectional area of the bed, u is the interstitial gas velocity (cm/sec), α is the void fraction, and N_0 represents the mass flux of the component from gas to outer surface of the particles. This flux may be described in terms of the external mass transfer coefficient k_f (cm/sec) or the intraparticle diffusivity D_e (cm^2/sec):

$$\begin{aligned} N_0 &= k_f[C - (C_i)_R] \\ &= D_e \left(\frac{\partial C_i}{\partial r}\right)_{r=R} \end{aligned} \tag{2}$$

where r is the radial coordinate in the particle and R its radius. For the particles, the conservation equation is

$$D_e \left(\frac{\partial^2 C_i}{\partial r^2} + \frac{2}{r}\frac{\partial C_i}{\partial r}\right) - N_i = \beta \frac{\partial C_i}{\partial t} \tag{3}$$

where N_i denotes the rate of disappearance of the tracer per unit volume of the particle, and C_i is the tracer concentration in the pores. For first-order adsorption kinetics, the expression for N_i is

$$N_i = \rho_p k_a \left(C_i - \frac{n_i}{K_a}\right) = \rho_p \frac{\partial n_i}{\partial t} \tag{4}$$

The coefficients k_a and K_a represent the adsorption rate constant $cm^3/(g)$(sec) and adsorption equilibrium constant (cm^3/g), respectively. In the above equations, ρ_p and β are the particle density and porosity of the particle, and C, C_i, and n_i denote concentrations of the component in the gas phase, in the pores of the particle, and adsorbed on the particle (moles/g), respectively.

The chromatographic experiment, based on introducing a pulse of tracer component of concentration C_i into the entrance of the bed, is illustrated in Fig. 1. For quantitative analysis of the effluent peak $C_e(t)$, the method of moments will be used. However, other methods, particularly the Fourier transform approach Silveston [31] described, can be used. Moments of the chromatographic peak are defined as

$$m_n = \int_0^\infty C_e t^n \, dt \tag{5}$$

Then the nth absolute moment is given as

$$\mu_n = \frac{m_r}{m_0} = \frac{\int_0^\infty C_e t^n \, dt}{\int_0^\infty C_e \, dt} \tag{6}$$

and the nth central moment as

$$\mu'_n = \frac{\int_0^\infty C_e (t-\mu_1)^n \, dt}{\int_0^\infty C_e \, dt} = \sum_{i=0}^{n} (-1)^{n-i} \mu_1^{n-i} \mu_i \binom{n}{i} \tag{7}$$

Since higher moments are subject to large experimental errors, first absolute and second central moments are of most significance. Moments can be related directly to the solution for C_e in the Laplace domain.

If the Laplace transform of C(t) is $\overline{C}(p)$, where

$$\overline{C}(p) = \int_0^\infty \exp(pt) C(t) \, dt \tag{8}$$

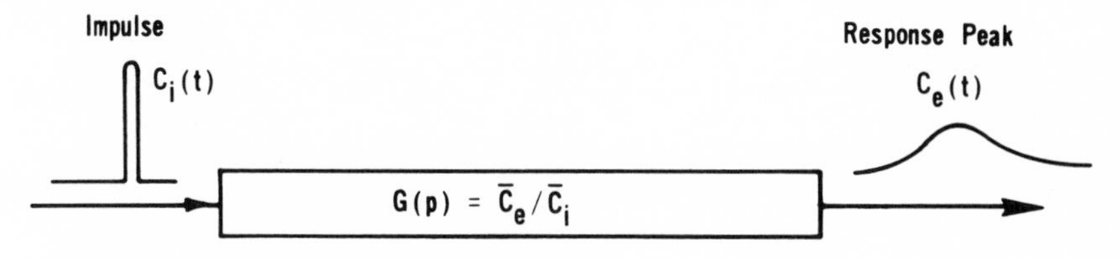

FIG. 1. Chromatographic experiment

Eqs. (1) to (4) can be transformed to a set of ordinary differential equations. If the input is a square pulse of injection time τ, that is,

$$z = 0;\ C = 0 \quad \text{for} \quad t < 0 \ \text{and} \ t > \tau$$
$$C = C_0 \quad \text{for} \quad 0 \le t \le \tau \tag{9}$$

the set of differential equations can be solved in the Laplace domain to give (Kubin [9]):

$$\overline{C}_e(p) = \frac{C_0}{p} \left[1 - \exp(-p\tau)\right] \exp(-\lambda z) \tag{10}$$

$$\lambda = \frac{u}{2E_z}\left[\sqrt{1 + \frac{4E_z}{u^2} G(p)} - 1\right] \tag{11}$$

$$G(p) = p + \frac{3(1-\alpha)}{\alpha} \frac{Bi}{R^2/D_e} \left[1 - \frac{Bi}{A_0(p) + Bi}\right] \tag{12}$$

$$A_0(p) = R\sqrt{\frac{\beta \cdot \kappa(p)}{D_e}} \tag{13}$$

$$\kappa(p) = p + \frac{\rho_p/\beta}{(1/k_a) + (1/pK_a)} \tag{14}$$

where Bi is the Biot number ($k_f R/D_e$). The moments of the chromatographic peak are related to the solution in the Laplace domain as follows:

$$m_n = (-1)^n \lim_{p \to 0} \left[\frac{d^n}{dp^n} \overline{C}\right] \tag{15}$$

By applying Eq. (15) and Eqs. (5) to (7) to Eq. (10), one can obtain the following expressions for the first absolute and second central moments:

$$\mu_1 = \frac{z}{u}\left[1 + \frac{1-\alpha}{\alpha}\beta\left(1 + \frac{\rho_p K_a}{\beta}\right)\right] \tag{16}$$

$$\mu_2' = \mu_2 - \mu_1^2 = \frac{2z}{u} \left\{ \frac{1-\alpha}{\alpha} \quad \frac{\rho_p K_a^2}{k_a} + \frac{R^2 \beta^2}{15} \left(1 + \frac{\rho_p}{\beta} K_a\right)^2 \right.$$

$$\left. \times \left(\frac{1}{D_e} + \frac{5}{k_f R}\right) + \frac{E_z}{u^2} \left[1 + \frac{1-\alpha}{\alpha} \left(1 + \frac{\rho_p}{\beta} K_a\right)\right]^2 \right\} \tag{17}$$

Equations (16) and (17) show that the first-moment expression includes only equilibrium parameters and that the contributions of axial dispersion, external mass transfer, intraparticle diffusion, and adsorption rate to the second central moment are additive.

B. First-Moment Analysis

For linear isotherm systems, first-moment data give reasonably accurate adsorption equilibrium contants. Typical plots of μ_1 versus z/u are shown in Fig. 2 for low concentration (1-3%) pulses of hydrocarbon in helium passed through beds of silica gel [13]. Slopes of the straight lines in Fig. 2 can be used with Eq. (16) to determine K_A. Such results from Fig. 2 are compared with direct equilibrium measurements (BET method) in Table 1.

C. Second-Moment Analysis

It is evident from Eq. (17) that each transport step gives a separate and additive contribution to the second moment. Then, for instance, from second moments for different gas velocities, one can separate the contribution of axial dispersion from the contribution of the other transport steps. Similarly, from data for different particle sizes, the intraparticle diffusion contribution can be separated from the contribution for adsorption. By properly choosing operating conditions, one can frequently maximize the contribution for the particluar step whose rate parameter ($E_A = \alpha E_z$, D_e, k_a) is desired. For example, large particle sizes increase the contribution due to intraparticle diffusion.

The following sections indicate one procedure for using the second-moment equation and μ_2' data for analyzing the various transport processes.

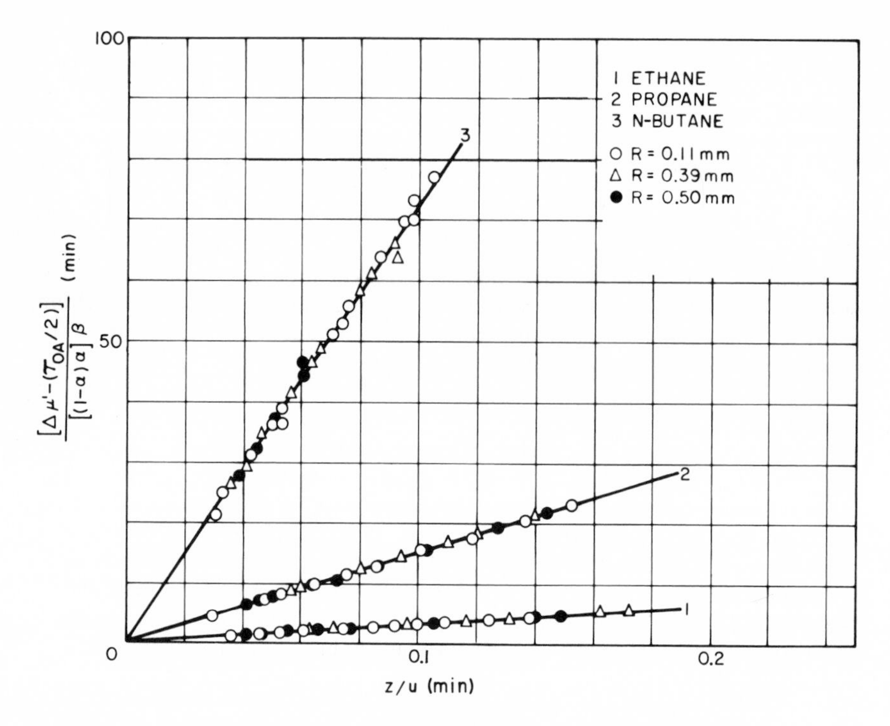

FIG. 2. Chromatography of hydrocarbons on silica gel (50° C). Dependence of the reduced first absolute moment on z/u. (From Ref. 13.)

TABLE 1

Adsorption Coefficients on Silica Gel at 50°C[a]

Hydrocarbon	Adsorption coefficient, K_A (ml/g SiO_2)	
	From equilibrium adsorption measurements	From evaluation of chromatographic peaks
Ethane	14.5	14.6
Propane	63	65.4
n-Butane	308	311

[a]From Ref. 13.

1. Axial dispersion in packed beds of small particles

Though axial dispersion data in a column of small particles are often needed for analyzing kinetic studies of gas-solid reactions, available data and correlations are scarce for particle sizes smaller than 1 mm. Also, as Eq. (17) indicates, the axial dispersion contribution to the second moment must be separated from the measured total contribution before other rate parameters can be evaluated by chromatographic means. It may be noted further than axial dispersion has a significant effect on the maximum separation gas velocity in a chromatographic separation column since the HETP is proportional to $\mu_2'/(\mu_1)^2$.

For use in chromatographic studies, the authors [14] measured axial dispersion coefficients by chromatographic methods in beds of small particles over a wide range of flow rates. Two distinct phenomena were manifested: the first was a finite contribution of intraparticle diffusion to axial dispersion in the low flow-rate range, and the second, very small Peclet numbers at high flow rates.

The apparatus used in this study consisted of a chromatograph in which the packed bed of particles replaced the separation column, and the outlet and inlet of the bed were connected directly to the detector (thermal conductivity cell) in order to minimize dead volumes. A small sample of inert gas (hydrogen or oxygen) was introduced by a seven-port sampling valve before the cell. In this way the stream carrying the inlet peak passed through one side of the cell and the stream carrying the effluent peak passed through the other side. Thus, the moment differences $\Delta\mu_i = (\mu_i)_{effluent} - (\mu_i)_{inlet}$ could be measured. Helium and nitrogen were used as carrier gases for hydrogen and oxygen pulses, respectively, in order to cover a wide range of $d_p u_0 / D_v$.

The lengths, particle sizes, and porosities of the packed columns and operating conditions are summarized in Table 2. The porous particles were prepared by crushing and sieving cylindrical pellets of Girdler Type G-66B unreduced catalyst (CuO-ZnO mixture), whose characteristics are shown in Table 3.

From the theoretical expressions for the first absolute moment μ_1 and for the second central moment μ_2' obtained for each peak, the differences $\Delta\mu_1$ and $\Delta\mu_2'$ were calculated. For a nonadsorbing tracer gas, these moment differences are related to the rate parameters such as axial dispersion coefficient E_z, gas-to-particle mass transfer coefficient k_f, and intraparticle diffusion coefficient D_e, as follows:

$$\mu_1 = \frac{z}{u}\left(1 + \frac{1-\alpha}{\alpha}\beta\right) \tag{18}$$

TABLE 2

Characteristics of Packed Columns[a]

Column no.	1A	B	2	3	4	5	6
Length, cm	200	100	100	100	100	100	100
Particles	Catalyst[b]		Catalyst[b]	Catalyst[b]	Catalyst[b]	Catalyst[b]	Glass beads
Mesh size	20–24		35–42	48–60	80–100	115–150	28–35
Average diameter, d_p, mm	0.767		0.384	0.271	0.161	0.114	0.51
Void fraction, α	0.439		0.454	0.466	0.450	0.449	0.377
Superficial velocity[c], u_0, cm/sec:							
Hydrogen	0.16–68.5		0.77–57.4	0.32–29.9	0.138–4.35[d]	0.254–3.11[d]	0.318–32.3
Nitrogen	3.2–24.1			1.09–2.14	0.438–4.60	0.470–3.48	0.412–21.4

[a] From Ref. 14.
[b] $CuO \cdot ZnO$, porous particles (properties given in Table 2).
[c] At bed conditions, 24°C and pressures from about 30 in. Hg to 54 in. Hg.
[d] Higher velocities were not possible because of excessive pressure drop.

TABLE 3

Properties of Catalyst Particles[a]

Particle density	= 2.38 g/cm^3
Particle porosity,	= 0.556
Most probable pore diameter	= 350 A
(Mono-disperse pore structure)	

[a]From Ref. 14.

$$\Delta\mu_2' = \frac{2z}{u} \left\{ \frac{1-\alpha}{\alpha} \beta^2 R^2 \left(\frac{1}{15D_e} + \frac{1}{3k_f R} \right) + \left[1 + \frac{(1-\alpha)\beta}{\alpha} \right]^2 \frac{E_A}{u^2\alpha} \right\} \tag{19}$$

or

$$H = H_0 + \frac{E_A}{u^2\alpha} \tag{20}$$

$$H_0 = \frac{[(1-\alpha)/\alpha]\beta^2 R^2}{\{1 + [(1-\alpha)/\alpha]\beta\}^2} \left(\frac{1}{15D_e} + \frac{1}{3k_f R} \right) \tag{21}$$

where H (= HETP/2u) is the following function of moments,

$$H = \frac{\Delta\mu_2'/(2z/u)}{\Delta\mu_1/(z/u)^2} \tag{22}$$

In most cases the contribution of gas-to-particle mass transfer resistance to the second moment is very small compared with the contribution of intraparticle diffusion resistance. Then Eq. (21) reduces to

$$H_0 = \frac{[(1-\alpha)/\alpha]\beta}{\{1 + [(1-\alpha)/\alpha]\beta\}^2} \frac{R^2}{15D_e} \tag{23}$$

Equations (20) and (23) indicate that H is the sum of two terms, one involving E_A and the other a function of D_e. In turn, H can be obtained from the experimentally determined $\Delta\mu_1$ and $\Delta\mu_2'$ using Eq. (22). For most accurate evaluation of E_A, the contribution of D_e to H should be minimized; that is, H_0 should be low. Fortunately, the objective of obtaining E_A for small particles agrees with this requirement; Eq. (23) shows that as R decreases, H_0 decreases rapidly.

The procedure used to evaluate E_A for the catalyst particles consisted first of calculating H_0 from runs at conditions for which intraparticle diffusion was significant, that is, for runs with Column 1A (largest particle size) at high velocities. At high velocities, the Peclet number, $d_p u_0/E_A$, is expected to be constant, so that E_A becomes proportional to u_0. According to Eq. (20), a plot of H vs 1/u should be a straight line with an intercept H_0 at 1/u = 0. The data showed a well-defined relation when plotted in this way.

The contribution of H_0 to H at these extreme conditions was as high as 50%. The intraparticle diffusivities calculated from the H_0 values using Eq. (23) were

$$D_e = 0.074 \text{ cm}^2/\text{sec for helium in hydrogen}$$
$$= 0.012 \text{ cm}^2/\text{sec for oxygen in nitrogen}$$

While these numerical values are reasonable, they must be considered as estimates since the contribution of intraparticle diffusion to H was never more than 50%. For obtaining an accurate diffusivity, the graphical method described in the next section is preferable.

Secondly, the values of H_0 for the other columns were obtained by using these diffusivity values.

Finally, E_A was obtained from H_0 using Eq. (20). Since H_0 decreases sharply when particle size decreases, the values H_0 for the other columns with smaller particle sizes were never greater than 10% of H for column 2, and less than 2% for columns 3-5, even at the highest velocity.

For nonporous glass beads, E_A was obtained directly from Eq. (20) by setting $H_0 = 0$.

The resultant E_A values are displayed for catalyst particles and for glass beads in Figs. 3 and 4, as E_A/D_V vs $d_p u_0/D_V$. This type of plot eliminates the effect of the molecular diffusivity D_V of the gas pair.

At low velocity, E_A/D_V ultimately reaches a constant value. In this region, the diffusibility η is defined as

$$E_A = \eta D_V \tag{24}$$

Since E_A is based on the cross-sectional area of the bed, η includes the void fraction α. The ratio of η and α is a measure of the tortuosity of the axial dispersion path. The data in Figs. 3 and 4 give

For glass beads, $\eta = 0.30$ and $\eta/\alpha \sim 0.79$

For catalyst particles, $\eta = 0.44$ and $\eta/\alpha \sim 0.97$

The unusually high value of η/α for catalyst particles, corresponding to small tortuosity factors α/η, suggests another contribution to axial dispersion. The data obtained agree with the possibility that this second contribution is due to intraparticle diffusion. Similar treatment to Burger's [15] derivation of the electrical conductivity in a heterogeneous system (4) leads to the following expression:

$$E_A = \frac{\alpha}{\alpha+k(1-\alpha)} D_V + \frac{k(1-\alpha)}{\alpha+k(1-\alpha)} D_e \tag{25}$$

FIG. 3. Axial dispersion coefficient for small catalyst particles. (From Ref. 14.)

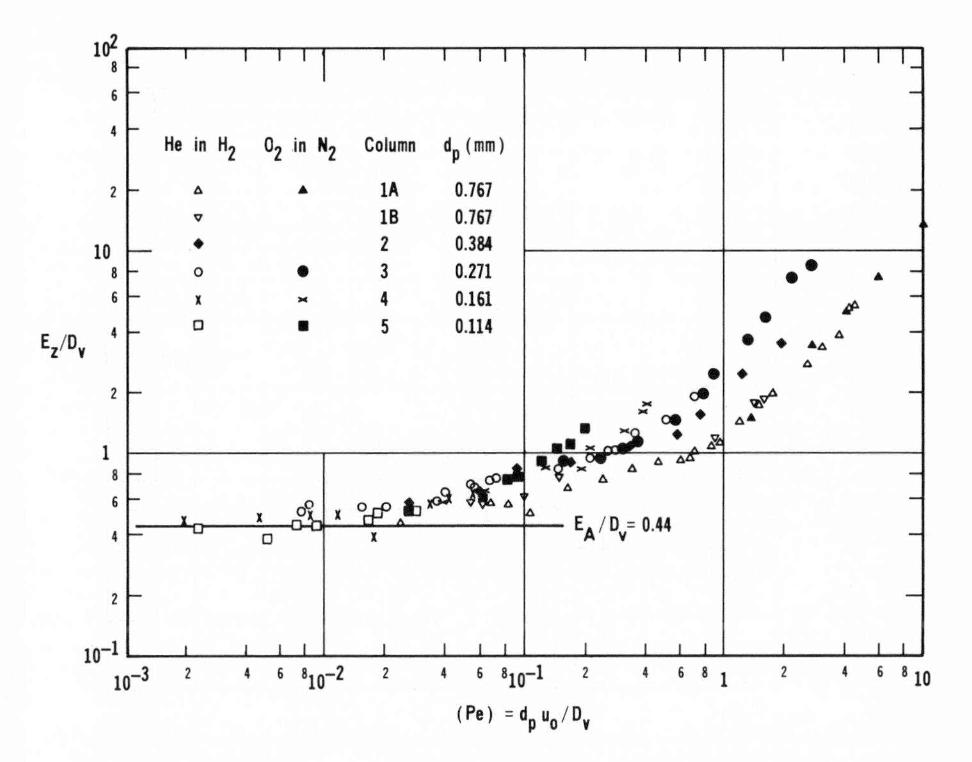

FIG. 3. Axial dispersion coefficient for small catalyst particles. (From Ref. 14.)

where k = 1.5, by assuming that particle shape is spherical. Consider, as an example, the data for column 3 ($\alpha = 0.466$, helium in hydrogen system; $D_v = 1.13\ \text{cm}^2/\text{sec}$ and $D_e = 0.074\ \text{cm}^2/\text{sec}$). Then the above equation gives the numerical value of E_z as

$$E_A = 0.36(1.13) + 0.64(0.074) = 0.46\ \text{cm}^2/\text{sec}$$

and from Eq. (24),

$$\eta = \frac{0.46}{1.13} = 0.41$$

This result agrees well with the experimental result of $\eta = 0.44$ and suggests that about 10% of E_A is due to the intraparticle diffusion contribution.

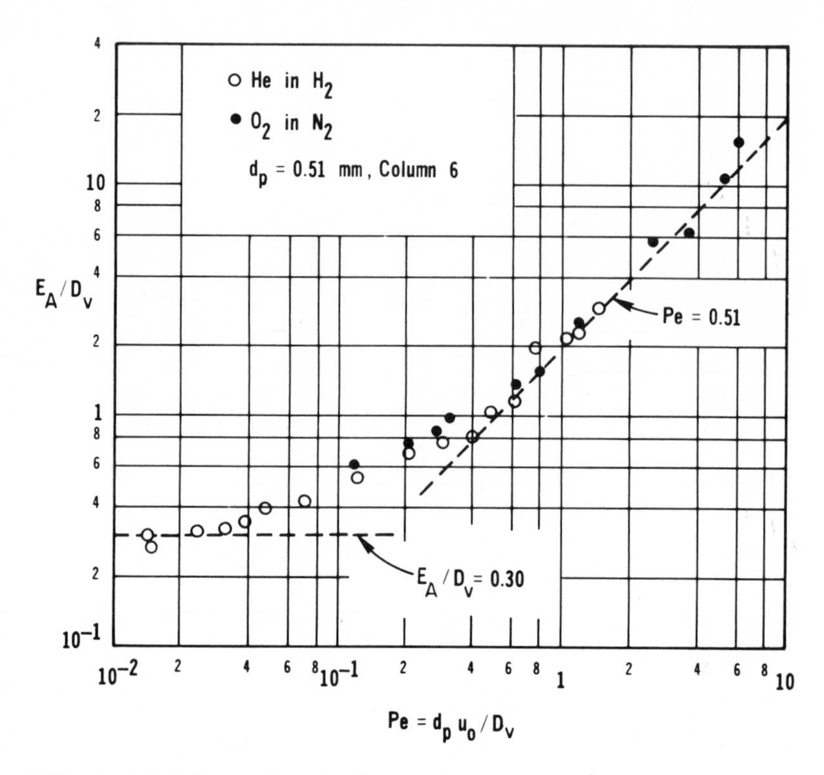

FIG. 4. Axial dispersion coefficient for glass beads. (From Ref. 14.)

At high velocities, the data points in Fig. 3 and 4 show that E_A is proportional to u_o, corresponding to a constant Peclet number. Both theoretical and experimental studies have shown that for larger particles the limiting value of Pe is 2 in the case of gaseous systems. For small particles, Pe still approaches a constant value, but it is smaller than 2 and is a function of particle diameter d_p. The constant values of Pe obtained at high velocities are plotted versus particle diameter in Fig. 5. Also included are all the other available data for small particles. The three low points (solid circle) are for beds of reduced particles of the same CuO-ZnO catalyst. As a result of reduction in situ, the particles shrank about 12% in volume and were composed of copper and zinc oxide [16]. The large difference in Peclet numbers between unreduced and reduced forms is striking. Apparently, shrinking caused a nonuniform distribution of void spaces in the bed and the resultant channeling causes a large increase in E_A. The data of Edwards and Richardson [16] show no trend with d_p. These authors suggested that this could be due to channels formed when the gas moved the smaller particles aside.

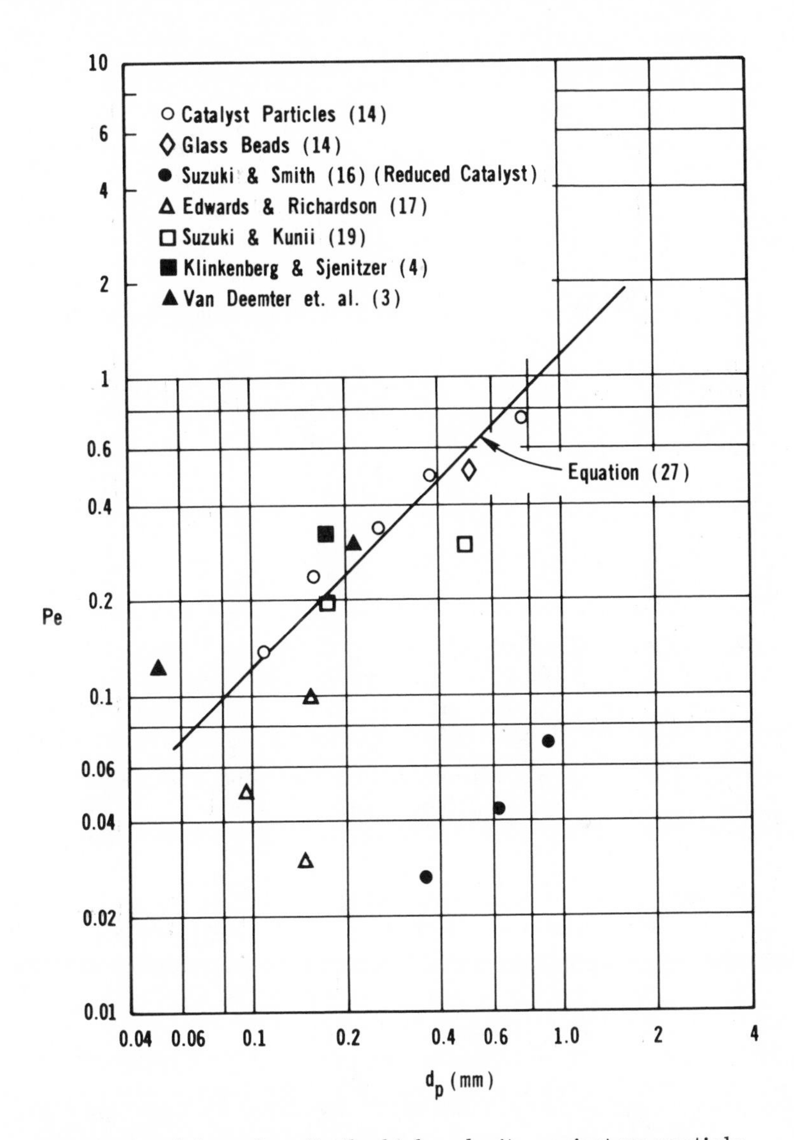

FIG. 5. Peclet number (in the high-velocity region) vs particle diameter. (From Ref. 14.)

Excluding the data for the reduced catalyst and that of Edwards and Richardson, a straight line of 45° slope was drawn through the data in Fig. 5 according to the equation

$$(26)$$

$$\mathrm{Pe} = \frac{d_p u_0}{E_A} = 1.2\, d_p \qquad d_p \text{ in mm} \tag{26}$$

or

$$\frac{E_A}{u_0} = 0.83 \text{ mm} \tag{27}$$

Equation (27) suggests that the scale of dispersion is independent of particle size and is about 1 mm.

An approximate method of predicting axial dispersion coefficients for small particles, over the entire range of velocities, can be obtained by summing the components due to molecular diffusion and to mixing at higher velocities. Then the final form is

$$\frac{E_A}{D_v} = \eta + \ell \frac{u_0}{D_v} \tag{28}$$

where ℓ is the scale of dispersion. Our results suggest that $\ell = 0.83$ mm (for $0.11 \text{ mm} < d_p < 0.77$ mm), $\eta = 0.44$ for CuO-ZnO catalyst particles and $\eta = 0.30$ for glass beads. Equation (28) is compared with experimental results for catalyst particles in Fig. 6.

The quantities ℓ and η are sensitive to bed packing methods and other ill-defined variables. Different investigations [18] can lead to substantially different values of η and ℓ. Hence, it is desirable to evaluate these constants from pulse tests at different velocities using Eq. (28) for each bed employed. Extreme care is necessary in the experimental tests, since the second moment accentuates small errors and leads to scatter in the results.

2. Intraparticle diffusion in porous particles

It is expected that at larger particle sizes, intraparticle diffusion would significantly affect the second moment, particularly at high flow rates.

Schneider and Smith [20] showed this clearly for silica gel-hydrocarbon systems by using three different sizes of particles. From Eq. (17), $[\Delta\mu_2'/(2z/u)]$ at $1/u \rightarrow 0$, or δ_1, should have a linear dependence on R^2. This was confirmed in Fig. 7 for silica hydrocarbons (ethane, propane, and n-butane) where δ_1 is plotted versus R^2. The slopes of the straight lines give the intraparticle diffusivities, and then tortuosity factors of the

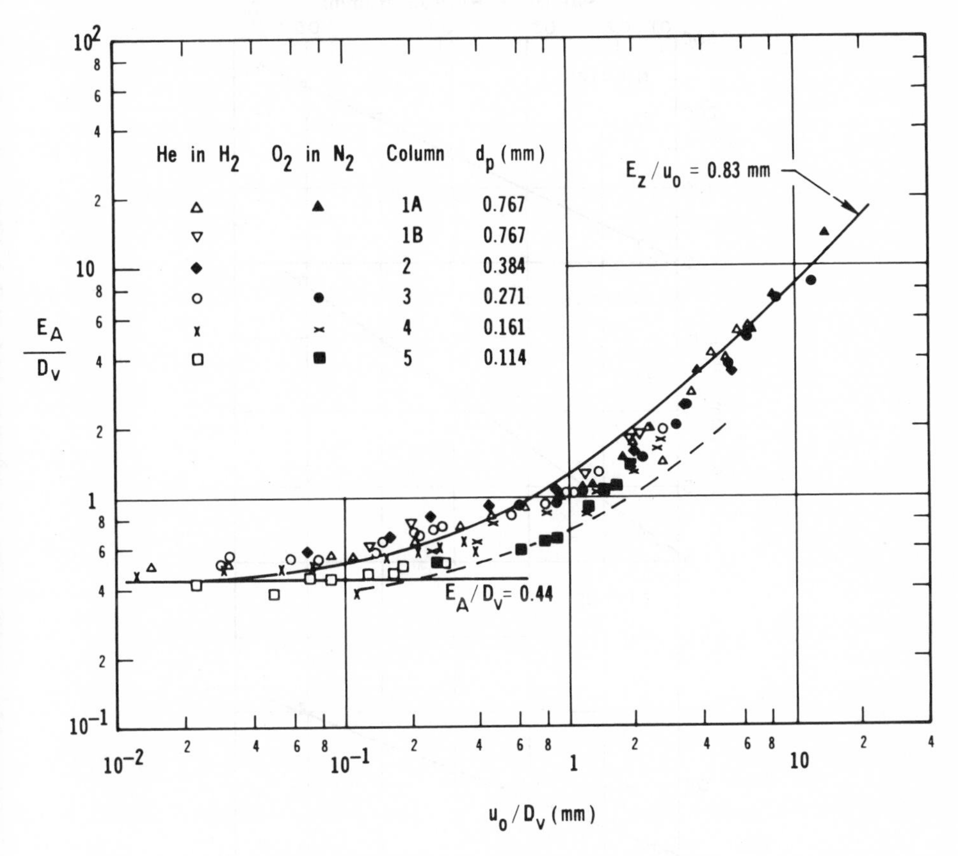

FIG. 6. Correlation of axial dispersion coefficients for small porous particles. (From Ref. 14.)

pores in the particle were obtained. The results are summarized in Table 4. If the porous gel is represented by an assembly of parallel cylindrical capillaries, the effective diffusivity is related to Knudsen diffusivity as

$$D_e = \frac{\beta}{q_{int}} D_K \tag{29}$$

In this expression, Knudsen diffusivity D_K is evaluated at $\overline{r} = 11$ Å for silica gel from the equation

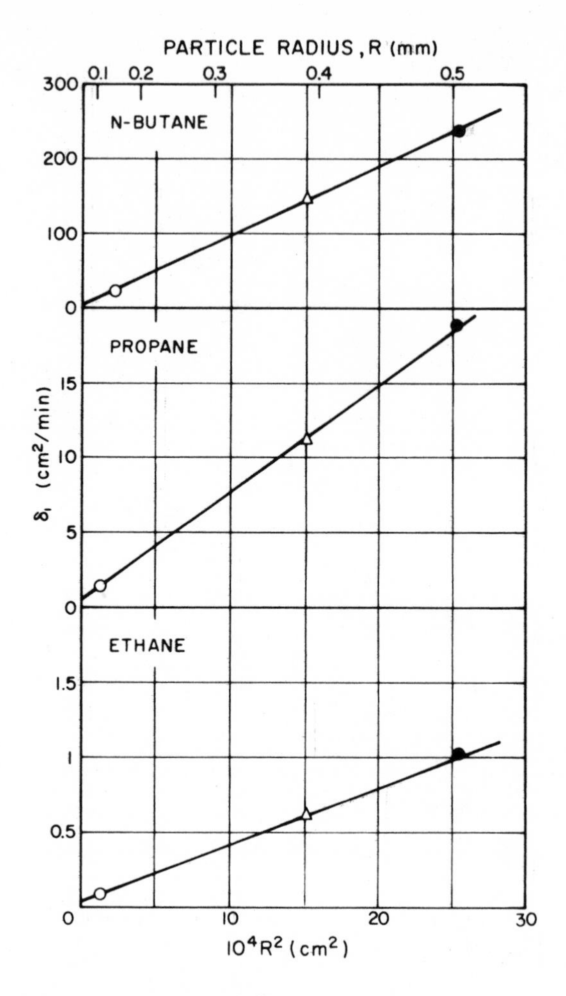

FIG. 7. Dependence of δ_1 on R^2.

$$D_K = \frac{4}{3}\,\bar{r}\left(\frac{2R_gT}{\pi M}\right)^{1/2} \tag{30}$$

The tortuosity factor q_{int} included in Table 4 is the one calculated from D_e and D_K by Eq. (29). If Eq. (29) is valid, q_{int} should be greater than unity and a constant, of the order of 2 to 3. However, if another diffusional process predominates, such as surface migration of the adsorbed molecules, the observed tortuosity would be much smaller. Since the values given in Table 4 are much less than 2, it is indicated that

TABLE 4

Effective Diffusion Coefficients (50°C)[a]

	Ethane	Propane	n-Butane
$D_e \times 10^3$ (sq cm/s)	1.41	1.54	2.93
$D_K \times 10^3$ (sq cm/s)	3.48	2.89	2.52
q_{int}	1.20	0.91	0.42

[a]From Ref. 20.

surface diffusion is occurring and that it becomes more significant in changing adsorbates from ethane to n-butane. The increase in significance with molecular weight of adsorbate is expected since the adsorbed concentration increases.

Since adsorption rate at the surface is very high in this system, as mentioned in the following section, the gas concentration in the pore can be assumed to be in equilibrium with the surface concentration of adsorbed species. Then surface diffusion can be accounted for in the intraparticle diffusivity expression as follows:

$$D_e = \frac{\beta}{q_{int}} D_K + \frac{\rho_p}{\beta} K_A D_{s,p} \tag{31}$$

where

$$D_{s,p} = \frac{\beta}{q_s} D_s \tag{32}$$

D_s is the true surface diffusion coefficient, and q_s denotes the tortuosity factor for surface diffusion.

Figure 8 shows the effective surface diffusion coefficient $D_{s,p}$ for ethane, propane, and n-butane. For the determination of q_{int}, high-temperature runs with ethane were made. At these conditions, the effect of $D_{s,p}$ was negligible. The value of q_{int} so determined was 3.35, which is reasonable as a tortuosity factor for silica gel.

The activation energies for surface diffusion can be obtained from Fig. 8. The results were 4.5 kcal/mole for propane and 4.4 kcal/mole for n-butane. These values are equal to 76% and 56% of the heats of adsorption of propane and n-butane, respectively.

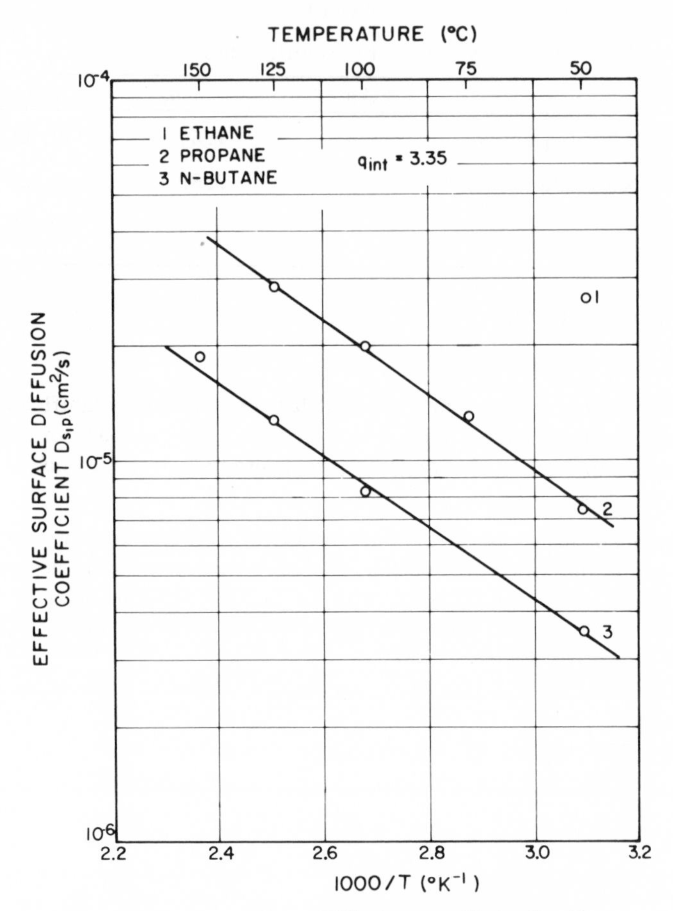

FIG. 8. Effective surface diffusion coefficients, $D_{s,p}$.

3. Adsorption rate at the surface.

From the plots shown in Fig. 7, Schneider and Smith [13] approximately determined adsorption rate constants. The external mass transfer coefficient k_f was determined by assuming that $Nu = k_f R/D_v = 2$. Then from the intercepts of the straight lines in Figure 7, k_a were determined by applying Eq. (17). The values of k_a for ethane, propane, and n-butane on silica gel are as follows:

Substance	k_a (50°C) $cm^3/(g\ SiO_2)(sec)$
Ethane	167.
Propane	255.
n-Butane	1500.

Such rate constant information is rather rare because of the rapidity of physical adsorption processes. Some uncertainties exist in the accuracy of these numerical values. However, they indicate clearly that k_a increases with molecular weight of adsorbate.

The analysis described illustrates how rate parameters for axial dispersion, intraparticle diffusion, and adsorption can be determined from pulse-response measurements for fixed beds of adsorbent particles. The same procedures can be used for nonadsorbing particles [21] ($K_a = 0$) or nonporous materials ($\beta = 0$). The equations given are applicable for monodispersed particles in a bed. Similar equations applicable to bidispersed particles such as commercial molecular sieve and alumina pellets have been developed [22-24].

In application of experimental data, the observed moments must be corrected for the retention time (first moment) and dispersion (second moment) in the dead volumes between pulse injection and entrance to the bed and between exit of the bed and the detector. Methods for making this correction are described in the literature [13, 14].

The procedure described, modifications of it, the plate method, and other methods have been used by many investigators for evaluating one or more of the rate parameters in a fixed bed with or without adsorption. These methods are described in the literature or discussed in the next section.

III. OTHER METHODS OF ANALYSIS

This brief survey of the literature is not intended to be complete but rather to point out specific examples of other methods of analysis. Mathematical derivations and equations appropriate to these methods are available in the references cited.

Following the early development [3, 4] of the plate theory, it has been used frequently to evaluate rates of adsorption, for example by Ozaki and his coworkers in hydrogen adsorption studies. One example of their work was concerned with the deuterium exchange on a nickel catalyst [25].

Numerous applications have been made to the determination of intraparticle diffusivities. Typical examples are those of Davis and Scott [26] for diffusing hydrogen and methane in alumina and of Ma and Mancel [27] for diffusing CO_2, NO, NO_2, and SO_2 in molecular sieves (zeolites). Researchers in Russia have also used the plate theory to examine rate parameters in combined adsorption and diffusion processes. An example is Denisova and Rozental's work [28] on diffusion and adsorption of n-butane in silica gel particles.

Rather than use moments, one can fit a Fourier series to the response peak. To predict the form of the peak, the coefficients in this series can be related to the rate parameters in the model used. The comparison between experimental and predicted peaks can be made either in the real time domain or in a transform domain. Clements and coworkers [29, 30] proposed a Fourier transform of the response peak followed by fitting in the frequency domain. Hudgins, Silveston, and colleagues [31] have shown that evaluation of the coefficients of the Fourier series representation of the peak, in essence, provides a transformation into the frequency domain. These authors then applied the method to evaluating mass transfer parameters in fixed beds. A potential advantage of the method is the nearly constant error in successive terms used in the series representation. This is in contrast to the moment method, in which errors build up very rapidly in successive moments due to the importance attached to the tail of the response curve [32]. Jefferson [33] describes the elements of the Fourier transform method.

A sinusoidal input function can be used instead of a single pulse input in chromatographic studies, for example, in a fixed bed. The analysis of the response function is then conducted by examining the change in frequency due to the bed. Jefferson [34] and Turner [35] discuss the procedure but examples of its use with experimental data in fixed beds do not seem to be available.

IV. ISOTOPE CHROMATOGRAPHY

Most adsorption equilibrium isotherms become nonlinear at moderate concentrations, and for chemisorption of gases the nonlinear region extends to a low concentration. Often this concentration is so low that accurate detection devices for the effluent concentration are not available. For nonlinear phenomena, the previously described procedure is not applicable. Also, direct numerical solution of the nonlinear equations for the concentration-vs-time relation of the effluent peak requires exorbitant amounts of computer time. Further, evaluating rate parameter by comparing such predicted peaks does not usually lead to unique sets of parameters.

These difficulties in studying intrinsically nonlinear adsorption phenomena can be circumvented in some cases by introducing a pulse of an isotope into a carrier of the adsorbing gas (e.g., D_2 in a carrier of H_2) at the entrance to the fixed bed. Then it is possible both to linearize the equation and to measure the rate constants at constant surface conditions corresponding to equilibrium with respect to the carrier.

A. Isotope Method for Chemisorption Studies

We shall describe our experiments with the H_2-D_2 system [16, 36-38]; other systems could be used, however. For example, chemisorption rates of CO on catalysts could be studied by using pulses of $C^{14}O$ with a suitable, small volume detector [39]. Consider a steady flow of hydrogen through a bed of catalyst particles at constant temperature and pressure. Under these equilibrium conditions, the surface coverage is constant, and the fluxes (designated as ϕ) of hydrogen from gas phase to catalyst surface and from catalyst surface to gas phase are equal. Suppose deuterium is introduced as a tracer to the hydrogen stream, with the total pressure (and hence concentration) of $H_2 + D_2$ kept constant. If the isotope effect is neglected, the total rates will not change, but it is possible to distinguish the molecules of D_2 and HD from molecules of H_2 in the detector (thermal conductivity cell) at the outlet of the bed. The rate of adsorption of deuterium equals ϕ multiplied by the mole fraction C_d/C_t of deuterium in the gas. Similarly, the rate of desorption of deuterium will be ϕ multiplied by the fraction of the adsorbed deuterium atoms. Therefore, the net rate of adsorption of the tracer is given by

$$\phi_D = \phi \frac{C_d}{C_t} - \phi \frac{n_d}{n_t} = \frac{\phi}{C_t}\left(C_d - \frac{1}{n_t/C_t} n_d\right) \tag{33}$$

It may be that near-equilibrium will be obtained for the exchange reaction $H_2 + D_2 = 2HD$. For the deuterium concentration C_d to be independent of whether the deuterium in the gas phase is present as D_2 or HD, C_d is defined as

$$C_d = \frac{1}{2} C_{HD} + C_{D_2} \tag{34}$$

Since the total concentration C_t of hydrogen plus deuterium and the total adsorbed concentration n_t are constant, Eq. (33) is linear in C_d and n_d.

Equation (33) can be rewritten in terms of quantities measured directly from the chromatographic data, as

$$\phi_D = k^* \left(C_d - \frac{n_d}{K^*}\right) \tag{35}$$

where

$$k^* = \frac{\phi}{C_t} \tag{36}$$

and

$$K^* = \frac{n_t}{C_t} \tag{37}$$

Once k^* and K^* are determined, ϕ can be calculated from Eq. (36). The value K^* is not the adsorption equilibrium constant in the usual sense but simply the ratio of n_t to C_t at a single point on the nonlinear isotherm. Experiments at different total concentrations (pressure of hydrogen) and temperature give different values of k^* and K^*. These results determine the effect of pressure and temperature on the rate and the adsorption isotherm.

When $N_i = \rho_p \phi_D$, Eqs. (1)-(3) and (35) give the same moment equations as Eqs. (16) and (17), except that k_a and K_a are replaced by k^* and K^*. Hence the moment method of analysis described in Section II can also be used for isotope chromatography.

We have carried out experiments for hydrogen on copper-zinc oxide catalyst [16] and on a nickel-kieselguhr catalyst [38] for variable hydrogen pressure, and on Ni-kieselguhr catalyst [36] and on cobalt catalyst [37] for constant hydrogen pressure.

It was possible to vary the hydrogen pressure by mixing H_2 and He to produce the carrier gas. Deuterium-helium pulses were then introduced to the bed and the effluent pulse analyzed as described. In this way, exchange rates could be measured as a function of hydrogen pressure for studies on the mechanism of hydrogen adsorption on these catalysts. Typical results are presented in the following paragraphs.

B. Applications

Typical results for effluent deuterium peaks (with pure hydrogen as carrier gas) on copper-zinc oxide catalysts (catalyst G-66B from Chemetron Corp.) at several temperatures and approximately the same flow rate are shown in Fig. 9. Also shown is the inlet peak (at 250°C) at an attenuation

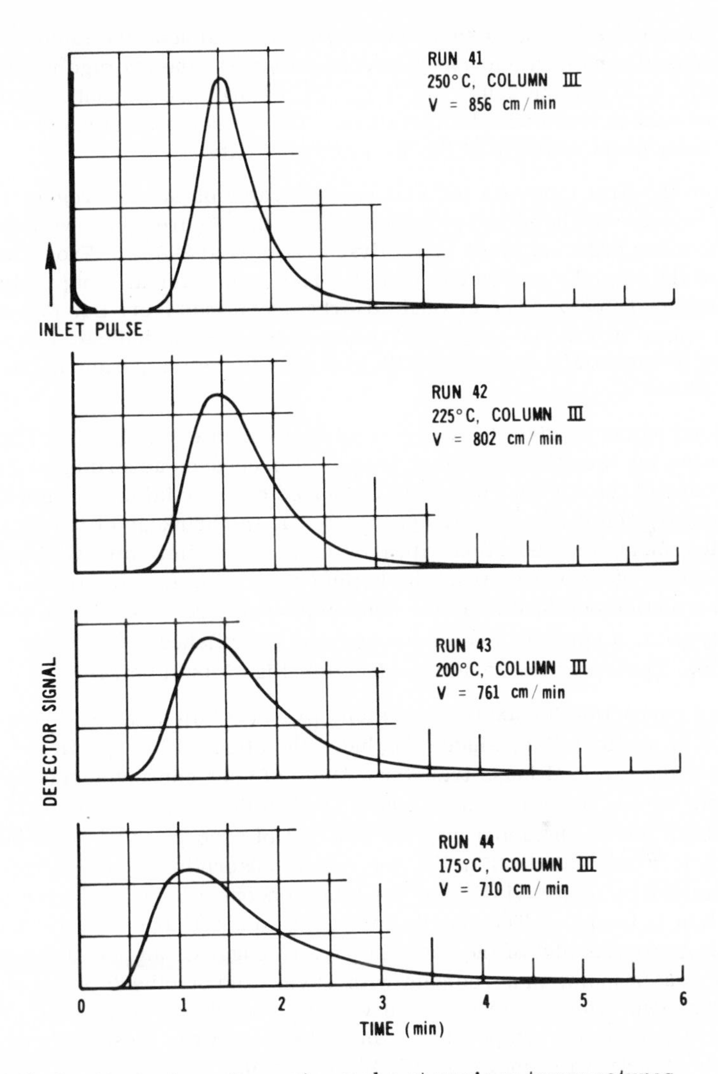

FIG. 9. Deuterium chromatographs at various temperatures.

32 times larger than that for the effluent peak. Comparing the two curves shows that the width of the inlet peak was negligible relative to the peak of the effluent chromatograph of deuterium.

The lower limit of temperatures studied was determined by the extent of tailing. The curves in Fig. 9 illustrate the difficulty in obtaining accurate values of the second moment for temperatures below 175°C.

While the curves in Fig. 9 show increasing dispersion as the temperature is decreased, and have increased second moments, the average retention time varies little from 250°C to 175°C. Therefore, the amount adsorbed does not change much with temperature. This is demonstrated quantitatively in the subsequent analysis of the first-moment data.

From the first moments for various flow rates of pure hydrogen, $\rho_p K^*z/u$ was calculated and plotted versus z/u. Figure 10 shows this plot for three different sizes of catalyst particles at 200°C. From the slope of the line, K* can be calculated. The results for all temperatures are given in Table 5. The amount adsorbed, obtained from Eq. (37) and K*, is shown in Fig. 11. The nearly constant n_t values indicate that the surface is essentially saturated with hydrogen in the temperature range 150°C to 250°C.

Second moments give the apparent adsorption rate constant k*, after correction for the effects of other transport processes according to Eq. (17). In the case of the Cu·ZnO catalysts employed here, axial dispersion showed an unusually significant effect, apparently due to shrinkage of the catalyst particles during the reduction with H_2. Axial dispersion was evaluated over a wide range of velocities from the helium-pulse data for comparison with the contribution obtained from the deuterium-pulse results. From the high velocity data, a constant Peclet number was established for each size of particle. These are the results included in Fig. 5 (solid circles).

After correction for axial dispersion, $[\Delta\mu_2/(2z/u)]_{1/u\to 0}$ could be a function of particle size, since it includes the effect of intraparticle diffusion and external mass transfer. Plots of such corrected second moments versus particle size at 200°C showed that these effects were negligible. Hence the apparent rate constant of adsorption k* could be immediately calculated. Finally, the rate of adsorption (or exchange) Φ was obtained by multiplying k* by the total concentration of hydrogen plus deuterium in the gas. The results for 150°C to 250°C are shown in Table 5. The Arrhenius plot of the rates in Fig. 12 suggests an activation energy of 14.6 kcal/g mole, in agreement with Taylor and Strother's study on hydrogen adsorption on zinc oxide at varying surface coverage [40]. For high coverages in the temperature range (184 ~ 218°C), these authors reported an activation energy of 15.3 kcal/g mole.

Results for rates of adsorption (exchange) of hydrogen on nickel [38] and on cobalt [37] catalysts supported on Kieselguhr were obtained at temperatures from -34 to 25°C. At 24°C, the rates were 60×10^{-2} and 9.6×10^{-2} g mole/g min. These results are about two orders of magnitude greater than the rates given in Table 5 for Cu·ZnO, despite the much lower temperature for the nickel and cobalt values.

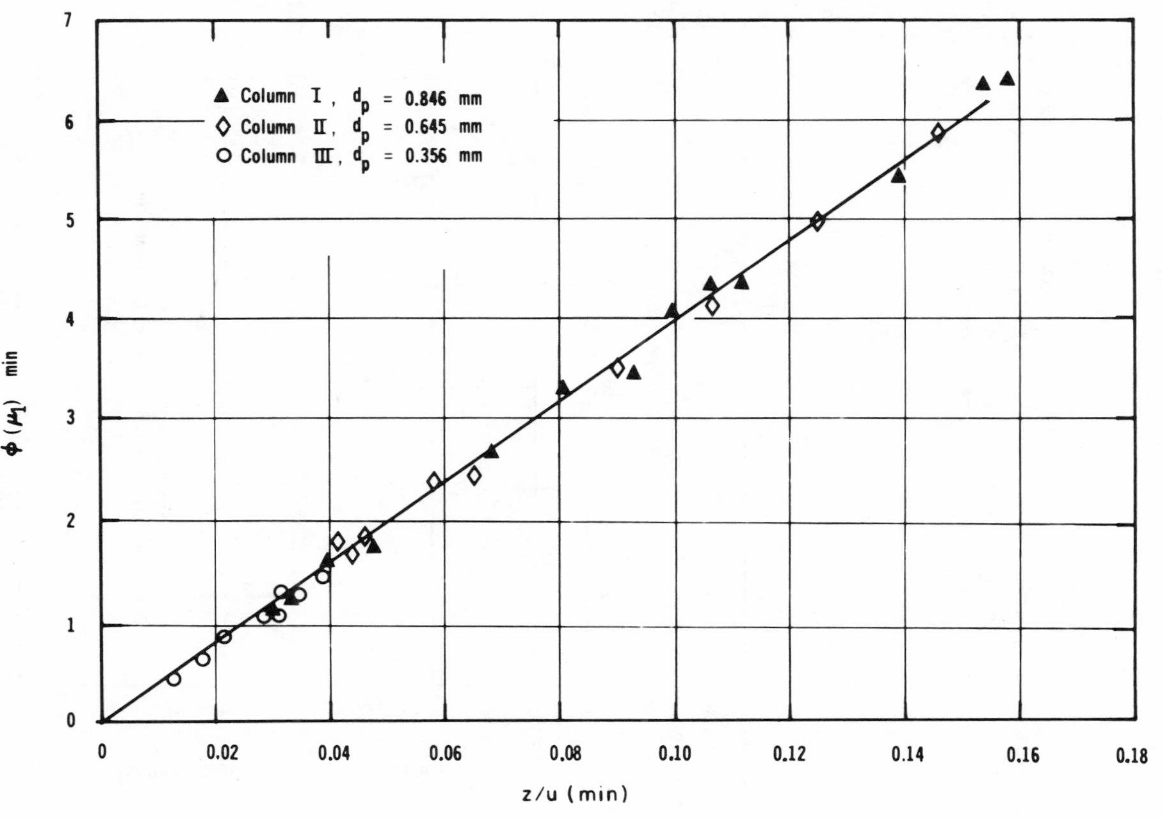

FIG. 10. First-moment data for deuterium.

TABLE 5

Rates of Adsorption at P_{H_2} = 780 mm Hg[a]

Temperature °C	Pseudo equilibrium constant K* (cm^3/g)	Amount adsorbed $n_t \times 10^4$ (g mole/g)	Adsorption rate constant $k^* \times 10^{-2}$ cm^3(g)(min)	Adsorption rate $\phi \times 10^2$ mole/(g)(min)
150	13.7	4.07	0.54	0.16
175	14.9	4.17	1.39	0.39
200	15.6	4.13	2.45	0.65
225	16.5	4.12	6.20	1.56
250	16.1	3.87	11.6	2.80

[a]From Ref. 16.

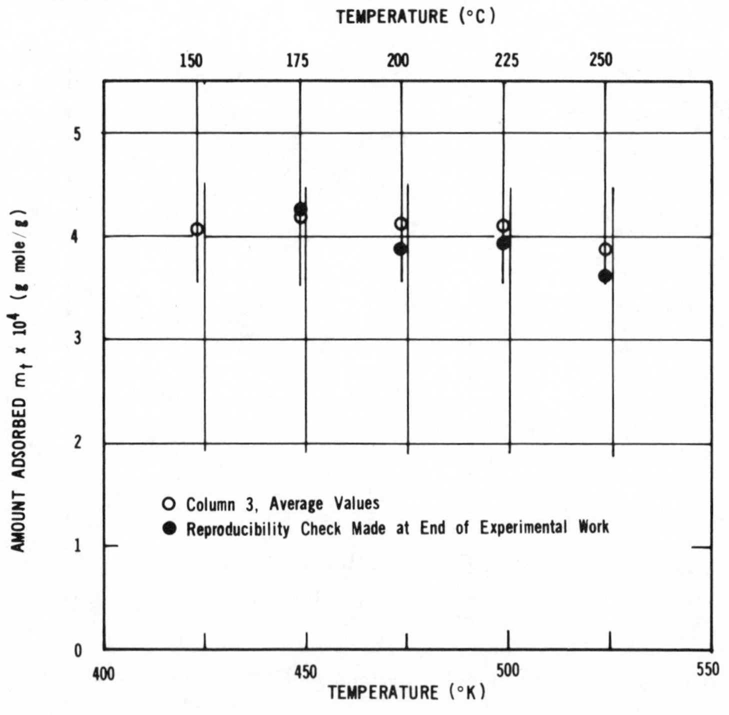

FIG. 11. Equilibrium adsorption for hydrogen on catalyst.

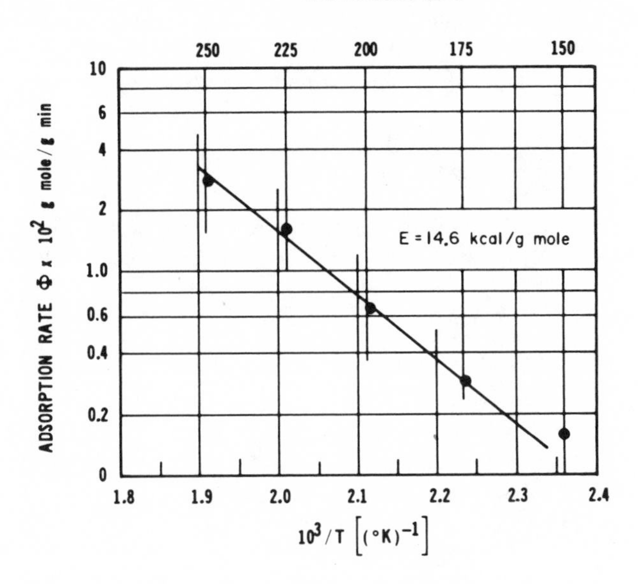

FIG. 12. Arrhenius plot for hydrogen adsorption.

The heat of adsorption deduced from the first-moment data [K_A vs temperature results] for the nickel catalyst was 12 kcal/g mole. Ozaki and colleagues [41] also used a chromatographic technique with deuterium to obtain a heat of adsorption of 12 to 15 kcal/g mole. Comparison of the activation energy with the heat of adsorption indicated that these chemisorptions were nonactivated chemisorption processes.

The rate data obtained for different mole fractions of hydrogen in the carrier gas show the effect of hydrogen partial pressure on the rate. Such information is valuable in evaluating proposed mechanisms for hydrogen adsorption on catalysts. Illustrations of this aspect of chromatographic investigation of rates is given in Ref. 38.

V. EXTENSION TO REACTING SYSTEMS

The purpose of this section is to present, for the reaction case, a solution in the Laplace domain for the response curve from a bed of catalyst particles. The model equations include axial dispersion, external mass transfer, intraparticle diffusion, reversible adsorption, and first-order surface reaction. This solution is then used to relate moments of the chromatographic curve to the transport and surface rate parameters. Finally, moment equations for some special cases are derived. A simple reaction involving but one reactant is used to illustrate the development. Details of the development are given in Ref. 42.

A. Modifications to Theory; Basic Equations

The mechanism of the surface processes for a reaction of the form $A \rightarrow S$ or $A + B \rightarrow C$ will be described by the mechanisms

$$\begin{aligned} &A + S \rightleftarrows A \cdot S, \qquad \text{rate} = N_i \\ &A \cdot S \rightarrow C \cdot S \text{ or } C + S, \qquad \text{rate} = r_c \end{aligned} \tag{38}$$

or

$$\begin{aligned} &A + S \rightleftarrows A \cdot S, \qquad \text{rate} = N_i \\ &B + S \rightleftarrows B \cdot S \\ &A \cdot S + B \cdot S \rightarrow C \cdot S \text{ or } C + S, \qquad \text{rate} = r_c \end{aligned} \tag{39}$$

It is further assumed that the rates of adsorption and surface reaction of A are both linear:

$$N_i = \rho_p k_a \left(C_i - \frac{n_i}{K_a} \right) = \rho_p \frac{\partial n_i}{\partial t} + \rho_p r_c \tag{40}$$

$$r_c = k_r n_i \tag{41}$$

where N_i is based upon a catalyst (particle) volume and r_c is based upon a unit of catalyst mass. Note that Eq. (40) accounts for the change in surface coverage (n_i) with time.

The conservation equations for reactant A in packed beds, at the surface of the catalyst and in the particle, are given by Eqs. (1), (2), and (3). If the Laplace transform of the variables with respect to time is applied to Eqs. (1) to (3), (40), and (41), then the final solution for the pulse input of injection time τ are the same as Eqs. (10) to (14), except that the following equation should be used for κ (p):

$$\kappa(p) = p + \frac{\rho_p/\beta}{(1/k_a) + [1/(p + k_r)K_a]} \tag{42}$$

B. Moments for a Reacting System

In the case of reaction chromatography, the zeroth moment is not conserved along the bed length coordinate z since there is a consumption of reactant A due to surface reaction. We define here a zeroth reduced moment μ° as

$$\mu^{\circ} = \frac{m_o}{(m_o)_{z=0}} \tag{43}$$

which is equal to the fraction of A unreacted in the effluent from the bed of catalyst particles. Then $1 - \mu^{\circ}$ is the conversion.

Equation (10), with Eqs. (11) to (13) and (42), is used in Eq. (15) to evaluate m_o, $(m_o)_{z=0}$, and m_n. Then zeroth reduced, first absolute, and second central moments are obtained in a manner similar to that described in Section II. The results are

1. Zeroth reduced moment:

$$\mu^{\circ} = \exp\left[-\frac{u}{2E_z}(q_o-1)z\right] \tag{44}$$

where

$$q_o = \sqrt{1 + \frac{4E_z}{u^2}\,G(0)} \tag{45}$$

$$G(0) = \frac{3(1-\alpha)Bi}{\alpha R^2/D_e}\left[1 - \frac{Bi}{A_o(0) + Bi}\right] \tag{46}$$

$$A_o(0) = \phi(0)\coth\phi(0) - 1 \tag{47}$$

$$\phi(0) = R\sqrt{\frac{\beta}{D_e K(0)}} \tag{48}$$

$$K(0) = \frac{\rho_p/\beta}{(1/k_a) + (1/k_r K_a)} \tag{49}$$

and the Biot number (Bi) is $k_f R/D_e$.

2. First absolute moment:

$$\frac{\mu_1 - t_o/2}{z/u} = \frac{1}{q_o} \left\{ 1 + \frac{3(1-\alpha)\beta}{2\alpha} (Bi)^2 \frac{A_1(0)(K')_o}{[A_o(0) + Bi]^2} \right\} \tag{50}$$

where

$$A_1(0) = \frac{1}{\phi(0)} \coth \phi(0) - \text{cosech}^2 \phi(0) \tag{51}$$

$$(K')_o = \left[\frac{d\kappa(p)}{dp}\right]_{p=0} = 1 + \frac{k_a{}^2 K_a (\rho_p/\beta)}{(k_a + k_r K_a)^2} \tag{52}$$

3. Second central moment:

$$\frac{\mu_2' - t_o^2/12}{z/u} = \frac{4E_z}{q_o^3 u^2} \left[1 + \frac{3(1-\alpha)\beta}{2\alpha} (Bi)^2 \frac{A_1(0)}{[A_o(0) + B_i]^2} (K')_o \right]^2$$

$$- \frac{1}{q_o} \cdot \frac{3(1-\alpha)\beta}{2\alpha} \cdot \frac{(Bi)^2}{[A_o(0) + Bi]^3} \left[- \frac{R^2\beta}{2D_e} \left\{ 2A_1(0)^2 \right. \right.$$

$$+ \left[A_o(0) + Bi\right] A_2(0) (K')_o^2$$

$$\left. + \left[A_o(0) + Bi\right] A_1(0) (K'')_o \right] \tag{53}$$

where

$$A_2(0) = \frac{1}{\phi(0)} \left[\frac{\coth \phi(0)}{[\phi(0)]^2} + \frac{\text{cosech}^2 \phi(0)}{\phi(0)} \right.$$

$$\left. - 2 \,\text{cosech}^2 \phi(0) \cdot \coth \phi(0) \right] \tag{54}$$

$$(K'')_o = \left[\frac{d^2\kappa(p)}{dp^2}\right]_{p=0} = - \frac{2(k_a K_a)^2 (\rho_p/\beta)}{(k_a + k_r K_a)^3} \tag{55}$$

If the zeroth, first absolute, and second central moments are evaluated from experimental chromatographic data, three equations --(44), (50), and (53)-- are available relating the five constants E_z, k_f, D_e, k_a, and k_r and the adsorption equilibrium constant K_a. If moments at various flow rates and particle sizes are obtained, it is possible to increase the number of relations, so that in principle, all the constants can be determined.

However, Eqs. (50) and (53) are too complicated to use effectively with experimental data. These equations are simplified when one or two of the three transport processes (axial dispersion, external mass transfer or intraparticle diffusion) offer negligible resistance to mass transport. Also, by proper choice of experimental conditions, it is possible to obtain chromatographic data for which simplified equations are applicable. Some of these special cases are considered in the next section.

Figure 13 is a plot of A_0, A_1, and A_2 versus $\phi(0)$ as determined from Eqs. (47), (51), and (54). This figure provides a simple way of estimating A_1 and A_2 from A_0 determined from zeroth reduced moment.

C. Special Cases

1. Axial dispersion and external mass transfer negligible.

It is sometimes possible to operate a laboratory, fixed-bed reactor at high enough gas velocities and with large enough catalyst pellets that only intraparticle diffusion and the surface processes are significant. For this case, Eqs. (44), (50) and (53) reduce to

$$\mu^{o} = \exp\left\{ -\frac{z}{u} \cdot \frac{3(1-\alpha)}{\alpha} \cdot \frac{D_e}{R^2} \cdot A_0(0) \right\} \tag{56}$$

$$\frac{\mu_1 - \tau/2}{z/u} = 1 + \frac{3(1-\alpha)\beta}{2\alpha} A_1(0) \cdot (K')_0 \tag{57}$$

$$\frac{\mu_2' - \tau^2/12}{z/u} = \frac{3(1-\alpha)\beta}{2\alpha} \left[\frac{R^2\beta}{2D_e} A_2(0) \cdot (K')_0^2 - A_1(0)\,(K'')_0 \right] \tag{58}$$

The above equations become identical to the results Denisova and Rozental [28] obtained, when the desorption rate is far larger than the surface reaction rate, that is $k_r << k_a/K_a$.

2. Axial dispersion significant

With small particles and low gas velocities, intraparticle diffusion and external mass transfer can be unimportant compared with axial dispersion. Then the moment equations are greatly simplified to give

$$\mu^{o} = \exp\left[-\frac{u}{2E_z} (q'-1)z \right] \tag{59}$$

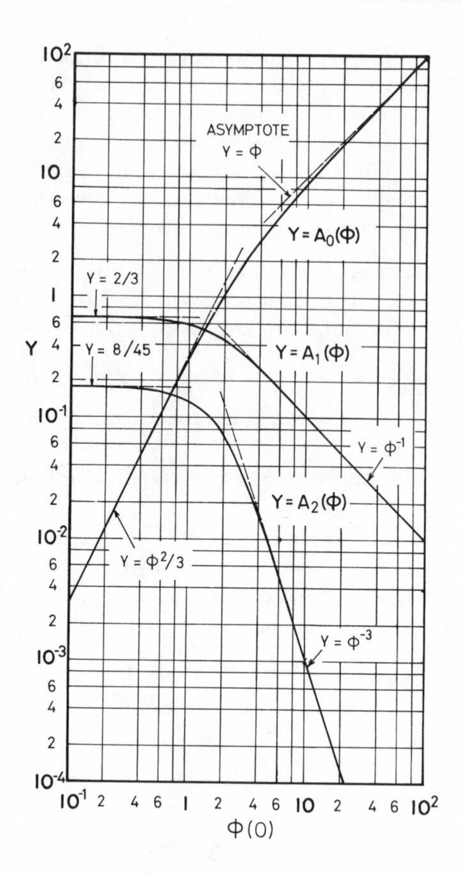

FIG. 13. Rate constant functions $A_0(\phi)$, $A_1(\phi)$, $A_2(\phi)$.

$$\frac{\mu_1 - (\tau/2)}{z/u} = \frac{1}{q'} \left\{ 1 + \frac{(1-\alpha)\beta}{\alpha} (K')_0 \right\} \tag{60}$$

$$\frac{\mu_2 - (\tau^2/12)}{z/u} = \frac{2E_z}{(q')^3 u^2} \left\{ 1 + \frac{(1-\alpha)\beta}{\alpha} (K')_0 \right\}^2 - \frac{1}{q'} \frac{(1-\alpha)}{\alpha}\beta \; (K'')_0 \tag{61}$$

where

$$q' = \sqrt{1 + \frac{4E_z}{u^2} \frac{(1-\alpha)}{\alpha}\beta \; K(0)} \tag{62}$$

These results are identical to those of Kocirik [43]. That is, they relate the moments to E_z, k_a, k_r, and K_a for the case of first-order adsorption and reaction with axial dispersion.

It is concluded that even for a first-order surface reaction, the equations for the moments that include all of the transport resistances appear too complicated for use with experimental data. However, for the case when only one transport process is significant, the results are tractable. Of special interest is the case when intraparticle diffusion becomes important compared with axial dispersion or external mass transfer. The effect of intraparticle diffusion on moments will be discussed in a later section.

Another attractive possibility is to obtain the surface constants k_a, K_a, and k_r from the moments of the chromatographic curve. A method by which separate values of adsorption and surface rate constants (k_a and k_r) could be evaluated would be very important in kinetics of heterogeneous catalysis. Hence, this possibility is examined in the following paragraphs.

D. Surface rate constants

Suppose chromatographic curves of a reactant are measured, and moments evaluated therefrom. When the resistance of each transport process can be established from independent measurements, such as adsorption studies, then $K(0)$, $(K')_0$, and $(K'')_0$ are determined from the zeroth reduced, first absolute, and second central moment by Eqs. (44), (50), and (53) for the general case, by Eqs. (56) to (58) when intraparticle diffusion is controlling, or by Eqs. (59) to (61) when axial dispersion is controlling. In principle, k_a, k_r, and K_a can be calculated from $K(0)$, $(K')_0$, and $(K'')_0$. Equations (49), (52), and (55) can be solved to give the explicit forms for the rate and equilibrium constants.

$$\frac{\rho_p}{\beta} k_a = K(0) - 2 \frac{[(K')_0 - 1]^2}{(K'')_0} \tag{63}$$

$$\frac{\rho_p}{\beta} K_a = \left[(K')_0 - 1\right] \left[1 - \frac{1}{2} \frac{K(0) \cdot (K'')_0}{[(K')_0 - 1]^2}\right]^2 \tag{64}$$

$$k_r = \left[\frac{(K')_0 - 1}{K(0)} - \frac{1}{2} \frac{(K'')_0}{(K')_0 - 1}\right]^{-1} \tag{65}$$

The challenge in using these equations to evaluate separate values for k_a and k_r is to find an experimental system meeting the restraints on the equation. Only one reactant can affect the rate in order to agree with

Eqs. (39) to (41). Furthermore, the rate should be first-order. Also, it must be possible to separate the effluent peak of A from the effluent peak of the other components B and C. Isomerization reactions, using the product as the carrier and introducing a pulse containing the reactant, might meet these requirements. Multicomponent systems could be used if they involved a detector sensitive only to the concentration of A in the effluent peak. Alternatively, detectors could be used that would separate the concentration-vs-time peaks for several components in the mixture (i.e., flow-type multipole mass spectrometers).

E. Effect of Intraparticle Diffusion on Moments

In this paragraph, the effect of intraparticle diffusion on zeroth, first absolute, and second central moments are discussed. For the zeroth moment, Eq. (56), with Eq. (47) for $A_o(0)$, indicates that the effect of intraparticle diffusion is well defined. Namely, from Eq. (56), the effective rate constant that includes intraparticle diffusion resistance is given by

$$k_{eff} = \frac{\ln(\mu_0)}{z/u} = \frac{3(1-\alpha)}{\alpha} \frac{D_e}{R^2} A_o(0) \tag{66}$$

For conditions of no intraparticle diffusion resistance, the rate constant can be written

$$k_{nd} = \frac{1-\alpha}{\alpha}\beta\ K(0) \tag{67}$$

The function of void fraction appears because K(0) is based on unit pore volume in the particle. From Eqs. (66) and (67), the ratio of k_{eff} and k_{nd}, which is the effectiveness factor, is

$$\text{E.F.} = \frac{3D_e}{\beta R^2} \frac{A_o(0)}{K(0)} \tag{68}$$

If Eqs. (47) and (48) are substituted, Eq. (68) reduces to the form

$$\text{E.F.} = \frac{3}{[\phi(0)]^2} [\phi(0) \coth \phi(0) - 1] \tag{69}$$

This relation is the conventional effectiveness factor form for a first-order reaction in a porous catalyst, where $\phi(0)$ is the Thiele modulus.

Figures 14 and 15 show the relations between the moments and the intraparticle diffusion resistance for constant values of the surface rate constants. The first absolute moments, calculated from Eq. (57), are plotted versus R^2/D_e; R^2/D_e is the time constant for intraparticle diffusion. The unity term in the right side of Eq. (57) is the value of $(\mu_1 - t_0/2)/z/u)]$ if there were no surface processes and no intraparticle diffusion. Hence $[(\mu_1 - t_0/2)/(z/u)] - 1$ represents the effect of surface processes and intraparticle diffusion. It is this term that is plotted as the ordinate in the upper part of Figs. 14 and 15. The curves for the second central moment were established from Eq. (58). All the curves are based n a bed porosity (α) of 0.4 and an intraparticle porosity (β) of 0.6.

Figure 14 shows the effect of R^2/D_e and k_r on the moments for $\rho_p k_a/\beta$ = 1000 sec^{-1} and $\rho_p K_a/\beta = 10$. The reaction rate constant k_r has a greater effect on the second central moment at higher diffusion times, i.e., larger intraparticle diffusion resistances. The level of the first and second moments at low diffusion times is determined primarily by the adsorption

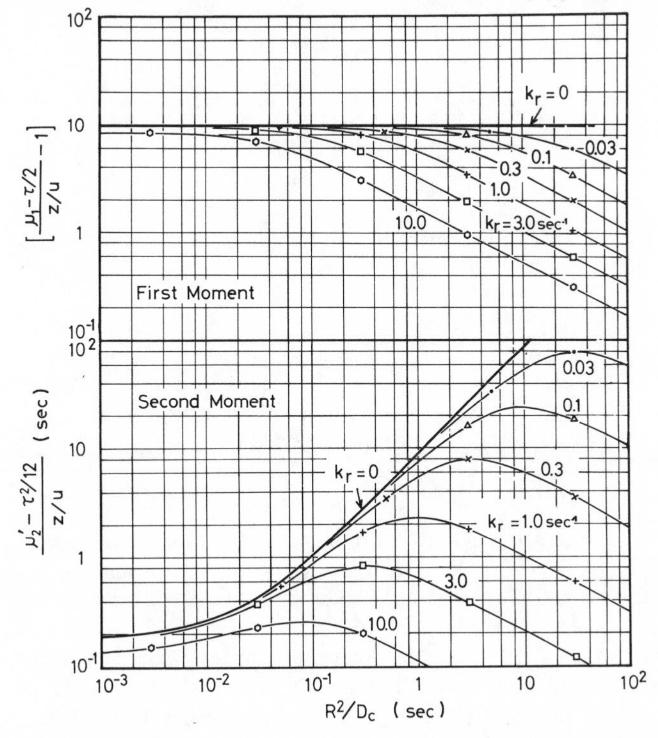

FIG. 14. Effect of surface rate constant on moments ($\rho_p k_a/\beta$ = 1000 sec^{-1}, $\rho_p K_a/\beta = 10$).

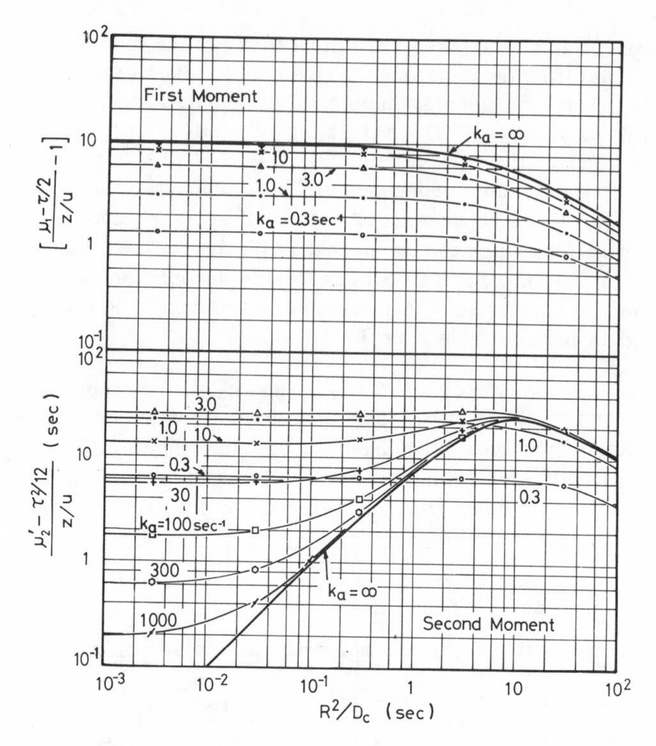

FIG. 15. Effect of adsorption rate constant on moments ($\rho_p K_a/\beta = 10$, $k_r = 0.1 \sec^{-1}$).

rate (k_a) and equilibrium (K_a) constants. However, when $k_r K_a$ becomes as large as k_a, it is true that k_r also has a significant effect at low diffusion times. It is interesting to note that the second central moment has a maximum value at an intermediate diffusion time. The solid lines in Fig. 14 correspond to no reaction. The equations for these solid curves are Eqs. (16) and (17), with $E_z = 0$ and $k_f = \infty$.

Figure 15 shows the effect of the adsorption rate constant for $k_r = 0.1 \sec^{-1}$, $\rho_p K_a/\beta = 10$. The solid lines apply for $k_a = \infty$ and were located from Eqs. (57) and (58) by using values of K(0), $(K')_0$, and $(K'')_0$ corresponding to surface reaction controlling the reaction steps. The figure shows that for $\rho_p k_a/\beta = 10$, the first moment is insensitive to k_a, while the second central moment still shows the effect of differences in k_a (except at high diffusion times). It is concluded that adsorption equilibrium can be safely assumed in evaluating rate processes when k_a is above a critical value determined by k_r, K_a, and the diffusion time.

It is worth while to show asymptotic form of Eqs. (56) to (58) for the cases when diffusion times are very high or very low. For high diffusion times, that is, when $\phi(0)$ is far greater than unity, A_0, A_1, and A_2 take simple forms, as the intercepts in Figure 13 show. Then the moment equations reduce to

$$\mu^{o} = \exp\left\{-\frac{z}{u}\,\frac{3(1-\alpha)}{\alpha}\,\frac{D_e}{R^2}\,\phi(0)\right\} \tag{70}$$

$$\frac{\mu_1 - (T/2)}{z/u} = 1 + \frac{3(1-\alpha)}{2\alpha}\beta\,\frac{K'(0)}{\phi(0)} \tag{71}$$

$$\frac{\mu_2' - (T^2/12)}{z/u} = \frac{3(1-\alpha)}{2\alpha}\beta\left\{\frac{(K')_o}{2K(0)} - (K'')_o\right\}\frac{1}{\phi(0)} \tag{72}$$

At very low diffusion time [$\phi(0) << 1$], moments will be determined solely by surface processes, and Eqs. (56) to (58) become

$$\mu^{o} = \exp\left\{-\frac{z}{u}\,\frac{(1-\alpha)}{\alpha}\,K(0)\right\} \tag{73}$$

$$\frac{\mu_1 - (T/2)}{z/u} = 1 + \frac{(1-\alpha)}{\alpha}\beta\,(K')_o \tag{74}$$

$$\frac{\mu_2' - (T^2/12)}{z/u} = -\frac{(1-\alpha)\beta}{\alpha}\,(K'')_o \tag{75}$$

These equations represent the most simple case of surface reaction chromatography.

F. Limiting Surface Processes

Moment equations developed thus far contain surface process functions $K(0)$, $(K')_o$, and $(K'')_o$. These are defined in general by Eqs. (49), (52), and (55), but take simplier forms for special surface reaction conditions. Of particular interest are cases where either adsorption or surface reaction controls the surface process.

1. Adsorption controlling

In this case, $k_a << k_r K_a$ so that Eqs. (49), (52), and (55) simplify to

$$K(0) = \frac{\rho_p}{\beta} k_a \tag{76}$$

$$(K')_0 = 1 \tag{77}$$

$$(K'')_0 = -2 \frac{\rho_p}{\beta} \frac{k_a^2}{k_r^3 K_a} \tag{78}$$

These special forms apply for any of the cases involving special mass transport situations.

2. Surface reaction controlling

For this case, $k_a >> k_r K_a$, and then

$$K(0) = \frac{\rho_p}{\beta} k_r K_a \tag{79}$$

$$(K')_0 = 1 + \frac{\rho_p}{\beta} K_a \tag{80}$$

$$(K'')_0 = -2 \frac{\rho_p}{\beta} \frac{K_a^2}{k_a} \tag{81}$$

Comparison of Eqs. (77) and (80) suggests that an estimate might be made of the rate-controlling step on the surface from the data for only the zeroth and the first absolute moments.

Although no applications to reaction systems of the chromatographic technique have apparently been published, the potential results appear to justify research in this direction. Appropriate experimental measurements coupled with an analysis such as described in this section could lead to important conclusions about the relative values of adsorption and surface rate constants for catalytic reactions.

VI. OTHER REACTOR FORMS

The foregoing sections have dealt with the effluent peaks from fixed beds of particles. It is possible to apply the chromatographic method of evaluating rate parameters in other reactor systems such as slurry and single-pellet reactors. This last section briefly treats methods and applications for these two systems. Impulse-response techniques have been used also for analyzing mass transport rates in fluidized beds [44] and for establishing residence-time distribution functions for a variety of types of equipment [45].

A. Slurry Reactor Chromatography

Consider a slurry adsorber containing a liquid phase and a dispersed phase of solid adsorbent particles; no gas phase is present, but there is a continuous flow of liquid in and out of the vessel, as depicted in Fig. 16. If a pulse of adsorbate is introduced into the feed stream, dispersion in the effluent will be due to mixing in the fluid phase in the vessel, to two mass transfer processes, transport of adsorbate from bulk fluid to particle surface and intraparticle diffusion, and to adsorption at an interior site in the particle. Conservation equations can be written for adsorbate in the liquid phase and within the particles. Solution of these equations for first-order adsorption in the Laplace domain can be combined with Eq. (15) to relate the several rate parameters to the moments of the effluent pulse.

Complete results for the first absolute and second central moments for such slurry adsorbers have been published [46]. To understand the type of equations involved, let us consider the special case where the resistance to intraparticle diffusion is negligible and the adsorption is irreversible. The moment equations then are a function of the resistances for but two steps, that for transport from bulk liquid to the surface of the particle and that for adsorption at a interior site. In slurry systems, this last mass transport resistance is often a principal determinant of the overall rate of adsorption. Under these restrictions, the chromatographic technique provides a method of measuring the transport coefficient k_f between fluid and particle surface. It is convenient experimentally to measure the moments with and without particles in the vessel. Then the difference

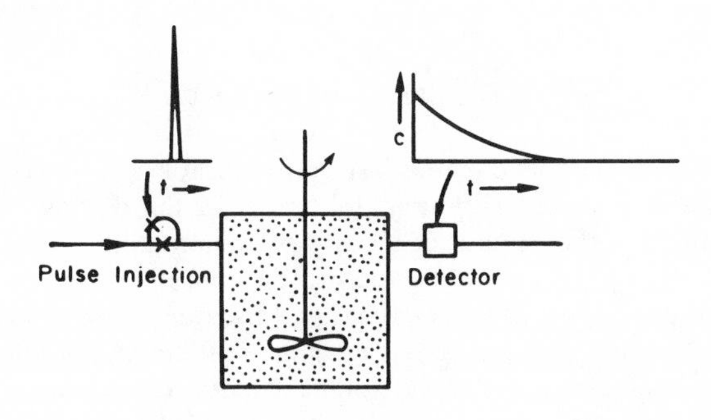

FIG. 16. Slurry chromatographic apparatus.

between the reciprocals of the two first moments, for example, is a function of the two rate constants, k_f and k_a, the latter applicable for the irreversible adsorption step. The resulting equation is

$$\left[\frac{1}{\mu_1} - \frac{1}{\tau'}\right]^{-1} = \frac{R\rho_p}{3m_s k_f} + \frac{1}{m_s k_a} \tag{82}$$

where m_s is the particle concentration (g/(cm^3) of particle-free liquid) and τ' is the average residence time V/Q of theliquid in the vessel (without particles). Furusawa [46] used Eq. (82) with data for the adsorption of benzaldehyde in an aqueous slurry of activated carbon particles to evaluate k_f. In principle, the adsorption rate constant k_a could also be determined; in the benzaldehyde system, however, the rate of this physical adsorption process was too rapid to make the second term on the right side of Eq. (82) significant.

It is interesting to note that for irreversible adsorption the first, as well as the second, moment is a function of the rate constants. For reversible adsorption, the first moment is independent of the rate constants, as shown, for example, for fixed beds by Eq. (16). In bed chromatography, this situation has been used advantageously to evaluate equilibrium constants (e.g., K_a) for a variety of systems. Kobayashi, Deans, and colleagues [47] have written a comprehensive review on using chromatographic techniques for evaluating equilibrium parameters.

While the treatment here has been restricted to adsorption in slurries, there is no inherent obstacle to extending equations such as (82) to apply to first-order reactions, just as illustrated in Section V for fixed beds. Similarly, the development could be enlarged to include the mass transport resistances associated with the gas phase where the adsorbate was introduced in a gas stream bubbled into the slurry. Here, the impulse would be supplied by introducing a pulse of adsorbate into the gas stream entering the vessel.

B. Single-Pellet Systems

When the purpose of chromatographic studies is to evaluate intraparticle rate parameters, fixed-bed experiments have the disadvantage that the moments include the effects of axial dispersion and fluid-to-particle mass transport. These two contributions to the moments can be eliminated by passing the pulse of tracer over one end face of a single cylindrical catalyst pellet and examining the response pulse on the other end face. If the thickness of the pellet is sufficient for the mass transport resistances in the fluid phase adjacent to the end faces to be negligible, the moment expressions are functions only of intraparticle rate parameters.

C. Closed End-Chamber Apparatus

Figure 17 is a schematic diagram of one type of single-pellet apparatus, in which the response pulse is measured in a small, closed chamber adjoining one end face of the cylindrical pellet. This scheme has been used [48] to measure diffusivities and hydrogen adsorption equilibrium constants for nickel catalysts. The first moment is a function of both the equilibrium constant and the diffusivity, and the second moment includes the adsorption rate constant. The contribution of the adsorption rate to the second moment seems likely to be small compared with the intrapellet diffusion contribution, so it is difficult to obtain accurate values of adsorption rate constants. However, the single-pellet experiment is particlularly well suited to determination of diffusivities since only first moments are needed. If the experiment is done with a nonadsorbing tracer, the first moment is a function only of D_e and the geometry of the pellet and the closed-end chamber.

The theory can be readily developed by reference to Fig. 17. Conservation of tracer with the pellet leads to the following differential equation:

$$D_e \frac{\partial^2 C_i}{\partial x^2} = \beta \frac{\partial C_i}{\partial t} + N_i \tag{83}$$

If an impulse function of tracer concentration is applied at the upper face (x = 0),

$$C = M \ \delta(t) \tag{84}$$

where M is the strength of the pulse. Assuming complete mixing in the detector volume of the lower, closed chamber, the boundary condition at x = L is

$$D_e \left(\frac{\partial C_i}{\partial x}\right)_{x=L} = -\frac{V_d}{S}\left(\frac{\partial C_L}{\partial t}\right) \tag{85}$$

where C_L is the concentration in the closed chamber, S is the cross-sectional area of the end face of the particle, and V_d is the volume of the closed chamber. For an adsorbing tracer, if the rate is linear, Eq. (4) gives N_i. The initial condition is C = 0 for x 0. Following the procedure described in Section II, the resulting expressions for the moments are

$$\mu_1 = \frac{L^2}{D_e}\left[\frac{\beta + \rho_p K_a}{2} + \frac{\gamma}{L}\right] \tag{86}$$

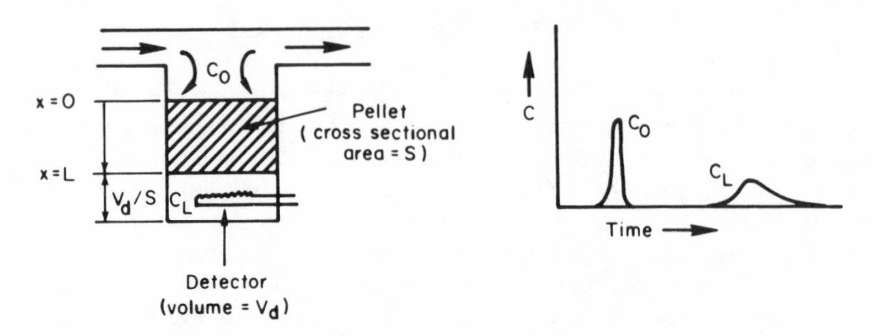

FIG. 17. Single-pellet chromatography.

$$\mu_2' = \left(\frac{L^2}{D_e}\right)^2 \left[\frac{2}{3}\left(\frac{\beta + \rho_p K_a}{2} + \frac{\gamma}{L}\right)^2 + \frac{1}{3}\left(\frac{\gamma}{L}\right)^2\right] + \frac{L^2}{D_e}\left(\frac{\rho_p K_a^2}{k_a}\right) \tag{87}$$

where γ is the length, V_d/S, of the closed chamber.

For a nonadsorbing tracer ($K_a = 0$), Eq. (86) shows that D_e can be calculated from the measured first moment, the porosity β of the pellet and the dimensions L and γ of the system. Suzuki [48] obtained reliable diffusivities for alumina catalyst pellets in this manner.

D. Dynamic Wicke-Kallenbach Apparatus

Another form of single-pellet experiment can be made by using the Wicke-Kallenbach apparatus (Fig. 18) for catalyst diffusivities. Instead of an operation in the conventional steady-state mode, a pulse of tracer is introduced in the stream fed to the upper ($x = 0$) face of the pellet. Then the analysis is conducted on the response pulse measured in a detector placed in the exit of the stream passing over the lower ($x = L$) face. Both streams consist of carrier gas.

For this case, Eqs. (83) and (84) are applicable, but instead of Eq. (85), the boundary condition at $x = L$ is

$$-S D_e \left(\frac{\partial C_i}{\partial x}\right)_{x=L} = F_1 C_L \tag{88}$$

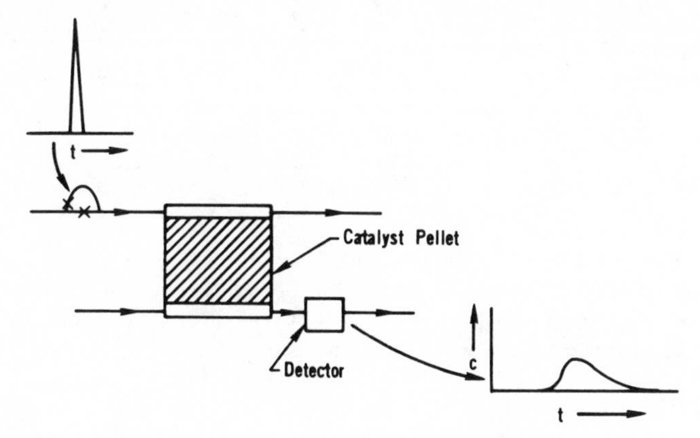

FIG. 18. Dynamic Wicke-Kallenbach experiment.

where F_1 is the flow rate of the stream passing over the face of the pellet at $x = L$.

Equations (83), (84), (88) and the initial condition can be solved for various expressions for N_i, the rate of disappearance of tracer per unit volume of pellet, including those for reaction. Dogu [49], among others, has considered the following possibilities:

1. Nonadsorbable tracer, $N_i = 0$.
2. Irreversible adsorption, $N_i = \rho_p k_a C_i$.
3. Reversible adsorption, N_i given by Eq. (4).
4. Irreversible reaction with equilibrium adsorption.

$$N_i = K_A \frac{dC_i}{dt} + k_r K_A C_A$$

We give here the moment expressions for case 3. This case is the one for which Eqs. (86) and (87) are applicable when Fig. 17 represents the single-pellet experiment. For the Wicke-Kallenbach apparatus (Fig. 18), the results are

$$\mu_1 = \left[\frac{L^2\beta}{D_e}\right] \frac{\left(3(S/L)D_e + F_1\right)}{6\left((S/L)D_e + F_1\right)} \left(1 + \frac{\rho_p K_a}{\beta}\right) \tag{89}$$

$$\mu_2' = \left[\frac{L^2}{D_e}(\beta + \rho_p K_A)\right]^2 \left\{\frac{[(S/L)D_e]^2 + (2/5)(S/L)D_e F_1 + (1/15)F_1^2}{6[(S/L)D_e + F_1]^2}\right\}$$

$$+ \frac{L^2 \rho_p K_A^2}{3D_e k_a} \frac{\left\{3[(S/L)D_e]^2 + 4(S/L)D_e F_1 + F_1^2\right\}}{\left[(S/L)D_e + F_1\right]^2} \tag{90}$$

For measurements with a nonadsorbable system, Eq. (89) provides an attractive method of measuring diffusivities of catalyst pellets. For example, at very high flow rates, $F_1 >> (S/L)D_e$, so that the equation reduces to the asymptotic form

$$\mu_1 = \frac{L^2\beta}{6D_e} \tag{91}$$

This means that runs can be made for one pellet at high enough flow rates that the first moment is constant. Then D_e can be calculated from μ_1 and the porosity β. For example, measurements could be carried out with pulses of helium, using nitrogen as a carrier gas to determine the diffusivity of helium in a helium-nitrogen system. This diffusivity could be converted to a value for any other gas in helium, using the molecular-weight ratio. Such results would be applicable for either Knudsen or bulk molecular diffusion as long as surface migration on the pore walls of the pellet is not significant.

For adsorbing gases, the first-moment equation involves both K_a and D_e. Dogu [49] has measured the diffusivity for adsorbing gases in this way by first obtaining K_a independently. The second-moment data can be used to establish the rate constant for adsorption k_a, but again the relatively small magnitude of the second term in Eq. (90) renders accurate evaluation difficult.

In summary, single-pellet chromatography offers a rapid, accurate method of measuring catalyst pellet diffusivities. Fewer data are required than for the fixed-bed method described in Section II. Of the two single-pellet procedures shown in Figs. 17 and 18, the dynamic Wicke-Kallenbach method is perhaps preferred, since more reliable peaks (more stable base line) are obtained. Single-pellet chromatography could be a more attractive method than bed chromatography for studying reactions, that is, for separate evaluation of adsorption and surface reaction rate constants. However, operating conditions would have to be chosen to enlarge the magnitude of the rate constant term compared with the diffusion term in the second-moment equation (e.g., Eq. (90)).

GLOSSARY

Bi	Biot number, $k_f R/D_e$.
C	Gas concentration; C_i = intraparticle concentration; $\bar{C}(p)$ = Laplace transform of C(t).
D_e	Effective intraparticle diffusivity; D_k = Knudsen diffusivity.
$D_{s,p}$	Effective surface diffusivity.
D_v	Molecular diffusivity.
D_s	Surface diffusivity.
d_p	Diameter of spherical particle.
E_z, E_A	Axial dispersion coefficient based on interstitial spaces and crossectional area of the bed, respectively.
F	Volumetric flow rate (Eq. (88).
G(p)	Defined by Eq. (12).
H	"Plate height," defined by Eq. (22).
K_a	Adsorption equilibrium constant.
k_a	Adsorption rate constant.
k_f	Mass transfer coefficient from bulk gas to particle surface.
k_r	Rate constant for surface reaction.
k_{eff}	Effective rate constant (including intraparticle diffusion effects).
L	Pellet thickness.
ℓ	Scale of dispersion, Eq. (28).
M	Molecular weight.
m	Moment contribution, defined by Eq. (5).
m_s	Mass concentration of particles in a slurry.
N_i	Rate of disappearance of tracer at an intraparticle radius r, per unit volume of particle.
N_o	Mass flux to particle surface, per unit volume of particle.
n_i	Concentration of adsorbed tracer, per unit mass of particle.
Pe	Peclet number, defined by Eq. (26).
p	Laplace variable.
Q	Volumetric flow rate.

q	Tortuosity factor.
R	Radius of particle.
R_g	Gas constant.
r	Radial coordinate in particle.
$\overline{r}$	Mean pore radius.
r_c	Rate of adsorption per unit mass of particle.
S	End-surface area of single pellet (Fig. 17).
T	Absolute temperature.
t	Time.
u	Interstitial velocity in bed; u_0 = superficial velocity.
V	Volume.
x	Coordinate of diffusion path through single pellet (Eq. (88)).
z	Bed length.

Greek

α	Bed void fraction.
β	Particle void fraction.
δ_1	Limiting value of $\mu_2'/(2\,z/u)$ at $1/u = 0$.
η	Diffusibility, defined by Eq. (24).
ϕ	Thiele-type modulus, defined by Eq. (48).
Φ	Adsorption rate per unit mass of particle.
ρ_p	Particle density.
τ	Pulse injection time.
τ'	Average residence time, V/Q.
μ_1	First absolute moment.
μ_2'	Second central moment.
μ^0	Zeroth reduced moment.

Subscripts

e	Effluent.
i	Interface between gas and particle; intraparticle property; inlet to bed.

n Identifies moment number (Eq. 5).

t Total.

REFERENCES

1. J. C. Giddings, "Dynamics of Chromatography," part 1, Chap. 4, Marcel Dekker, New York, 1965.
2. E. Glueckauf, Disc. Faraday Soc. 7, 12 (1949).
3. J. J. van Deemter, F. J. Zuiderweg, and A. Klinkenberg, Chem. Eng. Sci. 5, 271 (1956).
4. A. Klinkenberg, and F. Sjenitzer, Chem. Eng. Sci. 5, 258 (1961).
5. W. Z. Nusselt, Ver. Deut. Ing. 55, 2021 (1911).
6. A. Anzelius, Z. Angeu. Math. u. Mech. 6, 291 (1926).
7. H. C. Thomas, J. Chem. Phys. 19, 1213 (1951).
8. J. B. Rosen, J. Chem. Phys. 20, 387 (1952).
9. M. Kubin, Collection Czechoslov. Chem. Commun. 30, 1104 (1965).
10. Ibid., 30 2900 (1965).
11. E. Kucera, J. Chromatogr. 19, 237 (1965).
12. S. Masamune and J. M. Smith, A. I. Ch. E. J. 11, 34 (1965); Ind. and Eng. Chem. Fundam. 3, 179 (1964).
13. P. Schneider and J. M. Smith, A. I. Ch. E. J. 14, 762 (1968).
14. M. Suzuki and J. M. Smith, Chem. Eng. J. 3, 256 (1972).
15. H. C. Burger, cited from D. A. DeVries. Trans. IVth Congr. Intern. Assoc. Soil Science, vol. 2, 41 (1950).
16. M. Suzuki and J. M. Smith, J. of Catal. 21, 336 (1971).
17. M. F. Edwards and J. F. Richardson, Chem. Eng. Sci. 23, 109 (1968).
18. N. Hashimoto and J. M. Smith, Ind. Eng. Chem., Fundam. 12, 353 (1973).
19. M. Suzuki and D. Kunii, "On a Fluid Flow in Packed Beds of Fine Particles," presented at 34th Ann. Mtg., Japan Soc. Chem. Engrs., April 1969.
20. P. Schneider and J. M. Smith, A. I. Ch. E. J. 14, 886 (1968).

21. R. L. Cerro and J. M. Smith, A. I. Ch. E. J. 16, 1034 (1970).

22. N. Hashimoto and J. M. Smith, Ind. Eng. Chem., Fundam. 12, 353 (1973).

23. H. W. Haynes, Jr., and P. N. Sarma, A. I. Ch. E. Journal 19, 1044 (1973).

24. N. Hashimoto and J. M. Smith, Diffusion in bidisperse porous catalyst pellets, Ind. Eng. Chem., Fundam., to be published February, 1974.

25. Y. Shigehara and A. Ozaki, J. of Catal. 15, 224 (1969).

26. B. R. Davis and D. S. Scott, Measurements of the effective diffusivity of porous pellets, Paper 5A, IV International Congress on Catalysis, Symposium III, Novosibirsk, 1968.

27. Y. H. Ma and C. Mancel, A. I. Ch. E. J. 18, 1148 (1972).

28. T. A. Denisova and A. L. Rozental, Kinetika i Kataliz 8, 441 (1967).

29. J. R. Hays, W. C. Clements, Jr., and T. R. Harris, Prepr. Paper, Frequency domain evaluation of dynamic models, A. I. Ch. E. Nat. Mtg., Philadelphia, December 1965.

30. W. C. Clements, Jr., Chem. Eng. Sci. 24, 957 (1969).

31. S. K. Gangwal, R. R. Hudgins, A. W. Bryson, and P. L. Silveston, Can. J. Chem. Eng. 49, 113 (1971). Also see S. K. Gangwal, Ph.D. dissertation, University of Waterloo, 1974.

32. R. L. Curl and M. L. McMillan, A. I. Ch. E. J. 12, 819 (1966).

33. C. P. Jefferson, An approximation method for Fourier transform applied to distributed parameter systems, Paper #2507, IFAC Symposium Sydney, Australia, August 1968.

34. C. P. Jefferson, Chem. Eng. Sci. 25, 1319 (1970).

35. G. A. Turner, A. I. Ch. E. J. 13, 678 (1967).

36. G. Padberg and J. M. Smith, J. Catal. 12, 172 (1968).

37. J. C. Adrian and J. M. Smith, J. Catal. 18, 57 (1970).

38. M. Suzuki and J. M. Smith, J. Catal. 23, 321 (1971).

39. E. Wolf and J. M. Smith, Ind. Eng. Chem., Fundam. 11, 413 (1972).

40. H. S. Taylor and C. O. Strogher, J. A. C. S., 56, 586 (1934).

41. A. Ozaki, F. Nozaki, K. Maruya, and S. Ogasawara, J. Catal. 7, 234 (1964).

42. M. Suzuki and J. M. Smith, Chem. Eng. Sci. 26, 221 (1971).

43. M. Kocirik, J. Chromatogr. 30, 459 (1967).

44. S. Morooka and T. Miyauchi, Kagaku Kogaku 33, 569 (1969).

45. Octave Levenspiel, "Chemical Reaction Engineering," 2d ed., pp. 253-315, Wiley, New York (1972).

46. T. Furusawa and J. M. Smith, Ind. Eng. Chem., Fundam. 12, 360 (1973).

47. K. T. Koonce, H. A. Deans, and R. Kobayashi, A. I. Ch. E. J. 11, 259 (1965); R. Kobayashi, P. S. Chappelear, and H. A. Deans, Ind. Eng. Chem., Applied Thermodynamics 227 (1970).

48. M. Suzuki and J. M. Smith, A. I. Ch. E. J. 18, 326 (1972).

49. Gulsen Dogu, Ph. D. dissertation, Single-pellet chromatography, Univ. of Calif., Davis, June 1974.

Chapter 6

QUALITATIVE ANALYSIS BY GAS CHROMATOGRAPHY

David A. Leathard

Sheffield Polytechnic,
Sheffield, England

I. INTRODUCTION

This review has a twofold aim: to provide new investigators in this broad field with a brief conceptual guide to identifying "unknown" gas chromatographic peaks, and to bring experienced workers up to date on recent applications and current trends. Critical reviews of earlier work can be found in Refs. 1-5, and a recent book is devoted to practical aspects of identification, using the minimum of apparatus apart from a gas chromatograph [6].

The variety of approaches to the identification of GC peaks is summarized in Fig. 1, which is used as a convenient framework for the arrangement of material for this chapter. It is important to appreciate, however, that the boundaries between each section are not always distinct, and that the optimum solution to a particular identification problem is likely to involve several methods, which the analyst selects on the basis of experience and available resources. Thus in some specialized environments, frequent use of directly coupled gas chromatography-mass spectrometry (GC-MS) may be justified by the urgency of the identification problem. More commonly, a GC-MS service will be used only occasionally, as the last stage in a particularly important identification.

II. RETENTION BEHAVIOR

The most common methods of peak identification rely on characterizing a substance by its retention behavior on one or more stationary phases, although the limitations of this approach are now widely recognized. For reviews and assessments of this area, see Refs. 5-9.

Coincidence of the retention time of an unknown substance with the retention time of a standard substance is particularly useful for confirming an identity that other considerations strongly suggest, but this coincidence is not suitable for analyzing a completely uncharacterized sample. Statistical arguments [10] show that if retention times can be measured with a precision of 1%, then an unknown substance and a standard having indistinguishable retention times t_R can be said, with 99.9% confidence, to be identical only if no more than 4 substances could possibly elute between injection and time t_R. If a confidence level of 95% is acceptable, then the possible number of conceivable substances eluting with retention times between 0 and t_R is allowed to rise to 36. In practice, this means that reliable identification from retention is impossible with a random distribution of retention volumes. This stresses the importance of gathering as much additional information about the sample as possible. As an extreme example, the chance of mistakenly identifying from retention behavior a substance known to be an n-alkane would be negligible.

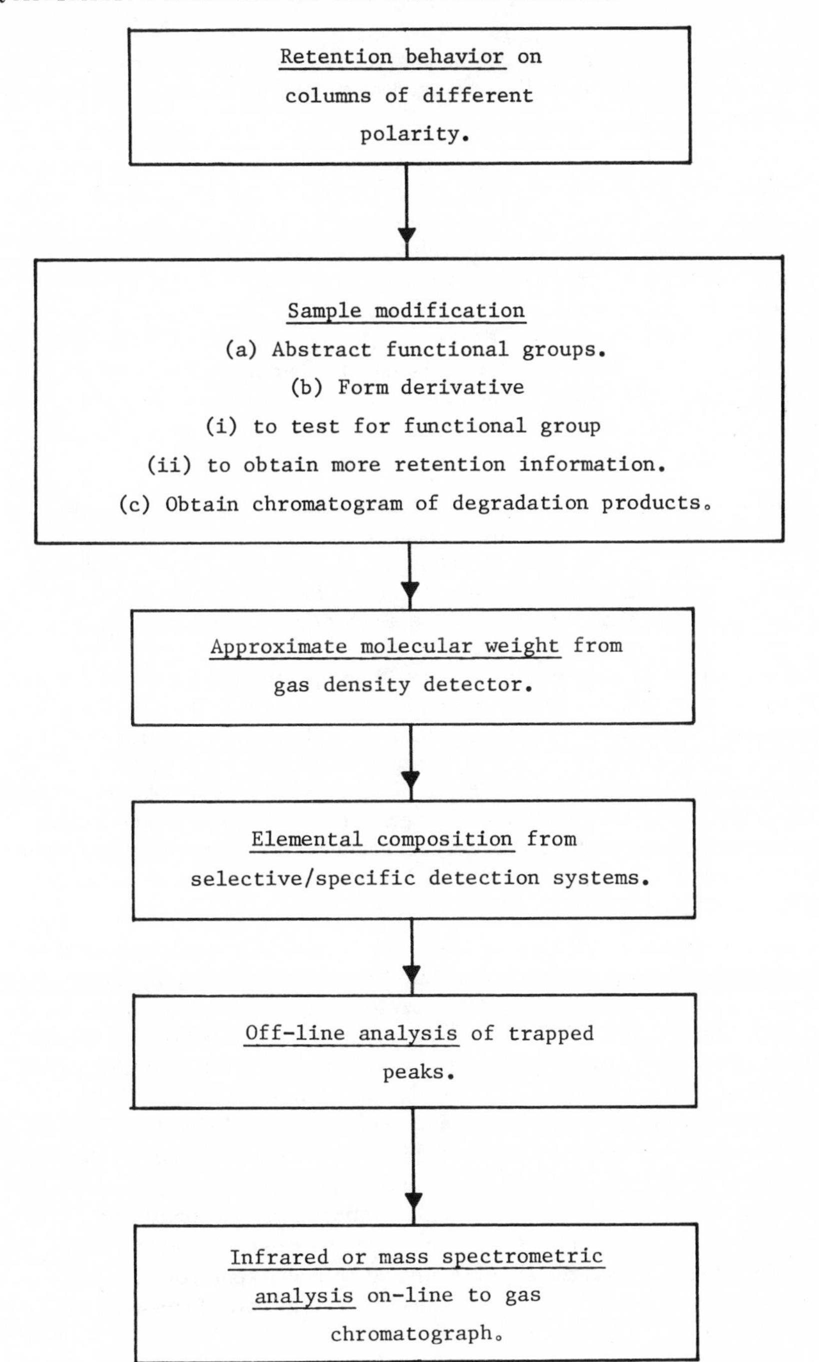

FIG. 1. Idealized scheme for identification by gas chromatography.

Such additional information can be provided by the various techniques indicated in Fig. 1, as well as by using two or more columns of significantly different "polarity."

One situation in which the identity of a peak is limited to only a few possibilities occurs in computerized interpretation of gas chromatograms of routine samples having approximately known composition. Over an extended period, factors such as column bleed, aging of the stationary phase, and fluctuations in inlet and outlet pressure cause the observed retention time of a particular component to vary. This leads to the possibility that a computer algorithm will confuse the identity of one peak with the identity of another in an adjacent timeband. Methods have been described [11, 12] by which such errors can be prevented, e.g., by injecting a standard mixture of n-paraffins from time to time, and programming the computer to identify the peaks by retention index.

Great interest still centers on correlating retention behavior with the molecular structure and physical properties of the solute and stationary phase [13-33]. In this context, predicting retention volumes, retention indices, equivalent chain lengths, etc. from the molecular structure of of the solute remains popular. It appears to the author that some workers in this field are in danger of attributing too much importance to retention data measured on a particular column in one particular laboratory, without any positive evidence that effects such as adsorption at the gas-liquid and liquid-solid interface can be neglected. It will always be better for identification purposes to directly compare retention of unknown solutes and standard solutes on two or more columns of different polarity, rather than rely on published retention indices and inexact correlation "rules" applied to a single column in a different laboratory. Nevertheless, such correlations can provide useful guidelines about the retention behavior to be expected on a particular stationary phase.

Recent studies of retention-structure correlation include applications to branched-chain paraffins [13], other hydrocarbons [14], linear chloro-, bromo-, and iodoalkanes [15], unsaturated fatty esters [16], alcohols, ketones, esters, amines, chlorosilanes, nitrogen heterocyclics, steroids, and halogenated hydrocarbons [17]. A rectilinear relation between the logarithm of the retention volume and of the molar volume has been demonstrated for a range of alkyl and aryl silanes and siloxanes and halogenated aromatic hydrocarbons [18]. A nice example of the utility of the linear relation between log retention and number of carbon atoms in a homologous series was reported in connection with analyzing a natural product extract [19]. The main components were separated by preparative GC of acetonide derivatives and identified by mass spectrometry as octadecan-1,2,3,4-tetrol and eicosan-1,2,3,4-tetrol. Four very minor constituents of the extract were then identified as the C_{16}, C_{17}, C_{21}, and

C_{22} homologs from a plot of the log of the retention times against the number of carbon atoms, when all the points were found to fall on a straight line.

In another recent study, C_{10} compounds containing oxygenated functional groups in the 1, 2, and 3 positions were distinguished by differences in retention index on polar and nonpolar stationary phases [20]. An analysis scheme for many of the individual components of a sample of naphtha involved preparative GC on a highly polar column (tetracyanoethylated pentaerythritol), followed by analysis of the collected fractions on an analytical squalane column [21]. Identification was helped by the retention index differences on the two stationary phases, by the temperature dependence of the retention indices, and by the retention indices themselves.

The characteristics of the various types of silicone stationary phase have usefully been reviewed [22-24].

The continued proliferation of new stationary phases, and problems of interlaboratory irreproducibility of retention data, have led to a renewed focusing of attention on the vexing problem of how a system of "preferred" stationary phases can be established and effectively promoted [25-28]. Rohrschneider made a significant early contribution to the characterization of stationary phases [34] by suggesting that the difference in retention index for a substance on a polar and nonpolar column could be represented as

$$\Delta I = ax + by + cz + du + es$$

The symbols a, b, c, d, e characterize the substance being chromatographed, and x, y, z, u, s characterize the stationary phase, as determined by

$$\Delta I_{\text{benzene}} = 100x, \quad \Delta I_{\text{ethanol}} = 100y, \quad \Delta I_{\text{2-butanone}} = 100z,$$

$$\Delta I_{\text{nitromethane}} = 100u, \quad \text{and} \; \Delta I_{\text{pyridine}} = 100s.$$

It has been pointed out [16] that although other workers have determined x, y, z, u, and s for columns (e.g., [35]), the use of a, b, c, d, and e for solutes appears to have been restricted to the original paper. This is probably because of the tedious least-squares procedure required to evaluate these constants. (A correction to this calculation method has been proposed [36]). Rohrschneider recently commented [37]: "Experience to date in the study of retention behavior indicates that simple models may be comprehended easily and used readily; however they provide no precise predictions of GC retention data. On the other hand, precise predictions require a system of equations which is not convenient to use and is too involved for laboratory purposes."

The direct use of ΔI values rather than x, y, z, etc. appears to be more attractive to gas chromatographers, and this approach was adopted by McReynolds, who published ΔI values for 120° (relative to squalane) for 10 selected compounds on over 200 liquid phases [38]. The ten components (benzene, butanol, 2-pentanone, nitropropane, pyridine, 2-methyl-2-pentanol, 1-iodobutane, 2-octyne, 1,4-dioxane, and cis-hydrindane) represent modifications and extensions to Rohrschneider's 5 substances, and were selected as those that could best be used to calculate the retention index of a set of 68 compounds on 25 liquid phases.

McReynolds's data [38] show clearly that many of the stationary phases studied have very similar characteristics. At the same time, however, several samples of stationary phase of nominally the same composition give significantly different values for ΔI. McReynolds's data are widely quoted and analyzed (e.g., [26, 39]) but it would be interesting to compare extensive data of this type obtained in several laboratories at different temperatures.

III. SAMPLE MODIFICATION

Modifying the sample before, during, or after separation can provide valuable information about the identity of individual components. Three alternative approaches may be adopted (cf. Fig. 1): one or more components may be completely or partially removed, converted by means of chemical reagents, or degraded by pyrolysis or other degradative techniques.

A. Abstraction

There is growing recognition of the usefulness of reactors that specifically remove one or more classes of compound. Such abstractors usually involve a physically or chemically active reagent, distributed on a convenient solid support material, which is placed "on stream" either before or after the separating column. Occasionally, however, reactions are carried out in solution prior to injection (e.g., [40]).

Abstraction has been applied to a wide variety of functional groups, including hydroxyl, carbonyl, carboxyl, olefinic, and sulfur compounds. Tables of suitable abstraction reagents are presented in earlier reviews [41, 42]. It should be stressed that abstraction of particular compounds is not always complete, that other functional groups are sometimes partially affected, and that reactor performance is often very dependent on the temperature of operation. The reader is strongly advised to refer to original references before attempting to implement an abstraction system.

It was reported some years ago that a commercial packing known as FFAP, a reaction product of polyethylene glycol and 2-nitroterephthalic acid, will selectively absorb aldehydes from mixtures of aldehydes, ketones, alcohols, and hydrocarbons [43]. Recent reports claim, however, that FFAP is extremely unreliable as an aldehyde abstractor, even when long precolumns are used [44, 45].

Other abstractors for carbonyl groups include o-dianisidine and benzidine [46], and these have been the subject of recent systematic studies [44, 47]. The results of these investigations conflict in many details with earlier findings and it appears that further work is needed before aromatic amine abstractors can be used with confidence. An abstractor for aldehydes and ketones with a bisulfite addition reagent has been evaluated and shown to be effective for compounds having boiling points up to 150° [47]. The abstraction behavior of several metal hydrides has been studied [48, 49] and it has been reported that sodium borohydride selectively reacts with carbonyl compounds, but not completely [49].

An example of the power of the abstraction approach to identification is shown in Fig. 2, which illustrates using 3-nitrophthalic anhydride and semicarbazide to remove alcohols and carbonyl compounds, respectively, from a mixture of saturated and unsaturated alcohols, ketones, and esters under temperature-programmed conditions [44]. Other abstractor systems that have recently been studied include boric acid for alcohols [44, 48], phosphoric acid for nitrogen bases [50], and PVC for aromatic substances [51].

B. Chemical Conversion

As indicated in Fig. 1, chemical modification of GC peaks to produce derivatives can assist in identification in two distinct ways. First, the fact that a GC peak undergoes a particular chemical reaction provides information about the functionality of the peak. Secondly, the retention volume of the product of reaction may help to confirm the identity of the reactant peak. Only the first type of information is obtained if the functional group analysis is carried out on separate peaks emerging from the column. To obtain both types of information, one must carry out the functional group reaction on the whole sample before it enters the column, and then compare the resulting chromatogram with the chromatogram obtained without precolumn reaction.

Apart from derivative formation in the normal sense, several other reactions have been used, including microozonolysis, to determine double bond positions, and hydrogenation, to allow materials to be classified as saturated or unsaturated. In this field, chemical insight and ingenuity are at a premium, and many novel techniques have been used. In this chapter,

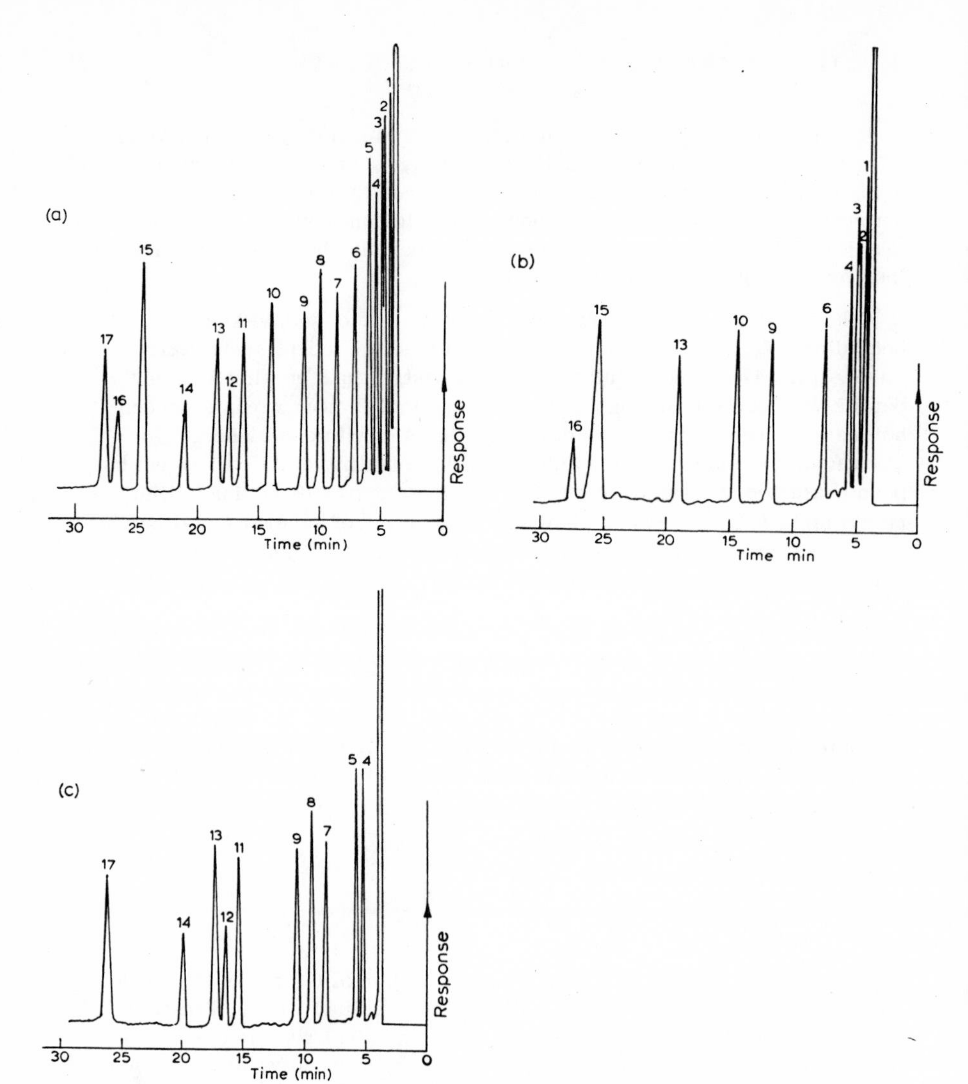

FIG. 2. Temperature-programmed analysis of a miscellaneous mixture with (a) no precolumn, (b) a 25-mm precolumn of 3-nitrophthalic anhydride (40% w/w) operated at an average temperature of 190°, and (c) a 25-mm precolumn of semicarbazide (40% w/w) operated at an average temperature of 115°. Column: 37 m x 0.5 mm I.D. PLOT column, coated with a 0.5% w/v solution of Carbowax 20M. Flow rate: 4 ml/min nitrogen, programmed from 70° at 2°/min from injection. Peak identity: 5 μg each in ether of (1) isobutanal, (2) 2-methylbutanal, (3) methyl-ethyl ketone, (4) n-propyl acetate, (5) n-propanol, (6) n-hexanal, (7) 3-buten-1-ol, (8) 2-methyl-butanol, (9) n-butyl butyrate, (10) 2-octanone, (11) n-hexanal, (12) cis-3-hexen-1-ol, (13) tridecane, (14) 1-octen-3-ol, (15) benzaldehyde, (16) 2-nonenal, and (17) n-octanol. (Reprinted from Ref. 44. Courtesy of the Journal of Chromatography.)

it is possible only to indicate some typical applications of pre- and post-column reactions to peak identification. References 2, 6, 52, and 53 review this area, and Ref. 54 emphasizes the utility of combining post column functional group classification reactions with retention volume data for the underivatized material. A recent book [55] about functional group analysis includes pre- and post column GC reactions in many of the chapters about particular functional groups. The proceedings of a "reaction GC" symposium have been published [56].

Many conventional organic qualitative analysis methods that rely on a color change or precipitation reaction to characterize a particular class of compound can be adapted for postcolumn use. For example [54, 57], if a five-way stream splitter at the column outlet was used, it was possible to classify peaks containing any one of eleven functional groups in two runs (at most) with the aid of a dozen reagents. Minimum detectable amounts were in the range 20-100 μg. In one application of this technique, it was found that a chromatogram that appeared to contain only 5 components in fact contained at least 10 components, because several peaks gave positive reactions for more than one type of functional group. Log retention volume vs carbon-number correlation plots enable bifunctional compounds to be distinguished from peaks due to a mixture of co-eluting compounds [54].

An interesting miniature technique involves trapping the peak in a short length of glass porous-layer open tubular capillary column, and then developing color reactions in situ [58]. A modification of this technique for nitrogen-containing compounds has been described [59], in which a small hydrogenolysis reactor is installed between the column exit and the trap. The packing in the trap is coated with either ninhydrin for offline development or phenolphthalein for continuous monitoring.

Another postcolumn technique uses the fact that GC separates the sample components in space. A horizontally advancing strip, of the type used in thin-layer chromatography, is positioned under the column outlet of the chromatograph. The strip moves at the same rate as the detector recorder chart. After the run is complete, the strip is wetted with a qualitative organic analysis reagent, thereby allowing peaks belonging to a given class to be identified. If a very large peak corresponds to a very light color spot, then it is probable that a minor component has been overlapped by a major component that does not contain the relevant functional group [60]. The moving strip system is not well suited for use with more than one reagent per run, but offers the additional attractive possibility of carrying out actual thin-layer chromatography on the trapped zones from the GC column. Examples of coupled GC and TLC have been reviewed [61, 62].

If a precolumn reaction produces an involatile derivative, then the effect is to abstract completely or partially the reactant peak from the chromatogram. Sometimes it is then helpful to chromatograph the derivative separately, either after decomposition, or by working at a higher column temperature. If a volatile derivative is formed, however, then "peak shifting" occurs, with the original peak replaced by one at a different retention volume, the value of which can be used to aid in identifying the shifted peak. This technique, first suggested for trimethylsilyl derivatives of alcohols [63], should be distinguished from derivative formation to increase volatility of materials such as amino acids and carbohydrates, or to increase the sensitivity of electron capture detection, e.g., by converting steroids to heptafluorobutyrate derivatives. Many examples of selective removal and peak shifting obtained by reacting vapor samples in a syringe prior to injection will be found in Ref. 64.

Since it is necessary to know the type of material being analyzed before a suitable peak-shift derivative can be prepared, it is best to use this technique after a reasonable amount of preliminary sample characterization has been carried out. The postcolumn functional group tests mentioned earlier are useful for this purpose.

The utility of chemical conversion techniques is well illustrated by a recent report that aldehydes and ketones have been distinguished by differences in reactivities with alcohols to form acetals and ketals [65]. For example, under acid-catalyzed conditions, aldehydes reacted completely with isopropanol, whereas ketones did not react. Thus, when a mixture of aldehydes and ketones was chromatographed after reaction with isopropanol, the aldehyde peaks disappeared and the ketone peaks were unaffected. Temperature programming after elution of the last carbonyl compound revealed new peaks representing the acetals. It appears that α,β-unsaturated aldehydes such as benzaldehyde are exceptions, because only partial conversion to the acetal occurred. In the same study [65], it was found that if methanol was used instead of propanol, then not only did all aldehydes react (including benzaldehyde) but cyclic ketones also partially reacted to form ketals. This enabled cyclic ketones to be distinguished from linear ketones, which were unaffected by methanol.

C. Degradative Methods

In pyrolysis gas chromatography (PGC), a sample for analysis is pyrolyzed and the resulting reaction products are separated in a concerted operation. The rapid growth of interest in PGC in recent years is mainly due to its usefulness in identifying solid materials such as plastics, polymers, and biological materials that frequently exhibit characteristic "pyrograms." Although PGC of solid samples is outside the scope of this review, the relatively neglected application of PGC to the identification of volatile materials, particularly of GC peaks, is directly relevant. This is an extreme example of chemical modification, where very many new pieces of information are provided by the retention volumes and relative amounts of the many pyrolysis fragments.

Sample handling and pyrolysis unit requirements for the PGC of volatile substances differ significantly from the requirements for solid samples. The chief difference is the great simplification of the problems associated with heat transfer and heating rate, so that a hot tube will suffice as a pyrolyzer. Further, the undecomposed portion of the sample can itself be chromatographed, which not only is useful in indicating the extent of pyrolysis but also means that the whole sample is removed from the reactor. This means that reproducibility is expected to be significantly better than for conventional PGC of solids [66, 67].

As with any postcolumn analytical technique, a choice must be made between trapping and direct coupling. One approach [68] is to trap the eluate from one column in a U-tube cold trap and subsequently sweep it into the pyrolyzer unit at the head of a second column. Direct coupling presents difficulties for two reasons. First, the flow rates through the first column, pyrolysis unit, and second column are unlikely to be optimal, and second, because the time taken to obtain a pyrogram is likely to be longer than the time between adjacent peaks from the first column. The first difficulty can be overcome by using a transfer system, incorporating a series of multiport taps and separate carrier gas supplies for each column and for the pyrolysis unit [69], so that all flows can be optimized independently. At the same time, subsequent peaks from the first column are diverted away from the pyrolysis unit. This apparatus has been used in a study of the effects of pyrolysis conditions and sample size on the PGC pyrogram "fingerprints" of n-alkanes, α-olefins, alcohols, mercaptans, and saturated and unsaturated methyl esters [69]. The second difficulty can be overcome by using an interrupted elution technique [70, 71] (to be considered later under infrared analysis). It should be stressed, however, that a well-resolved pyrogram can be obtained only if peaks entering the pyrolysis unit from the first column are sharp. This demands high-efficiency columns, or temperature programming, or both. At the same time, sample size, pyrolysis temperature, and flow rate must be carefully controlled and all traces of oxygen must be removed [72, 73].

An instrument is now available commercially with facilities for directly coupled interrupted elution PGC of chromatographic peaks [74-76]. The use of a split flow system and suitable catalysts in this instrument provides quantitative elemental C, H, N, and O analysis of the peak and also "functional-group" information from the pyrogram. These elemental analysis data should do much to simplify identifying an unknown peak from its pyrolysis fragments. Other approaches to tackling this problem have been suggested, including restriction of the pyrogram to small molecules such as CH_4, CO_2, and H_2O [77], and classifying the pyrograms by set theory [78].

Although photolysis has obvious similarities to pyrolysis as a tool for qualitative analysis, it has received much less attention in connection with GC. Reported applications include the mercury-photosensitized photolysis of milligram quantities of alcohols, esters, aldehydes, ketones, and ethers [79, 80], and also chlorinated insecticides [81], followed by chromatographic analysis of the products.

IV. GAS DENSITY DETECTOR

There is a steady trickle of interest in the gas density detector. This device ideally gives a response proportional to the difference between the density of the effluent gas stream and a reference stream of pure carrier gas. In turn, this difference depends on the molecular weights of the eluate (M_s) and the carrier gas (M_c), and also on the total mass of eluate W_s. It can be shown, e.g., [82], that the area of a peak is given approximately by

$$A = kW_s \frac{M_s - M_c}{M_s}$$

where k is constant for a given carrier gas in a given detector cell. It follows that measurement of the peak area for a known mass of substance leads directly to M_s, provided that k is known (e.g., from injection of a known amount of reference material). Such a procedure demands an exact knowledge of the amount injected. Alternatively, a known mass of reference material can be added to a known mass of the unknown compound and the ratio of peak areas determined from a single injection. An extension of this approach to multicomponent mixtures has been described [83]; this approach uses nitrogen carrier gas, and estimates of molecular weights between 46 and 74 were accurate to between 1.6 and 3.8 units.

When dealing with a mixture of unknown materials, however, one must determine the weight percentage of each component in the sample by a separate method. The absolute mass detector has been used for this purpose in conjunction with a gas density detector [84-86], but accuracy has been often rather poor.

Other workers have used two or more carrier gases, a procedure that eliminates the need for measuring the mass of the unknown and reference samples. When hydrogen and nitrogen were used to measure molecular weights above 100, an error of 4% was quoted [87], whereas errors of less than 3% were reported for molecular weights below 200 when nitrogen and argon were used [88]. Accuracies of between 1% and 2% were found for a method using several carrier gases [89]. For example, the molecular weight of CCl_4 was obtained to within 1% with CCl_2F_2 (M.W. 121) and C_4F_8 (M.W. 200) used as carrier gases.

A commercial molecular-weight, or "mass chromatography" system is now available [90-92]; it uses the two carrier gas approach (usually with SF_6 and CO_2), which is claimed to give molecular weights in the range 2 to 400 accurate to within 1%. There have been very few published statements from users that this application of the gas-density detector has helped them to identify an unknown substance. One report [93], however, demonstrates that the technique can usefully supplement GC-MS by suggesting to the mass spectroscopist where to look for the molecular ion. Thus, from "mass chromatograms," an unknown peak was calculated to have a molecular weight of 231.2 ± 4.9, while the highest mass ion observed on a GC-MS mass spectrum was 201. This suggested the possible loss of a methoxyl group (mass 31), and further study of the mass spectrum then led to a proposed structure (methyl-9, 9-dimethoxynonanoate), which in this case could be confirmed by examining a sample of the pure substance.

The precision indicated by the molecular weight estimate of 231.2 ± 4.9 seems typical of present practice. The workers [93] found significant variations in the instrument "constant" associated with different classes of compound, and also with day-to-day runs. They therefore recommend the inclusion of internal standards of various classes of compound if the functionality of the unknown peak is not known.

V. SELECTIVE DETECTION SYSTEMS

Most of the current interest in specific or selective detection systems concerns establishing optimal design and operating conditions so that precise quantitative analysis can be achieved for concentrations below the part per million level. Such interest appears to be most intense in flame photometric and microwave plasma detectors that measure the intensity of emitted light at a particular wavelength when the column effluent is excited in a flame or electrical discharge. (Atomic absorption has been employed for detection of mercury in GC effluents [94, 95]). Work also continues on obtaining a better understanding of the factors governing the performance of electron capture and thermionic detectors (which are in fairly common use but have acquired a reputation for unreliable operation). Several recent reports have appeared about less common electrochemical detectors, including electrical conductivity, coulometric, and ion-selective detectors. There has also been some interesting work on the development of piezoelectric sorption detectors.

A brief assessment of recent work on selective detection systems follows, with emphasis on applications to qualitative analysis. Earlier reviews and discussions of selective detector systems appear in book chapters and review articles devoted to selective detectors [96-100], and also in more general reviews of GC detectors [101, 102].

A. Flame Photometric Detectors

The principle of operating flame photometric detectors is that the intensity of molecular or atomic emission from a hydrogen-air flame is measured at a particular wavelength. The most common application is to the analysis of phosphorus- and sulfur-containing pesticides, for which it is usual to monitor the emission at 526 nm, due to HPO, or 394 nm, due to S_2, by a photomultiplier tube and interference filter transmiting at one of these wavelengths. Commercial systems incorporating two filters and two photomultiplier tubes are available for simultaneous operation in phosphorus and sulfur modes. Reference 103 reviews flame photometric detectors with emphasis on applications to analysis of substances containing sulfur and phosphorus.

The response to phosphorus is linear with sample size, while the response to sulfur has been reported to be roughly proportional to the square of the concentration (when the emitting species is S_2). As a result, the ratio of the P response to the square root of the S response has been used to determine the atomic ratio of phosphorus to sulfur. The ratio $R_P:\sqrt{R_S}$ was reported [104] to have values (in arbitrary units) in the range 5.2-6.0 for PS compounds such as $C_{10}H_{14}NO_5PS$, values between 2.8 and 3.3 for PS_2 compounds such as $C_{10}H_{19}O_6PS_2$ or $C_9H_{22}O_4P_2S_4$, and values between 1.7 and 2.3 for PS_3 compounds such as $C_7H_{17}O_2PS_3$.

The detailed behavior of the S_2 emission has recently been investigated in several laboratories [105-110]. It is clear that the presence of other closely eluting substances (not containing sulfur) can quench the emission [105, 107, 110]. One study has reported that depending on the flow rate of air and hydrogen, intensity of emission was proportional to concentration raised to a power that varied between 1.69 and 2.00 [109]. In a study of the analysis of coal sulfur, a <u>linear</u> response was observed [108]. Such reports indicate that the square relation cannot be assumed, and that caution should therefore be exercised in using the detector response to deduce the number of sulfur atoms in a molecule.

An interesting example of a useful "non-identification" is provided by the report that the odor of diesel exhaust fumes is <u>not</u> due to sulfur-containing compounds; this was concluded because the flame photometric detector gave no response [111].

There have been several recent reports about applying flame photometric detectors to the trace analysis of volatile metal compounds, including those of chromium [112], iron [113], lead [113, 114], and tin [113, 115]. Such studies usually use a monochromator rather than a filter, in order to take full advantage of the narrow atomic emission lines.

A fluorine-specific detector has been described, in which the column effluent is mixed with a stream of argon containing calcium vapor, before being fed to an oxyacetylene microburner. The intensity of emission from

CaF at 529.9 nm is monitored with a monochromator [116]. Alternatively, an interference filter can be used, which gives an improved detection limit [117].

An interesting variation of the conventional flame photometric detection system involves introducing solid sodium sulfate [118, 119] or indium metal [120-122] to achieve a highly selective, sensitive response to pesticides containing Cl, Br, or I (not F). Radiation at 589 nm (sodium D line) is monitored for the sodium sulfate system, and at 360 nm for indium-sensitized detectors. Such detectors are reported to be less sensitive than thermionic or electron-capture detectors for these applications; however, they offer better specificity and more nearly rectilinear response. A dual flame photometric detector has been reported [122], in which P and S are monitored conventionally in the lower flame while reaction with indium occurs in the upper flame, allowing simultaneous monitoring of halogens.

The introduction of indium into the flame recalls early reports of use of the Beilstein effect to detect halogens specifically [123-125]. Recently the utility of a simple, inexpensive "Beilstein detector" has been demonstrated in analyzing compounds of neutral oil from red alga [126]. The aim of the investigation was to locate and analyze halogen-containing organic compounds in natural products. A GC-MS system was available for the investigation, but the hundreds of fractions obtained precluded routine analysis of all peaks by the mass spectrometer. The Beilstein copper-sensitized flame detector provided a routine screening device: GC peaks corresponding to a green flame were marked on the chart. Subsequent GC-MS analysis showed that polychlorinated biphenyls produced these peaks.

B. Plasma Emission Detectors

The principles of plasma emission detectors are very similar to the principles of flame photometric detectors. Both types of device monitor the intensity of light of a particular wavelength that an excited species emits. Much more energy is available in a microwave-stimulated plasma than in a flame, however, so that the possibility of observing characteristic intense, narrow atomic emission lines is greater than in flame photometry. Reports of studies with microwave plasma detectors [127, 128] appeared in 1965, just before many of the early flame photometric studies. That microwave plasma detectors have not achieved the same modest popularity as the flame photometric detectors is mainly because no commercial detector has been available until recently; this in turn is probably because a large number of parameters have to be optimized before satisfactory routine operation of a microwave detector is possible. In particlular, there is a need for a high-quality monochromator because of the high spectral background; a narrow bandpass filter is adequate for most flame photometric work.

A fairly extensive literature about instrumentation and applications of microwave detectors exists [129-143] although large discrepancies in reported detection limits (in the ng to pg range) are common. Two types of detector have been used: argon carrier gas can sustain a discharge at atmospheric pressure [127-129, 138], and helium must be operated at a reduced pressure of a few mm Hg [127, 132, 133, 136, 141]. It has been reported [130] that even for the argon discharge operation at 200 mmHg improves sensitivity and selectivity by an order of magnitude compared with operating at atmospheric pressure.

A reflected power meter has been used in conjunction with an atmospheric pressure argon discharge detector to give a nonselective response [139], thus avoiding the need for a dual detector system. The attractions of a microwave plasma detector for determining empirical formulas are apparent from several publications [137, 140]. Other applications include the determination of organomercury compounds in fish and other biological material [136, 141]. In one inorganic application, detection limits for specific detection of acetylacetonate complexes were found to be 100 ng for Al, 1 ng for Cr, and 0.01 ng for Be, and this enabled beryllium in aluminum to be determined down to 10 ppm [142].

Persistent deposition of carbon within the plasma tube was reported as a problem with the low-pressure helium plasma detector [137]. The problem was solved by introducing a permanent bleed of air into the plasma to act as a carbon scavenger. This was reported to give an improved stability to the plasma and improved element selectivity (typically 1000:1 relative to a hydrocarbon matrix). The utility of a dual detector system using a flame ionization detector in conjunction with this microwave detector to unambiguously identify elements present in GC peaks is shown in Figure 3 for a synthetic mixture of 14 components. An additional carrier gas supply was added at the column exit to maintain the pressure above atomospheric. The combined flow was then split equally into two streams across suitable flow restrictors to the two detectors [137]. Figure 3a shows that only 2 of the 14 components contain chlorine, and that one of these contains twice as many chlorine atoms per carbon atom as the other. Figure 3b shows that the two components containing bromine had similar bromine-to-carbon ratios. Other reported applications of this dual detector system include the determination of the hydrogen:carbon ratios for paraffins, olefins, naphthenes, and aromatics to within 2% of theoretical values [137]. A similar system is now available commercially in either a single-channel or a direct-reading multichannel form [143].

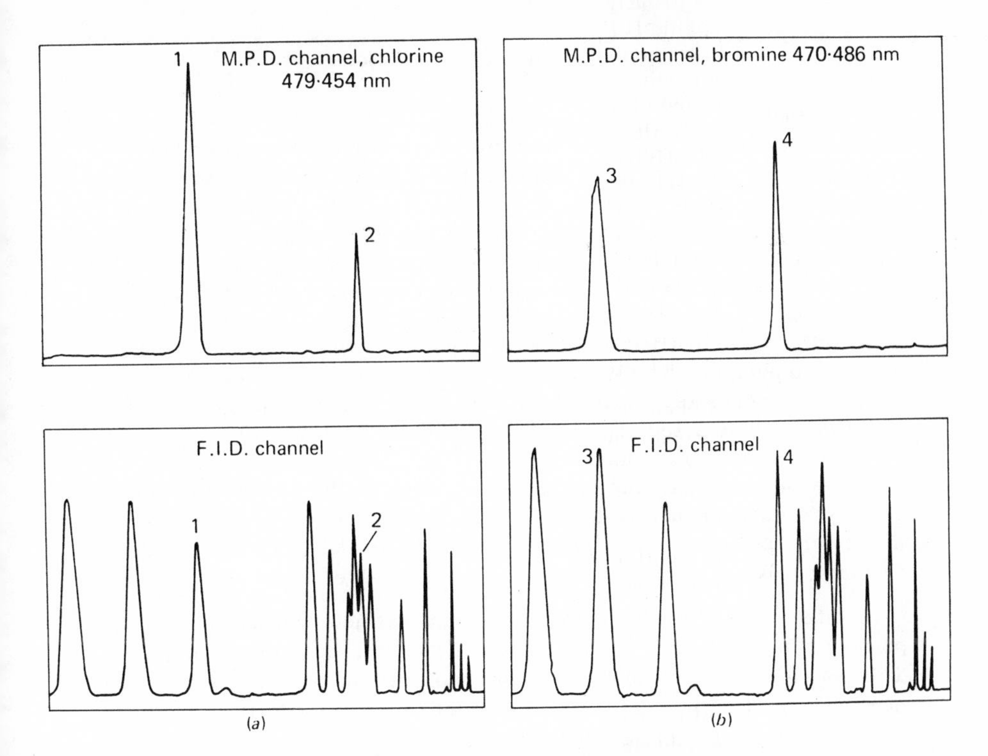

FIG. 3. Selective detection of (a) chloro compounds and (b) bromo compounds with a dual detection system using a microwave plasma detector in conjunction with a flame ionization detector. 1. o-Dichlorobenzene; 2. chlorocyclohexane; 3. o-bromotoluene; 4. bromobenzene. (Reprinted from Ref. 137. Courtesy of the Society for Analytical Chemistry.)

Most reports on using emission spectroscopy for GC detection concern ultraviolet/visible radiation from flames or microwave-stimulated plasmas. One recent exception was the application of vacuum ultraviolet atomic emission for qualitative analysis of C, N, and S compounds [144]. Another was a report on using a plasma emission device stimulated by 8-MHz radiofrequency radiation instead of the usual 2450-MHz microwave radiation [145]. The dc discharge emission detector first described some years ago [146] has recently been applied to detecting and identifying some air pollutants [147].

C. Alkali Flame Ionization Detectors

More than ten years have elapsed since the first reports that the response of a flame ionization detector to compounds containing halogens or phosphorus was greatly enhanced if the detector electrode was coated with sodium hydroxide [148, 149]. During that period very little basic understanding of the mechanisms responsible for the enhanced response has emerged. Nevertheless, several commercial alkali flame ionization detectors (sometimes called thermionic detectors) have been marketed, which usually use a salt tip to the jet made of CsBr or Rb_2SO_4. The high sensitivity and selectivity towards P, together with the low cost compared with flame photometric detectors, have established these detectors in fairly large numbers, especially for analyzing organophosphorus pesticide residues. (Some detectors have been specially designed for enhanced response to nitrogen.) Stability, selectivity, and sensitivity depend crucially on many operating parameters, such as the flow rates of carrier gas, hydrogen, and air, and also on the exact geometry of the jet, alkali metal salt, and electrodes. As a result, this is still very much a gas chromatographer's detector. Casual users beware!

A detector has been described containing two jets with facilities for passing sample through one jet and CsBr vapor through the other [150]. This was found to give enhanced reproducibility for phosphorus detection over salt-tip designs. Several descriptions of easily constructed alkali flame ionization detectors have appeared [151-153].

Apart from the usual applications to phosphorus pesticides, there have been recent studies of response to silicon, tin, and lead [154], to chlorinated materials [155, 156], and to compounds containing nitrogen [151, 157, 158]. In one study [158], the characteristics of an alkali FID and an electrolytic conductivity detector were compared. The conducitivity detector was preferred for compounds containing nitrogen and no phosphorus, because of the simplicity of operation and better specific response. The alkali FID was preferred, however, for compounds containing both P and N because of greater sensitivity, to a factor of 100.

A conventional flame ionization detector, without any added alkali metal salt, was found to respond selectively to metals when the air and hydrogen connections were reversed, so that air was mixed with the column effluent, which was then burned in an atmosphere of hydrogen. As little as 20 pg of substances such as ferrocene and tetraethyllead were detected, with selectivities of 10^4 compared with decane [159].

D. Electron Capture Detector

The electron capture detector is selective in the sense that it responds only to those molecules that readily capture electrons. These include many oxygen-, phosphorus-, sulfur-, and halogen-containing materials and also certain unsaturated compounds. Sensitivity to these electro-negative species is extremely high, and it is this fact that accounts for the widespread adoption of the detector for quantitative trace analysis. As with the alkali flame ionization detector, however, understanding of the best operating conditions and detector geometry is still far from complete, and problems of instability and irreproducibility can be severe. For a useful assessment of electron capture detector performance, the reader is referred to a recent review [160] that includes the statement "for better or for worse, in today's trace analysis [the electron capture detector] is indispensable and we may as well make the best of it."

Recent attempts to "make the best of it" include studies of the effect of oxygen on detector performance [161, 162], an important topic because of the difficulty of preventing some oxygen from entering the detector. It was found that under certain conditions it is possible for oxygen to increase the detector response to other electronegative components, through charge transfer processes [161]. Studies of radiation sources for the detector have included an evaluation of tritiated scandium [163] and a comparison of ^{147}Pm (half-life 2.7 years) with the usual ^{63}Ni source under identical geometrical conditions [164, 165]. A simplified electron capture detector made of glass, with a tritiated titanium foil source, has been described [166]. This detector is claimed to be especially suitable for chlorinated pesticides because it is easy to clean and thermal degradation of thermo-labile compounds is minimized. A constant-current system is used with a new commercial detector [167], and it is claimed that this gives improved linearity and is less susceptible to contamination.

E. Electrochemical Detectors

The <u>electrolytic conductivity</u> detector measures changes in the electrical conductivity of water, due to solution of simple molecules produced from eluates by processes such as pyrolysis, oxidation, and hydrogenolysis [168, 169]. A differential response can be obtained by circulating the contents of the conductivity cell continuously through an ion-exchange column. Hydrogenolysis of halogenated compounds at 800°C yields hydrogen halides that produce relatively large changes in conductivity when dissolved in water, so that the device becomes highly selective towards halogens (although nitrogen also gives a response under these conditions). Perhaps the greatest attraction of the detector is that it can specifically detect 100 pg of nitrogen in an eluate [100]. This is achieved by conversion into ammonia on a nickel catalyst, followed by absorption of any acids produced by $Sr(OH)_2$ [170]. A commercial electrolytic conductivity detector is available [168], the design and performance of which has been reviewed with examples of specificity towards sulfur, nitrogen, and halogens [171].

Recently, a method for enhancing the sensitivity and selectivity of the commercial detector towards chlorinated hydrocarbon pesticides has been described [172].

Chlorine, sulfur, and nitrogen can be determined absolutely, and in some cases specifically, by coulometric detectors. These devices therefore allow limited elemental and gravimetric analysis. The GC effluent is usually made to undergo a reaction such as oxidation or reduction (as with electrolytic conductivity detection) before being passed through a solution of a suitable electrolyte. Any reaction between eluate and electrolyte tends to change the potential difference of the cell, but a coulometric titration system maintains the concentration of electrolyte constant by electrolytic generation, thus restoring any loss due to reaction. The number of coulombs passed during the elution of a peak is an absolute measure of the amount of eluate. Microcoulometers suitable for use with GC are commercially available.

This type of detector was originally devised for chloride determination, by titration with Ag^+ [173]; it has also been used, however, for analyzing sulfur-containing compounds, which are converted to SO_2 before being titrated with I_3^- [174-176] or Br_2 [177]. Nitrogen-containing compounds can be detected by conversion to ammonia over a Ni-MgO catalyst, followed by titration with hydrogen ion [178]. Reported applications include the detection of trace sulfur-containing components in petroleum [175] and the identification of pesticides at concentrations of about 0.04 ppm. There have not been many recent reports, however, about microcoulometric detectors [100, 179-181].

Analytical chemists are showing a rapidly growing interest in ion-selective electrodes. Although many of these electrodes have only limited selectivity, they could obviously be used for GC detection after postcolumn reactions similar to reactions used in conjunction with electrolytic conductivity and microcoulometric detectors. This approach has been reported for chloride and sulfide, when hydrogen carrier gas and a platinum or nickel catalyst at 800°C [182] are used. Substances containing fluorine [183] and nitrogen [184] have also been detected when suitable ion-selective electrodes were used.

F. Piezoelectric Sorption Detector

The piezoelectric sorption detector is based on the fact that the oscillation frequency of a quartz-crystal oscillator is reduced if a substance is adsorbed on its surface. Such devices have been used commercially for many years for detecting water vapor. Of particular interest for qualitative analysis, however, is that the detector selectivity can be modified at will by altering the adsorbent layer, with the result that a greater or lesser amount of the GC solute is reversibly or irreversibly held by the detector. For example, if the response of benzene relative to the response of cyclohexane is taken as unity with a layer of nonpolar silicone oil on the crystal, then the corresponding value for a polar adsorbent layer of 1,2,3-tris-(2-cyanoethoxy)propane is approximately 8. The possibility therefore arises of being able to obtain the same information from the detector that the retention behavior on a second column [185, 186] would provide. The detector could also be used as a very sensitive means of ascertaining whether an unknown eluate has reacted with a specific and fast-acting reagent coated on the crystal [186].

It still remains for such possibilities to be fully exploited, but interest in the detector has grown in recent years [187-193]. Examples of applications include the specific detection of sulfur dioxide in the atomsphere [188, 190] and of part-per-million levels of organophosphorous pesticides using metal chloride coatings [189]. A portable chromatograph incorporating a piezoelectric detector is now commercially available.

VI. PEAK TRAPPING

In most laboratories it is often necessary to trap and store chromatographic fractions for analysis at some later date. Even if ancillary equipment is available that can be used on line to GC, access to such equipment is usually limited, and the scan rate is often slow compared with the rate of elution of the GC peak. When several columns are used in attempts to identify samples by retention behavior, single peaks from one column are trapped and reinjected on to another. Alternatively, the trapped material can be subjected to some reaction before being reinjected. Examples of such reactions that have been applied to microgram quantities of material include silylation [194], ozonolysis [195], and hydrogenation [195]. Similarly, a technique has been described for the hydrogenation or hydrogenolysis in a trap of as little as 0.01 μg of material prior to analysis by GC-MS [196].

Many different trapping systems have been used in conjunction with GC, and the reader is referred to an earlier review of these systems for consideration of the physical factors that limit trapping efficiency [197]. The main factors involved are the equilibrium vapor pressure of the eluate in the trap, and the tendency for the eluate to pass straight through the trap, either as a supersaturated vapor or as an aerosol fog composed of small liquid droplets.

Methods of improving trapping efficiency include [197]:

1. The use of temperature gradient traps.
2. The creation of turbulence in a double-walled trap by maintaining the the inner and outer wall at different timperatures.

3. The use of electrostatic precipitators.
4. Entrainment techniques using either a condensible carrier gas or a suitable solvent mixed as a vapor with the effluent at the column outlet.

Most interest today, however, centers on the use of simple miniature trapping systems, most of which use glass or metal capillary tubing that is cooled or packed, or both, with a suitable sorbent material. A feature of all these systems is the high surface-to-volume ratio that facilitates collision of aerosol droplets with the surface.

Cooled, empty stainless steel hypodermic needles or glass capillary tubes have been found to be effective for trapping trace components prior to analysis by infrared [198] or mass spectrometry [199-202]. Empty glass capillary tubes, 0.3 mm internal diameter, were used to trap various phenylpropanes and styrenes. The inside of the capillary was then "swept out" with a narrower tube (0.1 mm i.d.) to provide approximately 5 nanoliters of sample for Raman spectroscopy [203].

A tube 10 in. long, 1/4 in. o.d., packed with glass wool, which could replace the detector, was found to be useful for collecting labile organic phosphorus compounds [204]. The compounds were recovered from the glass wool by elution with a suitable solvent. It was reported that, on a smaller scale, glass tubes of 2 mm i.d. packed with quartz wool and cooled in solid CO_2 gave more than 80% recoveries for sample sizes in the range 10 to 30 μg [205]. Small borosilicate glass tubes containing a coating of porous glass have been reported to trap 10 μg of material with almost 100% efficiency, with no coolant required [206]. The trapped material is desorbed by heating slowly under vacuum in the sampling system of a mass spectrometer. Activated charcoal has been used in a similar way [207].

Gas chromatography support material coated with typical liquid phases has also been used for trapping. Even without refrigeration, this allows considerable reduction of the vapor pressure of the sample. The method is equivalent to isolating a peak in a portion of the column. A 4-in. packed column, cooled with solid CO_2, was found to give good recoveries even below the microgram level [208], and this trap was used to transfer as little as 0.1 μg to a mass spectrometer. Miniature packed columns made of capillary tubes have also been used successfully [209].

Recently, the use of XAD-4 resin loosely packed in borosilicate capillary tubes has been recommended as an effective simple device for collecting samples for infrared, ultraviolet, NMR, and MS analysis [210]. Trapping recoveries for unspecified amounts of a range of model compounds were between 60% and 95%.

Several of these techniques allow the trap to take the place of the usual sample holder in a direct insertion probe for mass spectrometry. For example, empty steel tubes [201], and glass capillaries packed with activated charcoal [207], with support and stationary phase [209] or with resin [210] have been used in this way.

VII. COMBINING INFRARED SPECTROSCOPY WITH GAS CHROMATOGRAPHY

For infrared analysis, a choice must be made between trapping the eluate in a gas cell or attempting to condense it. Despite the difficulties inherent in condensation, this approach is the most common for all but the lowest-boiling eluates. Among the reasons for this is that liquid-phase spectra are better documented than spectra in the gas phase. The short pathlength of liquid samples is also an advantage, since it allows the whole sample to be held close to the focus of the sample beam. The main drawback of gas-phase trapping is the need for a long, narrow cell. On the other hand, a gas cell offers the attractive possibilities of 100% trapping efficiency and direct on-line coupling of GC and IR. Reviews of methods of trapping GC fractions for infrared analysis are to be found in Refs. 211-214. In particular, Ref. 214 provides a thorough discussion of coupling GC with vapor-phase IR, including spectra of over 300 substances in the vapor phase at elevated temperatures and an account of frequency correlations.

The potassium bromide microdisk technique is particularly attractive for small sample sizes of fairly involatile material. Samples can be trapped in KBr minicolumns attached to the exit port of the gas chromatograph. The powder is then pressed into a disk in the usual fashion [215]. There have also been reports of condensing the column eluate directly onto a KBr disc [216] or forming a thin film in a KBr disc "sandwich" [217]. Infrared microcells have internal volumes that range from a few microliters down to 0.2 μl. Corresponding pathlengths are from 0.1 to 0.01 mm. The cells are usually filled by centrifuging trapped material from a larger "reception" volume into the microvolume. In favorable cases, the cells themselves can be used as traps [218-220]. Alternatively, trapped material can be transferred to the microcells from some other trap (e.g., [221]).

Although infrared spectroscopy of trapped eluates has been used as a means of identification almost since the beginning of GC, the continuous monitoring of GC effluents by IR is a relatively recent development and no entirely satisfactory solutions to the problems involved in this combination have yet emerged. There have been two main difficulties: the relative insensitivity of infrared spectroscopy, and the fact that conventional infrared instruments scan a spectrum in minutes rather than seconds. Various types of fast scanning instruments are now available, but the sensitivity problem still remains.

In connection with sensitivity, there appear to be some common misconceptions about the relative importance of pathlength, cell volume, and ratio of pathlength to cell volume of cells for vapor samples collected from gas chromatographs. If an unlimited amount of sample is available, then ideally absorbance depends only on pathlength. For a fixed sample size, however, the total volume of the cell is also important, because this determines the sample concentration: maximum sensitivity is then obtained for a cell with a high pathlength-to-volume ratio. This is illustrated in Table 1 on the assumption that the eluted GC peak just fills the cell and is entirely contained within the beam [213]. It may be thought that cell C, with a pathlength-to-volume ratio of 10 cm^{-2}, would be the obvious choice, and so it would--provided that the GC peak was contained in only 0.6 cm^3 of carrier gas. Such small peak volumes are found only in highly efficient columns. For example, a column of 300,000 theoretical plates that is operated at a flow rate of 2 cm^3 min^{-1} produces peaks with volumes of less than 0.6 cm^3, provided that the retention time is less than 30 minutes. However, the minimum size of sample required to produce a weak band in this cell (40 μg) would grossly overload such a capillary column. Realistic sample sizes of eluates emerging from this column

TABLE 1

Characteristics of Typical GC-IR Gas Cells

	Cell A	Cell B	Cell C
Volume (cm^3)	20	2.5	0.6
Path length (cm)	30	6	6
Path length/volume (cm^{-2})	1.5	2.4	10.0
Approximate detection limit for a weak band (μg)[a]	270	170	40
Corresponding concentration ($\mu g\ cm^{-3}$)[a]	13	67	67

[a]These theoretical figures assume that the eluted peak just fills the cell and is completely contained within the beam, giving 99% transmittance for a solute of molecular weight 100 and an extinction coefficient of 1 l $mole^{-1}$ cm^{-1} [213].

could not be readily detected by IR. However, the increasingly popular support coated or porous layer open tubular columns may be able to cope with 50-μg samples in peak widths of about 1 cm^3, and thus provide sufficient material for the production of a spectrum using cells such as C of Table 1.

It has been pointed out [213, 214] that for conventional packed columns, volumes of 50 or 100 cm^3 would have to be accommodated if the whole peak were to be trapped. For a pathlength-to-volume ratio of 10 cm^{-2}, this corresponds to a pathlength of 5 or 10 m, which is impracticable. In this connection, it has recently been reported that a conventional multireflection cell of 40 cm^3 volume offers advantages over a minimum volume cell for packed column work [222]. Although approximately one-half of the total cell volume was dead space, not traversed by the beam of radiation, the total pathlength was 76 cm (corresponding to 5 traversals of 12 cm between mirrors, plus entrance and exit paths of 8 cm each). Under these conditions, a light pipe such as cell C of Table 1 would give only one-quarter of the absorbance observed in the multireflection cell, even when the light pipe contained the central portion of the peak. It was concluded that the overall sensitivity of cell C, whether used in connection with a packed column or support coated capillary column, is appreciably worse than the sensitivity of a multiple reflection cell used in conjunction with packed columns. Useful spectra were obtained with the heated multireflection cell for peaks containing 20-100 μg of sample, with chromatographic conditions adjusted to confine peak volumes to within 40 cm^3 of carrier gas [222].

If a conventional slow-scanning infrared spectrometer is to be used, then heated gas cells can be used to trap individual GC peaks. In such a system, any peak not trapped is lost. By halting the carrier gas flow, however, one can store peaks in the GC column itself while the IR spectrum of the peak that has just emerged is taken. Such a stop-go or "interrupted elution" system has been used in connection with mass spectrometry [223, 224], and pyrolysis GC [70, 71] as well as infrared spectroscopy [222, 224]. It appears [71, 225] that venting of the pressure at the inlet to the column serves no useful purpose, and that support coated open tubular columns retain better peak resolution during extended period of interruption than packed columns do [225]. Interrupted-elution GC-IR has recently been used to identify the main components in a mixture of unsaturated C_4 hydrocarbons that could not be completely characterized by GC-MS [222].

Rapid scanning infrared spectrometers, discussed in several earlier reviews [211, 213, 214], have not achieved widespread use in connection with continuous-flow gas chromatography. A scan time of 20 sec, typical of so called fast-scanning dispersive instruments, is not really fast in the context of GC, yet it has proved difficult to achieve really fast scanning of 1 sec or so without unacceptable loss of resolution. There is therefore considerable interest in the use of interferometry, which has an inherently faster scan rate and better signal-to-noise ratio than dispersive techniques. Such instruments require high technology, including computer processing, and are therefore unlikely to become as readily available as GC-MS.

Two double-beam dispersive instruments specially designed for GC-IR that are claimed to overcome the sensitivity, resolution, and scan rate limitations of earlier systems have recently been described. One [226] involves a 1-cm^3 flow-through multiple internal reflectance cell of 10-cm pathlength, which can be heated to 250°C. A fast response time is obtained by electronically ratioing the reference and sample beams using a 90° phase null technique in conjunction with a mercury-doped germanium detector cooled in liquid helium. The region from 3700 to 750 cm^{-1} can be scanned in 5 to 20 seconds, using a single grating, and it is reported that good spectra are obtained for sample sizes of 20 μg, with functional group bands visible for samples of 3 μg.

The second system, which is commercially available, incorporates 5 cm^3 flow-through cells of 30-cm pathlength [227, 228]. The region from 2.5 to 15 μm (4000 to 670 cm^{-1}) can be scanned linearly in wavelength in 6 seconds, using a fast-response pyroelectric detector that does not require cooling. In common with the instrument described above, a single grating, and electronic ratioing of reference and sample beams are employed. Identifiable spectra have been reported for sample sizes in the range 22 to 150 μg [227].

Despite the attractions of vapor phase spectra, discussed in detail in Ref. 214, many investigators prefer to use liquid phase spectra. Such workers will be interested in an interface that condenses the GC eluate while retaining on-stream operation [229]. The principle (which is essentially similar to that involved in the piezoelectric sorption detector discussed above) is that the eluate dissolves in and then is desorbed from a thin liquid film as the column effluent flows over the film. The liquid film is deposited on the cell window of a fixed-pathlength liquid cell held in the beam of a fast-scanning infrared spectrometer. The cell is provided with facilities for heating and for flowthrough connection to the gas chromatograph. Spectra of 50-μg samples have been obtained. A cholesteric liquid crystal was actually employed [229], the intention being eventually to use the change in birefringence color associated with solution of the eluate to trigger the scan--but many GC liquid phases would also form suitable liquid films.

A novel molecular beam technique has been proposed [230] to increase the sensitivity of GC-IR so that it becomes comparable with the sensitivity of GC-MS. The proposal is that the column effluent, in helium carrier gas, should enter a molecular beam source, where it would be irradiated by a monochromatic beam from a fast-scanning infrared source. The vibrational energy gained by the eluate molecules would be transferred by collisions to the atoms of helium carrier gas. The helium would then be ionized by low-energy electron impact, and an electrostatic analyzer would allow only ions with energies above the thermal level to proceed to the detector. The infrared spectrum could then be obtained as a plot of observed ion current against frequency of infrared radiation.

VIII. COMBINED GAS CHROMATOGRAPHY AND MASS SPECTROMETRY (GC-MS)

In this section it seems appropriate to give a brief survey of GC-MS, restricting references in the main to review articles and books that have been published recently. These include a book of 448 pages devoted to techniques of GC-MS [231], and an excellent review article that explains and surveys current GC-MS instrumentation and also provides a comprehensive citation list for the period 1957-1970 indexed into various application categories with over 700 references [232]. Less comprehensive reviews are to be found as chapters in books [233-235]. The first biennial review of GC-MS covered the literature of June 1968 to June 1970, together with important earlier work [236], while the second review in that series covered the literature to June 1972 [237]. The proceedings of an international GC-MS symposium have been published [238]. Reference 239 stresses the synergistic effects involved in computerized GC-MS systems.

This phenomenal recent growth in the literature about directly coupled GC-MS is in marked contrast to the situation with GC-IR, which is still very much a novelty. Among the reasons for the spectacular development of on-line GC-MS are the following facts:

1. Useful mass spectra can be obtained only with pure samples, so there is an obvious attraction in using gas chromatography to provide samples for analysis by mass spectrometry.

2. Both gas chromatography and mass spectrometry can readily be used for the study of submicrogram quantities of material.

3. Scanning either is not necessary or can now be carried out very rapidly. For example, scan times of less than 1 sec are now routine on magnetic deflection instruments.

4. Several small, mobile mass spectrometers designed specifically for GC are now available. In addition, several firms now offer highly developed GC-MS units employing deflection, quadrupole, or time-of-flight instruments. Combined electron-impact/chemical ionization sources are now available for GC-MS.

5. Computerized systems have been developed to aid acquisition and presentation of large quantities of mass spectral data. For example, each mass spectrum can be stored and correlated with the corresponding part of the gas chromatogram, then made available later, on request. Experimental spectra can automatically be compared with library spectra. In the case of high-resolution spectra, element maps can be determined.

Although mass spectrometry cannot always provide an unambiguous structure for an unknown compound, it probably provides useful information to a greater degree than any other single analytical technique does, particularly for submicrogram samples. Nevertheless, it is important to remember that quite apart from economic considerations there are several fundamental limitations to the usefulness of mass spectrometry as an analytical tool, especially when it is used in a fast-scan mode on small samples heavily diluted with carrier gas. In general, therefore, it is necessary to supplement the mass spectrum with other information about the eluate, before any confidence can be placed in an identification.

The main difficulty that has to be overcome when directly coupling GC and MS is the limited rate at which it is possible to admit gas to the source chamber while maintaining a sufficiently low pressure in the spectrometer. The actual value of the limiting flow rate for a particular spectrometer depends on the detailed design of the vacuum system, including factors such as the pumping rate and whether differential pumping of the source and analyzer is provided. Typical values are between 0.2 and 2 $cm^3\ min^{-1}$ at atmospheric pressure, compared with carrier gas flow rates of about 30 $cm^3\ min^{-1}$ for "conventional" GC packed columns. It follows that at least 90% of the maximum possible MS response to a particular GC peak will be lost. This explains the great interest in molecular separators, which by concentrating the eluate allow a greater total percentage of the sample to enter the source. Two useful reviews about GC-MS interface systems have been published recently [240, 241]. With many separators, the enrichment factor is only about 100, and as a result carrier gas molecules entering the source chamber still commonly outnumber eluate molecules by at least 1000:1. However, this modest enrichment can in principle allow almost all the eluate from packed columns to enter the mass spectrometer. In practice, however, the efficiency of the separator may be rather poor, so that despite a high enrichment factor, only a small percentage of the sample leaving the column actually reaches the spectrometer. There may then be little advantage in using the separator compared with simply splitting the effluent from the GC column so that a suitable flow rate enters the source.

Many types of molecular separator have been designed. Perhaps the most widely used is the glass frit type in which hydrogen or helium carrier gas is removed by preferential effusion through a porous glass wall. Similar devices use porous metal or Teflon tubing.

The commercial LKB GC-MS system uses a molecular jet design that also depends on the more rapid diffusion of a carrier gas that is "light" compared with the "heavy" organic molecules. The high permeability of palladium to hydrogen, and total impermeability to other molecules, is used in another type of separator.

Recently, increasing use has been made of a completely different method of enrichment, in which organic molecules dissolve in and then diffuse through a silicone rubber membrane. The membrane is highly permeable to organic molecules, but almost impermeable to the carrier gas, which can be of high molecular weight, in contrast to the situation with most other types of separator.

A molecular separator may not be necessary if the chromatographic column is of the open tubular "capillary" type, operating at flow rates of about 1 $cm^3 min^{-1}$. However, the amount of sample that can be injected onto such a column is usually less than 1% of that which could be injected onto a packed column. As a result, there is usually less material in a capillary column peak than in that fraction of a packed column peak that can be sent directly to the spectrometer without a separator. As a rule therefore, capillary columns should be used only if the chromatographic separation cannot be achieved with a packed column. There is some evidence that support coated open tubular (SCOT) columns may offer a suitable compromise between sample size and carrier gas flow rate for direct coupling without a splitter or separator.

It is sometimes desirable to use the mass spectrometer as a GC detector to produce a chromatogram. This is usually achieved by monitoring the total ion beam leaving the source. Such a system can be used to trigger the production of mass spectra when the total ion current rises above a certain threshold level, thereby removing the need for constant inspection of the detector signal by the operator. However, a great many spectra are likely to be produced by this automated approach, so that computer processing becomes almost essential. Another use for the total ion monitor is to allow the intensity of a recorded mass spectrum to be automatically corrected for the change in component concentration occurring during the scan, using ratio recording.

Despite the growing use of directly coupled GC-MS, there is still a considerable need for trapping methods. Indeed, it is sometimes argued that directly coupling produces bad gas chromatograms and bad mass spectra [242, 243], and there is some substance in such arguments, particularly concerning memory effects and background due to the stationary phase.

REFERENCES

1. S. G. Perry, Chromatog. Rev., 9, 1 (1967).
2. V. G. Berezkin, "Analytical Reaction Gas Chromatography," Plenum Press, New York, 1968.
3. D. A. Leathard and B. C. Shurlock, "Advances in Analytical Chemistry and Instrumentation," vol. 6, ed. J. H. Purnell, Wiley, New York, 1968, p. 1.
4. L. S. Ettre and W. H. McFadden, eds. "Ancillary Techniques of Gas Chromatography," Wiley, New York, 1969.
5. D. A. Leathard and B. C. Shurlock, "Identification Techniques in Gas Chromatography," Wiley, London, 1970.
6. R. C. Crippen, "Identification of Organic Compounds with the Aid of Gas Chromatography," McGraw-Hill, New York, 1973.
7. V. G. Arakelyan and K. I. Sakodynskii, Chromatogr. Rev. 15, 93 (1971).
8. R. A. Keller, J. Chromatogr. Sci. 11, 49 (1973).
9. G. Schomburg and G. Dielmann, J. Chromatogr. Sci. 11, 151 (1973).
10. Pp. 41-46 of Ref. 5.
11. F. Caesar, Chromatographia 5, 173 (1972).
12. N. Guichard-Loudet, Analusis 2, 247 (1973).
13. G. Castello, M. Lundarelli, and M. Berg, J. Chromatogr. 76, 31 (1973).
14. A. W. Ladon and S. Sandler, Analyt. Chem. 45, 921 (1973).
15. G. Castello and G. D'Amato, J. Chromatogr. 76, 293 (1973).
16. J. K. Haken, J. Chromatogr. Sci. 11, 144 (1973).
17. J. M. Takacs et al., J. Chromatogr. 81, 1 (1973).
18. M. Wurst and J. Churacek, J. Chromatogr. 70, 1 (1972).
19. V. D. Patil, V. R. Nayak, and S. Dev, Tetrahedron 29, 1595 (1973).
20. B. A. Bierl, M. Beroza, and M. H. Aldridge, J. Chromatogr. Sci. 10, 712 (1972).
21. N. C. Saha and G. D. Mitra, J. Chromatogr. Sci. 11, 419 (1973).
22. J. K. Haken, Chromatogr. Rev. 16, 419 (J. Chromatogr. 73) (1972).
23. C. R. Trash, J. Chromatogr. Sci. 11, 196 (1973).

24. A. E. Coleman, J. Chromatogr. Sci. 11, 198 (1973).

25. R. A. Keller, J. Chromatogr. Sci. 11, 188 (1973).

26. J. J. Leary et al., J. Chromatogr. Sci. 11, 201 (1973).

27. J. R. Mann and S. T. Preston, J. Chromatogr. Sci. 11, 216 (1973).

28. R. S. Henly, J. Chromatogr. Sci. 11, 221 (1973).

29. P. H. Weiner and D. G. Howery, Analyt. Chem. 44, 1189 (1972).

30. P. H. Weiner, C. J. Dack, and D. G. Howery, J. Chromatogr. 69, 249 (1972).

31. P. H. Weiner and J. F. Parcher, J. Chromatogr. Sci. 10, 612 (1972).

32. P. H. Weiner and J. F. Parcher, Analyt. Chem. 45, 302 (1973).

33. J. M. Takacs, J. Chromatogr. Sci. 11, 210 (1973).

34. L. Rohrschneider, J. Chromatogr. 22, 6 (1966).

35. W. R. Supina and L. P. Rose, J. Chromatogr. Sci. 8, 214 (1970).

36. J. J. Leary, S. Tsuge, and T. L. Isenhour, J. Chromatogr. 82, 366 (1973).

37. L. Rohrschneider, J. Chromatogr. Sci. 11, 160 (1973).

38. W. O. McReynolds, J. Chromatogr. Sci. 8, 685 (1970).

39. S. Wold and K. Andersson, J. Chromatogr. 80, 43 (1973).

40. P. A. Hedin, R. C. Gueldner, and A. C. Thompson, Analyt. Chem. 42, 403 (1970).

41. P. 67 of Ref. 5.

42. P. 128 of Ref. 4.

43. R. R. Allen, Analyt. Chem. 38, 1287 (1966).

44. D. A. Cronin, J. Chromatogr. 64, 25 (1972).

45. M. K. Withers, J. Chromatogr. 66, 249 (1972).

46. B. A. Bierl, M. Beroza, and W. T. Ashton, Mikrochim. Acta. 1969, 637.

47. J. K. Haken, D. K. M. Ho, and M. K. Withers, J. Chromatogr. Sci. 10, 566 (1972).

48. N. A. Prokopenko, et al., J. Chromatogr. 69, 47 (1972).

49. F. A. Reginier and J. C. Huang, J. Chromatogr. Sci. 8, 267 (1970).

50. J. Frycka and J. Pospisil, J. Chromatogr. 67, 366 (1972).

51. I. L. Jamieson, J. Appl. Chem. Biotechnol. 22, 1157 (1972).

52. Pp. 86-110 of Ref. 5.

53. M. Beroza and M. N. Inscoe in Ref. 4, pp. 89-144.

54. C. Merritt in Ref. 4, pp. 325-340.

55. M. Beroza and M. N. Inscoe in various sections of "Instrumental Methods of Organic Functional Group Analysis" ed. S. G. Siggia, Wiley, New York, 1972.

56. First Symposium on Reaction Gas Chromatography (Tallinn, July, 1971), J. Chromatogr., 69, 1-86 (1972).

57. J. T. Walsh and C. Merritt, Analyt. Chem. 32, 1378 (1960).

58. D. A. Cronin and J. Gilbert, J. Chromatogr. 71, 251 (1972).

59. D. A. Cronin and J. Gilbert, J. Chromatogr. 89, 209 (1974).

60. B. Casu and L. Cavallotti, Analyt. Chem. 34, 1514 (1962).

61. R. Kaiser in Ref. 4, pp. 300-323.

62. R. Kaiser, Chem. in Britain 5, 54 (1969).

63. S. H. Langer and P. Pantages, Nature 191, 141 (1961).

64. J. E. Hoff and E. D. Feit, Analyt. Chem. 36, 1002 (1964).

65. P. E. Manni and T. D. Streif, J. Chromatogr. Sci. 10, 178 (1972).

66. Pp. 111-134 of Ref. 5.

67. Pp. 55-88 of Ref. 4.

68. J. H. Dhont, Analyst 89, 71 (1964).

69. E. J. Levy and D. G. Paul, J. Gas Chromatogr. 5, 136 (1967).

70. J. Q. Walker and C. J. Wolf, Analyt. Chem. 40, 711 (1968).

71. C. J. Wolf and J. Q. Walker, in "Gas Chromatography 1968," ed. C. L. A. Harbourn, Institute of Petroleum, London, 1969, p. 385.

72. P. 120 of Ref. 5.

73. D. A. Leathard and J. H. Purnell, Ann. Rev. Phys. Chem. 21, 197 (1970).

74. S. A. Liebman et al., Thermochimica. Acta. 5, 403 (1973).

75. S. A. Liebman et al., Research/Development 23, (12), 24 (1972).

76. S. A. Liebman et al., Analyt. Chem. 45, 1360 (1973).

77. C. Merritt and C. DiPietro, Analyt. Chem. 44, 57 (1972).

78. C. Merritt and D. H. Robertson, Analyt. Chem. 44, 60 (1972).

79. R. S. Juvet and L. P. Turner, Analyt. Chem. 37, 1464 (1965).

80. R. S. Juvet, R. L. Tanner, and J. C. Y. Tsao, J. Gas. Chromatogr. 5, 15 (1967).

81. K. A. Banks and D. D. Bills, J. Chromatogr. 33, 450 (1968).

82. P. 145 of Ref. 5.

83. J. Hetper and A. Janik, J. Chromatogr. 64, 162 (1972).

84. S. C. Bevan and S. Thorburn, Chem. in Britain 1, 206 (1965).

85. S. C. Bevan, T. A. Gough, and S. Thorburn, J. Chromatogr. 44, 241 (1969).

86. C. F. Simpson, Column 3, (4), 2 (1970), Pye Unicam.

87. A. Liberti, L. Conti, and V. Crescenzi, Nature 178, 1067 (1956).

88. A. G. Vitenberg, S. K. Pospelova, and B. V. Ioffe, Neftekhimya 12, 623 (1972).

89. J. S. Parsons, Analyt. Chem. 36, 1849 (1964).

90. D. G. Paul and G. E. Umbreit, Research/Development 21 (May), 18 (1970).

91. C. E. Bennett et al., Amer. Lab., May 1971.

92. R. S. Swingle, Ind. Res. 14 (Feb.), 40 (1972).

93. A. C. Lanser et al., Analyt. Chem. 45, 2344 (1973).

94. J. G. Gonzalez and R. T. Ross, Anal. Lett. 5, 683 (1972).

95. J. E. Longbottom, Analyt. Chem. 44, 1111 (1972).

96. Pp. 152-190 of Ref. 5.

97. Pp. 341-373 of Ref. 4.

98. J. D. Winefordner and T. H. Glenn, Adv. Chromatogr. 5, 263 (1968).

99. M. Krejci and M. Dressler, Chromatogr. Rev. 13, 1 (1970).

100. D. F. S. Nautsch and T. M. Thorpe, Analyt. Chem. 45, 1185A (1973).

101. T. A. Gough and E. A. Walker, Analyst 95, 1 (1970).

102. G. H. Hartmann, Analyt. Chem. 43 (2), 113A (1971).

103. M. L. Selucky, Chromatographia 4, 425 (1971).

104. M. C. Bowman and M. Beroza, Analyt. Chem. 40, 1448 (1968).

105. W. E. Rupprecht and T. R. Phillips, Anal. Chim. Acta. 47, 439 (1969).

106. A. I. Mizany, J. Chromatogr. Sci. 8, 151 (1970).

107. S. G. Perry and F. W. G. Carter, "Gas Chromatography, 1970," ed. R. Stock, Institute of Petroleum, London, 1971, p. 381.

108. L. J. Darlage, S. S. Block and J. P. Weidner, J. Chromatogr. Sci. 11, 272 (1973).

109. T. Sugiyama, Y. Suzuki, and T. Takeuchi, J. Chromatogr. 77, 309 (1973).

110. T. Sugiyama, Y. Suzuki, and T. Takeuchi, J. Chromatogr. 80, 61 (1973).

111. G. Goretti and M. Possanzini, J. Chromatogr. 77, 317 (1973).

112. R. Ross and T. Shafik, J. Chromatogr. Sci. 11, 46 (1973).

113. H. H. Hill and W. A. Aue, J. Chromatogr. 74, 311 (1972).

114. P. M. Mutsaars and J. E. Van Steen, J. Inst. Petrol. London 58, 102 (1972).

115. W. A. Aue and H. H. Hill, J. Chromatogr. 70, 158 (1972).

116. B. Gutsche, R. Herrmann, and K. Ruediger, Z. Analyt. Chem. 258, 273 (1972).

117. B. Gutsche and R. Herrmann, Z. Analyt. Chem. 259, 126 (1972).

118. A. V. Nowak and H. V. Malmstadt, Analyt. Chem. 40, 1108 (1968).

119. M. C. Bowman, M. Beroza, and K. R. Hill, J. Chromatogr. Sci. 9, 162 (1971).

120. M. C. Bowman, M. Beroza, and G. Nickless, J. Chromatogr. Sci. 9, 44 (1971).

121. R. F. Moseman and W. A. Aue, J. Chromatogr. 63, 229 (1971).

122. B. Versino and G. Rossi, Chromatographia 4, 331 (1971).

123. J. L. Monkman and L. Dubois, in "Gas Chromatography," ed. H. J. Noebels, R. F. Wall, and N. Brenner, Academic Press, New York (1961), p. 333.

124. F. A. Blunther, R. C. Blinn, and D. E. Ott, Analyt. Chem. 34, 302 (1962).

125. M. C. Bowman and M. Beroza, J. Chromatogr. Sci. 7, 484 (1968).

126. J. F. Siuda, J. F. De Bernardis, and R. C. Cavestri, J. Chromatogr. 75, 298 (1973).

127. A. J. McCormack, S. C. Tong, and W. D. Cooke, Analyt. Chem. 37, 1470 (1965).

128. C. A. Bache and D. J. Lisk, Analyt. Chem. 37, 1477 (1965).

129. C. A. Bache and D. J. Lisk, Analyt. Chem. 38, 783 (1966).

130. C. A. Bache and D. J. Lisk, Analyt. Chem. 38, 1757 (1966).

131. H. A. Moye, Analyt. Chem. 39, 1441 (1967).

132. C. A. Bache and D. J. Lisk, Analyt. Chem. 39, 786 (1967).

133. C. A. Bache, L. E. St. John, and D. J. Lisk, Analyt. Chem. 40, 1241 (1968).

134. R. M. Dagnall, S. J. Pratt, T. S. West, and D. R. Deans, Talanta 16, 797 (1969).

135. R. M. Dagnall, S. J. Pratt, T. S. West, and D. R. Deans, Talanta 17, 1009 (1970).

136. C. A. Bache and D. J. Lisk, Analyt. Chem. 43, 950 (1971).

137. W. R. McLean, D. L. Stanton, and G. E. Penketh, Proc. Soc. Analyt. Chem. 1972, 296, and Analyst 98, 432 (1973).

138. R. M. Dagnall, T. S. West, and P. Whitehead, Anal. Chim. Acta 60, 25 (1972).

139. R. M. Dagnall, M. D. Silvester, T. S. West, and P. Whitehead, Talanta 19, 1226 (1972).

140. R. M. Dagnall, T. S. West, and P. Whitehead, Analyt. Chem. 44, 2074 (1972).

141. W. E. L. Grossman, J. Eng, and Y. C. Tong, Anal. Chim. Acta 60, 447 (1972).

142. H. Kawaguchi, T. Sakamoto, and A. Mizuike, Talanta 20, 321 (1973).

143. B. J. Lowings, Source 2 (2) 1972, Applied Research Lab.

144. W. Braun et al., J. Chromatogr. 55, 237 (1971).

145. L. E. Boos and J. D. Winefordner, Analyt. Chem. 44, 1020 (1972).

146. R. S. Braman and A. Dynako, Analyt. Chem. 40, 95 (1968).

147. R. S. Braman, Atmosph. Environ 5, 669 (1971).

148. A. Karmen and L. Giuffrida, Nature 201, 1204 (1964).

149. L. Giuffrida, N. F. Ives, and D. C. Bostwick, J. Assoc. Offic. Agric. Chemists 49, 8 (1966).

150. H. Fischer, M. Nevfelder, and D. Pruggmayer, Chromatographia 5, 613 (1972).

151. D. A. Craven, Analyt. Chem. 42, 1679 (1970).

152. W. H. Stewart, Analyt. Chem. 44, 1547 (1972).

153. A. Karmen and H. Haut, Analyt. Chem. 45, 822 (1973).

154. M. Dressler, V. Martinu, and J. Janak, J. Chromatogr. 59, 429 (1971).

155. W. A. Aue and R. F. Moseman, J. Chromatogr. 61, 35 (1971).

156. W. A. Aue et al., J. Chromatogr. 63, 237 (1971).

157. R. F. Coward and P. Smith, J. Chromatogr. 61, 329 (1971).

158. R. Greenhalgh and W. P. Cochrane, J. Chromatogr. 70, 37 (1972).

159. W. A. Aue and H. H. Hill, J. Chromatogr. 74, 319 (1972).

160. W. A. Aue and S. Kapila, J. Chromatogr. Sci. 11, 255 (1973).

161. H. J. Van de Wiel and P. Tommassen, J. Chromatogr. 71, 1 (1972).

162. F. W. Karasek and D. M. Kane, Analyt. Chem. 45, 576 (1973).

163. C. H. Hartmann, Analyt. Chem, 45, 733 (1973).

164. J. A. Lubkowitz and W. C. Parker, J. Chromatogr. 62, 53 (1971).

165. J. A. Lubkowitz, D. Montoloy, and W. C. Parker, J. Chromatogr. 76, 21 (1973).

166. A. Soedergen, J. Chromatogr. 71, 532 (1972).

167. R. J. Maggs, Analyt. Chem. 43, 1966 (1971).

168. D. M. Coulson, J. Gas Chromatogr. 3, 134 (1965).

169. P. Jones and G. Nickless, J. Chromatogr. 73, 19 (1972).

170. D. M. Coulson, J. Gas Chromatogr. 4, 285 (1966).

171. M. L. Selucky, Chromatographia 5, 359 (1972).

172. J. W. Dolan and R. C. Hall, Analyt. Chem. 45, 2198 (1973).

173. D. M. Coulson and L. A. Cavanagh, Analyt. Chem. 32, 1245 (1960).

174. R. L. Martin and J. L. Grant, Analyt. Chem. 37, 644 (1965).

175. E. M. Fredericks and G. A. Harlow, Analyt. Chem. 36, 263 (1964).

176. H. V. Drushel, Analyt. Chem. 41, 569 (1969).

177. P. J. Klaas, Analyt. Chem. 33, 1851 (1961).

178. R. L. Martin, Analyt. Chem. 38, 1209 (1966).

179. F. W. Williams, Analyt. Chem. 44, 1317 (1972).

180. H. Schulz and M. Muniv, Erdoel Kohle, Erdgas, Petrochem. 25, 14 (1972).

181. V. E. Stapanenko and S. I. Krichmar, Zh. Analit. Khim. 26, 147 (1971).

182. T. Kojima, M. Ichise, and Y. Seo, Jap. Analyst 20, 20 (1971).

183. T. Kojima, M. Ichise, and Y. Seo, Talanta 19, 539 (1972).

184. T. Kojima, M. Ichise, and Y. Seo, Jap. Analyst 22, 208 (1973).

185. W. H. King, Analyt. Chem. 36, 1735 (1964).

186. P. 152 of Ref. 5.

187. F. W. Karasek and K. R. Gibbins, J. Chromatogr. Sci. 9, 535 (1971).

188. A. Lopez Roman and G. G. Guilbault, Analyt. Lett. 5, 225 (1972).

189. E. P. Scheide and G. G. Guilbault, Analyt. Chem. 44, 1764 (1972); and correction in Analyt. Chem. 45, 975 (1973).

190. M. W. Frechette, J. L. Fasching, and D. M. Rosie, Analyt. Chem. 45, 1765 (1973).

191. G. A. Ivanov, V. G. Guglya, and A. A. Zhukhovitskii, Zavod. Lab. 39, 15 (1973).

192. M. Janghorbani and H. Freund, Analyt. Chem. 45, 325 (1973).

193. F. W. Karasek, J. Chromatogr., in press.

194. W. A. McGugan and S. G. Howsam, J. Chromatogr. 82, 370 (1973).

195. G. Stanley and B. H. Kennett, J. Chromatogr. 75, 304 (1973).

196. G. Stanley and K. E. Murray, J. Chromatogr. 60, 345 (1971).

197. Pp. 191-209 of Ref. 5.

198. R. F. Kendall, Appl. Spectr. 21 (1); 31 (1967).

199. K. R. Burson and C. T. Kenner, J. Chromatogr. Sci. 7, 63 (1969).

200. F. Armitage, J. Chromatogr. Sci. 7, 190 (1969).

201. K. E. Murray, J. Shipton, A. V. Robertson, and M. P. Smyth, Chem. Ind. 1971, 401.

202. R. E. Snyder, J. Chromatogr. Sci. 9, 638 (1971).

203. B. J. Bulkin, K. Dill, and J. J. Dannenberg, Analyt. Chem. 43, 974 (1971).

204. A. F. Machin and C. R. Morris, Analyst 97, 289 (1972).

205. H. Copier and L. Schutte, J. Chromatogr. 47, 464 (1970).

206. D. M. Kane and F. W. Karasek, J. Chromatogr. Sci. 10, 501 (1972).

207. J. N. Damico, N. P. Wong, and J. A. Sphon, Analyt. Chem. 39, 1045 (1967).

208. B. A. Bierl, M. Beroza, and J. M. Ruth, J. Gas Chromatogr. 6, 286 (1968).

209. J. W. Amy, E. M. Chait, W. E. Baitinger, and F. W. McLafferty, Analyt. Chem. 37, 1265 (1965).

210. J. L. Witiak, G. A. Junk, G. A. Calder, J. S. Fritz, and H. J. Svec, J. Org. Chem. 38, 3066 (1973).

211. S. K. Freeman, Ref. 4., pp. 227-241.

212. J. G. Grasselli and M. K. Snavely, Progress in Infrared Spectroscopy 3, 55 (1967).

213. Pp. 211-217 of Ref. 5.

214. D. Welti, "Infrared Vapor Spectra," Heyden, London, 1970.

215. A. S. Curry, J. F. Read, C. Brown, and R. W. Jenkins, J. Chromatogr. 38, 200 (1969).

216. A. G. Osborne, Lab. Pract. 20, 579 (1971).

217. E. I. M. Bergstedt, Chromatographia 2, 545 (1969).

218. N. W. R. Daniels, Process Biochem. 3, 34 (1968), and Column 2, (1); 2 (1967).

219. H. Gunther, Z. Anal. Chem. 252, 145 (1970).

220. M. G. Matas and A. P. Sauro, Afinidad 27, 171 (1970).

221. I. A. Fowlis and D. Welti, Analyst 92, 639 (1967).

222. J. E. Crooks, D. L. Gerrard, and W. F. Maddams, Analyt. Chem. 45, 1823 (1973).

223. J. H. Beynon, R. A. Saunders, and A. E. Willisms, J. Sci. Instr. 36, 375 (1959).

224. R. P. W. Scott, I. A. Fowlis, D. Welti, and T. Wilkins, Gas Chromatography 1966, ed. A. B. Littlewood, Institute of Petroleum, London, 1967, p. 318.

225. J. Q. Walker and C. J. Wolf, Analyt. Chem. 45, 2263 (1973).

226. R. A. Brown, J. M. Kelliher, J. J. Heigl, and C. W. Warren, Analyt. Chem. 43, 353 (1971).

227. G. J. Penzias, Analyt. Chem. 45, 890 (1973).

228. G. J. Penzias and M. J. Boyle, Int. Lab., Nov/Dec. 1973, p. 49.

229. L. O. Lephardt and B. J. Bulkin, Analyt. Chem. 45, 706 (1973).

230. S. Behrendt, Sepn. Sci. 6, 479 (1971).

231. W. H. McFadden, "Techniques of Combined Gas Chromatography/Mass Spectrometry," Wiley, New York, 1973.

232. G. A. Junk, Int. J. Mass Spectr. Ion Physics 8, 1 (1972).

233. J. T. Watson in Ref. 4, pp. 146-225.

234. Pp. 229-263 of Ref. 5.

235. R. Ryhage and S. Wikström, "Mass Spectrometry--Techniques and Applications," ed. Milne, Wiley, New York, 1971, p. 91.

236. C. J. W. Brooks, Mass Spectrometry (D. H. Williams, Senior reporter), Chemical Society, London, vol. 1, 1971, p. 288.

237. C. J. W. Brooks and B. S. Middleditch, Mass Spectrometry (D. H. Williams, Senior reporter), Chemical Society, London, vol. 2, 1973, p. 302

238. "Proceedings of the Int. Symp. on GC-MS, Isle of Elba, May 1972," Tamburini Editore, Milan, 1972.

239. F. W. Karasek, Analyt. Chem. 44, (4), 32A (1972).

240. A. N. Freedman, Anal. Chim. Acta 59, 19 (1972).

241. C. F. Simpson, CRC Crit. Rev. Analyt. Chem. 3, 1 (1972).

242. K. E. Murray, J. Shipton, A. V. Robertson, and M. P. Smyth, Chem. and Ind. 1971, 395.

243. J. Dandoy and J. Delvaux, Chem. and Ind. 1971, 592.

AUTHOR INDEX

Numbers in parentheses are reference numbers and indicate that an author's work is referred to although his name is not cited in the text. Underlined numbers give the page on which the complete reference is listed.

H

I

J

L

M

N

O

T

SUBJECT INDEX

H

I

T

V

W